Building Sustainable Cities

Aldo Alvarez-Risco • Marc A. Rosen
Shyla Del-Aguila-Arcentales • Dora Marinova
Editors

Building Sustainable Cities

Social, Economic and Environmental Factors

Springer

Editors
Aldo Alvarez-Risco
Facultad de Ciencias Administrativas y Recursos Humanos
Universidad de San Martín de Porres
Lima, Peru

Shyla Del-Aguila-Arcentales
Escuela Nacional de Marina Mercante "Almirante Miguel Grau"
Callao, Peru

Universidad Nacional de la Amazonia Peruana
Iquitos, Peru

Marc A. Rosen
Faculty of Engineering and Applied Science
University of Ontario Institute of Technology
Oshawa, ON, Canada

Dora Marinova
Curtin University Sustainability Policy (CUSP) Institute
Curtin University
Perth, WA, Australia

ISBN 978-3-030-45532-3 ISBN 978-3-030-45533-0 (eBook)
https://doi.org/10.1007/978-3-030-45533-0

This Springer imprint is published by the registered company Springer Nature Switzerland AG
The registered company address is: Gewerbestrasse 11, 6330 Cham, Switzerland

Preface

Building sustainable cities is a global challenge for governments, and needs to be done in coordination with public and private organizations and, importantly, their citizens. When working towards the objective of achieving sustainable development worldwide, especially in light of the UN Sustainable Development Goals (SDGs) for 2015–2030, it is important to have a reference source to guide research and teaching efforts, as well as practical applications. This is needed in universities and other educational institutions focused on teaching sustainable development as well as to support R & D activities in academe, business, government, and other organizations. Achieving sustainable cities is a mission and challenge involving professionals with a wide variety of backgrounds and training, and requires multidisciplinary approaches in practice, teaching, and research.

The purpose of this book is to provide the theoretical and practical support needed to build sustainable cities, in a comprehensive manner. As a consequence, the book provides support for various areas, including the implementation of social enterprises, qualitative and quantitative research activities, the development of sustainable institutional strategies, education related to sustainability, and the realization of applied industrial activities on various facets of sustainability. Covering the social, environmental, and economic components of sustainability allows a global vision of what is necessary for sustainable development. It is emphasized that we include social aspects, which are usually less addressed.

In Part I of the book, general material is provided on sustainability, cities, and sustainable cities. In Chaps. 1 and 2, introductory aspects of the book and sustainability are reviewed, including concepts, definitions, and common applications. In Chap. 3, issues about energy sustainability are described. Chapter 4 is focused on cities and their characteristics, while in Chap. 5 the concept of sustainable cities is discussed, along with its centrality to the UN Sustainable Development Goals (especially SDG 11).

Part II is centered on social sustainability and comprises five chapters. Chapter 6 describes aspects related to formal jobs, identifying problems that exist in various societies and showing the deep impact generated by the persistence of informal work in the development of a country. Tax, organizational, and behavioral aspects of informal work are described. Chapter 7 addresses a sensitive issue in people's lives: health. It also covers how health systems, to be sustainable, need to ensure that clinical, economic, and humane results are achieved. The chapter shows global evidence of the interventions that are needed to achieve these results. In an increasingly demanding world where job opportunities are forged from the first years of school, it is possible that the need for survival may lead to a lack of equity, which is expressed in business environments as well as in the daily coexistence of citizens. Hence, the various efforts that are made to promote equity are presented in Chap. 8. Food has a central role in citizens' health, leading to pressure to have increasing quantities of food. However, equitable supplies of food are not available in all areas of the world. Also, as described in Chap. 9, the population's eating habits have a direct impact on the sustainable development of families. In Chap. 10, the significant efforts made in recent years by institutions and companies to increase development opportunities for women are described. This implies both the stopping violence against women and a greater appreciation of their many roles in society, including as workers, mothers, and transformational leaders of society.

Part III is centered on environmental sustainability and comprises seven chapters. Cities require sustainable urban designs as part of their contribution to sustainable development, and more and more citizens are demanding ecological construction. Sustainable urban forms and designs are thus described in Chap. 11. All countries have transport systems for both people and goods. The transition to sustainable and ecological means of transport requires governments to have a clear vision of sustainable development, so that changes made in a society can be within reach of the great majority. Current sustainable transport alternatives are presented in Chap. 12. In Chap. 13, aspects related to the development of urban infrastructure are described that can make it more sustainable; likewise, green building certifications are reviewed. The sustainable use of energy in cities, described in Chap. 14, is one of the most important challenges facing the world and requires a behavioral change in planning by authorities and the behavior of citizens. Efforts made globally to improve and optimize the use of energy at company, individual, and government levels are detailed. Chapter 15 discusses waste management and its regulation in various countries, including the obligations it generates on institutions and citizens. This coincides with an important part of the annual planning of the management of a city, which is the management of waste, especially hazardous types. Chapter 16 addresses a crucial issue in the contamination of water resources: the management of medical waste, which involves medicines and medical devices. There are still many countries that do not have specific regulations, and continue to permit patients to decide where to discard expired medicines. Donations are an example of sources of medicines with upcoming expiration dates whose management needs to be regulated so as not to generate excessive contamination. Since the first world summits to address the issue of climate change, various agreements on goals have been reached and still others must be achieved. This activity has required the joint efforts of countries, for example, through world meetings such as the Conference of the Parties (COP) as well as the signing of agreements such as the Paris Agreement. Chapter 17 describes government, business, and organized citizen efforts to mitigate and adapt to climate change, including data related to global initiatives to contribute to climate change mitigation.

Part IV is centered on economic sustainability and comprises four chapters. In Chap. 18, a crucial issue in people's lives is addressed: debt. There is ample evidence of the personal finance problems that many people in the world have and that impacts their quality of life. Also, strategies to improve the personal financial management of consumers are described. Chapter 19 describes one of the most emerging thematic areas: entrepreneurship. The concepts and applications of entrepreneurship and related ideas such as social entrepreneurship, green entrepreneurship, and sustainable entrepreneurship are addressed. Results of various measurements of the enterprise at a global level are presented. Chapter 20 reviews the green consumption habits of consumers as well as the planning of companies to offer products and services in a manner that is environmentally responsible; in addition, the normative measures that help to promote responsible consumption are described. Chapter 21 describes in detail the 17 UN Sustainable Development Goals (SDGs), barriers to achieving them, and the social, environmental, and economic impacts expected from each. The urgent need is demonstrated for multidisciplinary planning, articulated between governments, institutions, companies, universities, communities, families, and individuals.

Finally, Chap. 22 provides closing remarks and proposes the necessary research for sustainable cities. This includes improving the knowledge base for businesses, universities, and the general population.

All chapters were reviewed. In many cases, contributors assisted in reviewing chapters written by others.

This book shows the passion of authors and editors in the most relevant aspects of social, economic, and environmental sustainability. It highlights the most select of literature and provides professional experiences. We hope that the book forms a lasting guide for higher education and school activities. We also hope that this work can help companies exploit the many benefits that can be achieved through the implementation of business models that ensure sustainable development. Communities are the cornerstone of society and for this reason this book includes a strong focus on contributing to their sustainable and harmonious development.

Lima, Peru — Aldo Alvarez-Risco
Oshawa, ON, Canada — Marc A. Rosen
Callao, Peru — Shyla Del-Aguila-Arcentales
Perth, WA, Australia — Dora Marinova

Contents

About the Editors

Aldo Alvarez-Risco is a full-time researcher at the Universidad de San Martín de Porres in Perú in Sustainability Unit Research. He also holds PhD from Universidad Autónoma de Nuevo León, Doctorate in Pharmacy and Biochemistry, Master in Pharmacology and Pharmacist at the Universidad Nacional Mayor de San Marcos and Master Pharmaceutical Care at Universidad de Granada. He is the coordinator of the South American Network of Pharmaceutical Care (REDSAF). He has been speaker at events in 22 countries. He has experience as a speaker for radio and TV appearances.

Marc A. Rosen is a professor at the University of Ontario Institute of Technology in Oshawa, Canada, where he served as founding Dean of the Faculty of Engineering and Applied Science. Dr. Rosen has served as President of the Engineering Institute of Canada and of the Canadian Society for Mechanical Engineering. He has many professional roles, including editor-in-chief of various journals and Director of Oshawa Power and Utilities Corporation. With over 70 research grants and contracts and 900 technical publications, Dr. Rosen is an active teacher and researcher in energy technology, sustainable energy, and the environmental impact of energy systems. Much of his research has been carried out for industry. Dr. Rosen has worked for Imatra Power Company in Finland, Argonne National Laboratory near Chicago, the Institute for Hydrogen Systems near Toronto, and Ryerson University in Toronto. He has received numerous awards and honors.

Shyla Del-Aguila-Arcentales is a researcher in Sustainability area. She completed a doctoral program. She is Master in Pharmaceutical Sciences at the Universidad Nacional Mayor de San Marcos and she is Pharmacist at Universidad Nacional de la Amazonia Peruana. She has experience in bio-business, audit, environmental management, import–export management. It is linked to the ancestral knowledge of medicinal plants in the jungle regarding their traditional uses.

Dora Marinova has over 400 refereed publications and has supervised 50 PhD students to successful completion. She is a member of the National Health and

Medical Research Council's Panel on Centres of Research Excellence in Population Health. Her research interests cover innovation models, including the evolving global green system of innovation and the emerging area of sustainometrics. Dora is Editorial Board member of the International Journal of Education Economics and Development (published by Inderscience, Switzerland) and Transformations: An Interdisciplinary Journal (published by EBSCO, USA). She is Elected Fellow of the prestigious Modelling and Simulation Society of Australia and New Zealand (MSSANZ) and International Environmental Modelling and Software Society (iEMSs).

Contributors

Aldo Alvarez-Risco Facultad de Ciencias Administrativas y Recursos Humanos, Universidad de San Martín de Porres, Lima, Peru

Raquel Chafloque-Cespedes Universidad de San Martín de Porres, Lima, Peru

Andrew J. Chapman International Institute for Carbon Neutral Energy Research, Kyushu University, Fukuoka, Japan

Shyla Del-Aguila-Arcentales Escuela Nacional de Marina Mercante "Almirante Miguel Grau", Lima, Peru

Universidad Nacional de la Amazonía Peruana, Iquitos, Peru

Eduardo De-La-Torre-Jave Ciudad Saludable, Lima, Peru

Mori-Pelaez Demostenez Pontificia Universidad Católica del Perú, Lima, Peru

Santiago Diaz-Risco Centro de Fertilidad Cajamarca, Cajamarca, Peru

Timothy Fraser Department of Political Science, Northeastern University, Boston, MA, USA

Alex Harras Center for MSP Research, Lima, Peru

Sandra Huamán-Pastorelli Universidad de San Martín de Porres, Lima, Peru

Dennis Lopez-Odar Universidad de San Martín de Porres, Lima, Peru

Dora Marinova Curtin University Sustainability Policy (CUSP) Institute, Curtin University, Perth, WA, Australia

Coralia Mesa-Gomez Ministerio de Salud Pública de Cuba, La Habana, Cuba

Marc A. Rosen Faculty of Engineering and Applied Science, University of Ontario Institute of Technology, Oshawa, ON, Canada

Marté Sánchez-Villagómez Universidad de San Martín de Porres, Lima, Peru

M. Chandra Sekar University of Findlay, Findlay, OH, USA

Arístides Vara-Horna Universidad de San Martin de Porres, Lima, Peru

Silvia Vásquez-Sánchez Pontificia Universidad Católica del Perú, Lima, Peru

About the Authors, Reviewers, Editors, and Their Contributions

This book would not have been possible without the incredible and dedicated efforts and contributions of its many authors, who not only wrote chapters but also often assisted in reviewing the chapters of others.

- **Aldo Alvarez-Risco** is coauthor of Chaps. 1, 4, 5, 7, 9, 11, 12, 13, 15, 16, 17, 18, 19, 20, and 21. Also, he is one of the editors and reviewers of the book.
- **Andrew J Chapman** is author of Chap. 8 and coauthor of Chap. 14.
- **Raquel Chafloque-Céspedes** is coauthor of Chap. 19.
- **Shyla Del-Aguila-Arcentales** is coauthor of Chaps. 4, 5, 7, 11, 12, 13, 15, 16, 17, 18, and 21. Also, she is one of the editors and reviewers of the book.
- **Eduardo De-La-Torre-Jave** is coauthor of Chap. 15.
- **Santiago Diaz-Risco** is coauthor of Chap. 7. Also, he is one of the reviewers of the book.
- **Timothy Fraser** is coauthor of Chap. 14.
- **Alex Harras** is coauthor of Chaps. 15 and 20. Also, he is one of the reviewers of the book.
- **Sandra Huamán-Pastorelli** is coauthor of Chap. 20.
- **Dennis López-Odar** is coauthor of Chap. 19.
- **Dora Marinova** is coauthor of Chap. 4. Also, she is one of the editors and reviewers of the book.
- **Coralia Mesa-Gómez** is coauthor of Chap. 7.
- **Demóstenez Mori-Pelaez** is coauthor of Chap. 13.
- **Marc A. Rosen** is author or coauthor of Chaps. 1, 2, 3, 5, 12, 16, 17, 21, and 22. Also, he is one of the editors and reviewers of the book.
- **Marte Sanchez-Villagomez** is author of Chap. 6.
- **M. Chandra Sekar** is coauthor of Chaps. 4, 7, 18, and 19.
- **Aristides Vara-Horna** is coauthor of Chap. 9 and author of Chap. 10.
- **Sylvia Vásquez-Sánchez** is coauthor of Chap. 11.

Part I
Sustainability, Cities, and Sustainable Cities

Chapter 1
Introduction

Aldo Alvarez-Risco and Marc A. Rosen

1.1 Practicing Sustainability and its Importance

People and families are the main nucleus of society and their optimal functioning ensures that cities, countries, and finally the world move towards social, environmental, economic, and cultural growth. Consequently, governments need to plan and manage bearing in mind the well-being of families and individuals. But we are far from achieving sustainability, and examining what is happening in the world in relation to the lives of individuals can be instructive.

Global statistics indicate that many people lose their lives daily due to causes linked to sustainability issues:

- The 2018 State of Global Air Report published by the Health Effects Institute (HEI) in Boston found that in 2016 over 95% of the world's population lived in areas that exceeded the World Health Organization (WHO) Guideline for ambient particulate matter less than or equal to 2.5 micrometers in aerodynamic diameter (PM2.5). Also, worldwide exposure to PM2.5 contributed to 4.1 million deaths from lung cancer, heart disease and stroke, respiratory infections, and chronic lung disease in 2016. Air pollution is currently the fourth-highest cause of death around the world, trailing smoking, hypertension, and diet, with the majority of deaths recorded in poorer nations (HEI, 2018).

A. Alvarez-Risco (✉)
Facultad de Ciencias Administrativas y Recursos Humanos, Universidad de San Martín de Porres, Lima, Peru
e-mail: aldo.alvarez@redsaf.org

M. A. Rosen
Faculty of Engineering and Applied Science, University of Ontario Institute of Technology, Oshawa, ON, Canada
e-mail: marc.rosen@uoit.ca

A. Alvarez-Risco et al. (eds.), *Building Sustainable Cities*,
https://doi.org/10.1007/978-3-030-45533-0_1

- Worldwide, at least 2 billion people use drinking water sources contaminated with feces; contaminated water transmitted diseases such as dysentery, cholera, polio, and typhoid. Also, contaminated drinking water causes 502,000 diarrheal deaths each year (WHO, 2018a).
- In 2015, there were an estimated 921 foodborne illness outbreaks in the USA and in 2017 American citizens reported foodborne illness from bacteria as the most important food safety issue. Also, CDC (2011) estimates 48 million people get sick, 128,000 are hospitalized, and 3000 die from foodborne diseases each year.
- Noncommunicable diseases (NCDs) kill 41 million people annually, equivalent to 71% of all deaths worldwide. 15 million people die from a NCD between the ages of 30 and 69, and over 85% of these "premature" deaths occur in low- and middle-income countries. Furthermore, cardiovascular diseases account for most NCD deaths, at 17.9 million people annually, followed by cancers (9.0 million), respiratory diseases (3.9 million), and diabetes (1.6 million) (WHO, 2018b).

When governments establish global agreements for the care of the planet, as in the case of the Aichi Targets (CBD, 2011) or the Paris Agreement (UNFCCC, 2018), much of the responsibility in a country rests with companies. Hence, it is important that they can carry out their activities taking into account sustainability approaches. In the world, several companies have exhibited a strong commitment to sustainability according to Forbes (2018), including Siemens AG, Storebrand ASA, Cisco Systems Inc., Danske Bank A/S, Ing Group, Commonwealth Bank of Australia, Koninklijke Philips NV, Johnson & Johnson, Koninklijke DSM NV, and Enagas.

The practices of sustainability by different organizations have different impacts and are interlinked. During primary and secondary school education, children and adolescents learn and perform sustainability practices such as recycling, including campaigns that are organized in conjunction with parents and teachers, and preparing foods based on natural products, among other activities. In the case of universities, sustainability is often undertaken through volunteer activities to help vulnerable populations and learn about social responsibility. The latter has recently been approved in Peru as mandatory in universities, and universities receive at least 2% of their budget for its execution. For companies, corporate social responsibility (CSR) efforts are growing, and having an impact on society from economic, social, and environmental points of view (or a combination thereof). At the same time, productive and commercial processes of companies are advancing. CSR also contributes to the transformation of company workers, who see how a company can operate and at the same time generate benefits for third parties, especially marginal populations.

1.2 Teaching Sustainability

The education component of efforts for sustainable development has been clearly evident since the meeting in Rio 1992 (UNESCO, 1992). Aspects of this include improving access to the education for the poor, increasing energy efficiency education, providing better access to education (primary, secondary, vocational, and training), incorporating ethics into education and research priorities, and introducing environment and development concepts into educational programs. One may wonder what has happened since then. In 2002, the report "Education for sustainability: from Rio to Johannesburg lessons learned from a decade of commitment" (UNESCO, 2002) stated that seeking sustainable development through education requires educators to:

- Instill in students an ethic for living sustainably, based on principles of social justice, democracy, peace, and ecological integrity, which represent core issues at the center of society's concerns.
- Encourage a meeting of disciplines, through a linking of knowledge and of expertise to create understanding that is more integrated and contextualized.
- Encourage lifelong learning, starting at the beginning of life and grounded in life experiences, based on improvement and transformation of the moral character of society.
- Develop to the fullest the potential of all human beings, throughout their lives, to facilitate the collective achievement of a viable future.
- Value aesthetics, the creative use of the imagination, an openness to take risks and exhibit flexibility, and a willingness to explore new options.
- Encourage new alliances between the state and civil society in promoting citizens' freedom and the practice of democratic principles.
- Mobilize society in a concerted effort to eliminate poverty and all forms of violence and injustice.
- Encourage a commitment to the values of peace in such a way as to promote the creation of new lifestyles and living patterns.
- Identify and pursue new human projects in the context of local sustainability within a planetary consciousness and a personal and communal awareness of global responsibility.
- Create realistic hope in which the possibility of change and the desire for change are accompanied by a concerted, active participation in change, at the appropriate time, in favor of a sustainable future for all.

In the same report, global initiatives that were carried out to promote sustainability education in teachers were described:

- A Toolbox in-service education project conducted by the National Consortium for Environmental Education and Training in the USA.
- Diverse initiatives in the United Kingdom sponsored by WWF, Forum for the Future and the UK Panel on Sustainable Development, and several local education authorities.
- The Environmental Education Initiative in Teacher Education in Europe.
- The UNESCO Learning for a Sustainable Environment: Innovation in Teacher Education project in Asia and the Pacific.
- A professional development program for over 70 teacher education colleges in the province of Karnataka in India.

- A network of teacher education and resource centers in China sponsored by WWF.
- A national teacher education program in New Zealand that has trained over 40 people to provide in-service training for teachers in their local regions.
- A national teacher education program in South Africa that has appointed a coordinator in each province, established a range of curriculum and resource development projects, and is developing a national structure for the accreditation of teachers who complete the courses.
- An international network of about 50 teacher education institutions, each conducting a project to reorient all or part of its pre-service curriculum towards sustainability, facilitated by UNESCO Chair for the Reorientation of Teacher Training to Address Sustainability at York University, Toronto, Canada.
- The on-line and CD-ROM-based multimedia teacher education program, Teaching and Learning for a Sustainable Future, developed by UNESCO as a demonstration project for adaptation and translation to suit local educational and cultural contexts.

More recently, at Rio + 20 (UN, 2012), educational commitments were presented at several levels:

- Information, education, and training on sustainability at all levels, including in the workplace, are key to strengthening the capacity of workers and trade unions to support sustainable development.
- Each country is encouraged to consider the implementation of green economy policies in the context of sustainable development and poverty eradication, ensuring that workers are equipped with the necessary skills through education and capacity building, and is provided with the necessary social and health protections.
- Action is required to enhance agricultural research, extension services, training, and education to improve agricultural productivity and sustainability through the voluntary sharing of knowledge and good practices.
- Workers should have access to education, skills, health care, social security, fundamental rights at work, social and legal protections, including occupational safety and health, and decent work opportunities.
- Commitments are needed to the right to education and a commitment to strengthen international cooperation to achieve universal access to primary education, particularly for developing countries. Access to quality education at all levels is an essential condition for achieving sustainable development, poverty eradication, gender equality and the empowerment of women, as well as human development, for the attainment of internationally agreed development goals, and for the full participation of both women and men, in particular young people. In this regard, the needs for equal access to education for persons with disabilities, indigenous peoples, local communities, ethnic minorities, and people living in rural areas are vital.
- As younger generations are the custodians of the future, the need for better quality and access to education beyond the primary level is important.

- Cooperation among schools, communities, and authorities needs to be enhanced to promote access to quality education at all levels.
- States need to be encouraged to promote sustainable development awareness among youth, in part by promoting programs for non-formal education in accordance with the goals of the UN.
- Greater international cooperation is required to improve access to education, including through building and strengthening of educational infrastructure and increasing investment in education, particularly investments to improve the quality of education for developing countries.
- Education for sustainable development needs to be promoted to integrate sustainable development more actively into education.
- Educational institutions need to be encouraged to adopt good practices in sustainability management on their campuses and in their communities, with the active participation of students, teachers, and local partners, thereby teaching sustainable development as an integrated component across disciplines.
- Educational institutions, especially higher educational institutions in developing countries, need to be supported in carrying out research and innovation for sustainable development, to develop quality and innovative programs (including entrepreneurship and business skills training, professional, technical and vocational training, and lifelong learning), geared to bridging skills gaps for advancing national sustainable development objectives.

All these reports are meaningful only when they are taken into account during operational and strategic planning in education ministries, public and private schools, universities, and higher education institutes as well as in companies and businesses.

A fundamental aspect for teaching sustainability is its multidisciplinary nature (Pullen & Brinkert, 2014; Rogers et al., 2015; Christie & Miller, 2015; Wolsko et al., 2016) which supports the need to develop work teams with professionals from different fields. For example, the 2030 Agenda for Sustainable Development (UN, 2015), adopted by all United Nations Member States in 2015, provides a shared blueprint for peace and prosperity for people and the planet, now and into the future. At its heart are the 17 Sustainable Development Goals (SDGs), which are an urgent call for action by all countries—developed and developing—in a global partnership. Specifically, SDG 3: "Ensure healthy lives and promote wellbeing for all at all ages" requires:

- Economists who can evaluate the costs associated with the sustainability of the health system (Mani, Gunasekaran, & Delgado, 2018).
- Pharmacists who can ensure the effectiveness and safety of pharmacological treatments through pharmaceutical care and impact economically on the sustainability of the health system (Alves da Costa, Van Mil, & Alvarez-Risco, 2019).
- Anthropologists who can provide a deeper insight into the individual to understand the behavior of health system actors and their impact on sustainability (American Anthropological Association, 2014).

This multidisciplinary approach must also be present in research activities, in the social, economic, and environmental contexts of sustainability. Within the competences that must be developed in the improvement of human capital with an SDG focus, the development of research competencies in sustainability should have a central focus. Thus, Leal Filho (2018) established a set of research needs that are urgent to contribute to the SDGs:

1. Increase the interdisciplinary and transdisciplinary nature of sustainability research in order to be more oriented towards solving the needs of society.
2. Further develop research on sustainability at the local level, in order to understand and adequately manage the impacts of local decisions on a broader scale.
3. Bring research on sustainability closer to society, so it is incorporated into defining directions and the sustainability research agenda.
4. Intensify the communication of scientific results to different interest groups, and share knowledge with them. This requires a change in the way that sustainability researchers offer value for the non-academic community.
5. Promote governance and provide better means to link science with policy formulation. Ideally, decisions should be based on good research that emphasizes trade-offs and multiple possibilities for action.

1.3 What is the Trend for Sustainability Research?

To examine the trend of sustainability research, it is useful to use macro-analysis techniques of research publications.

According to Alvarez-Risco et al. (2018), an effective way to learn more about a discipline or subject area is to identify the patterns that organize the topics addressed in those disciplines or areas. In general, such patterns are expressed in conglomerates of authors or themes, which are structured according to the links that exist between the members of each conglomerate. In the case of conglomerates of authors, the links or nexuses are established from the simultaneous mentions to two authors by a third researcher. Thus, if a researcher A cites two academics B and C simultaneously, it can be assumed that such a citation or mention of both authors is due to the fact that researchers B and C are related because they share a line of research or an approach of analysis, or because they belong to the same school of thought. It could even be that authors B and C are cited because researcher A disagrees with them or disagrees with the results presented by both researchers. Even in the case of negative quotes, the fact of quoting reveals that the cited work has a minimum level of importance for the author making the appointment. That is, whatever the reason for citing a publication, the fact that authors B and C are cited at the same time provides an idea of the affinities or similarities between the scholars and how they are organized relative to these affinities. When a joint mention of authors B and C is repeated many times in a large set of publications, that grouping based on conceptual or theoretical affinity is known as conglomerates of invisible authors or colleges. The technique for detecting these groupings is the co-citation analysis of authors. In the case of conglomerates of topics, groupings are formed from the joint appearance of specific terms in key areas of academic documents, for example, the title or summary of the text analyzed. When two terms (where each term can consist of two or more words) are mentioned simultaneously, it can be understood that there is a thematic link or affinity between them. To the extent that simultaneous mentions are repeated several times within a set of texts, thematic conglomerates are formed. The technique that produces sets of associated terms is known as co-word analysis. A complementary approach to the review of thematic conglomerates is the ordered list obtained through the automatic extraction of terms, a text mining technique that uses linguistic and statistical analysis to generate a ranking of terms with greater semantic potential. In this way, the automatic extraction of terms produces a list of terms ordered according to the contribution of each term to better understand the set of texts analyzed.

To use the three techniques mentioned above, it is necessary to have two fundamental elements: a) a large number of texts to be analyzed, b) the possibility of obtaining such information from academic databases. To date, this availability is offered by two multidisciplinary databases: Scopus and Web of Science. The total of documents indexed in both databases is known as the mainstream of science, or in practical terms, scientific knowledge. Both are the largest multidisciplinary databases that exist to date (Scopus with 65 million records and Web of Science with 70 million records), whose documents have gone through peer review, unlike what happens with other search engines of free access academics. It should be added that these two databases are commercial products, published by the companies Elsevier for the case of Scopus and Clarivate Analytics for the Web of Science. Since they not only record the bibliographic information of the studies published in the most prestigious academic journals, but also the number of times the studies and authors are cited (individual or collective, in the case of research groups), both Scopus and Web of Science are known as citation databases. It is estimated that Google Scholar has between 180 and 200 million indexed documents, which are automatically analyzed by Google robots that examine academic websites on a permanent basis (Alvarez-Risco et al. 2018).

Figure 1.1 shows the publications on sustainability indexed in Scopus and Web of Science from 2010 to 2018. The data show increases in the number of publications in both databases each year.

To evaluate the publications on sustainability that have more citations, Google Scholar is used for the years between 2010 and 2018. These results are presented in Table 1.1.

The first step to obtain conglomerates of authors is to describe the results in general. This means identifying journals where many works on sustainability are published and authors whose works received the highest number of citations. For that reason, we present the journals with the highest number of studies and the most cited authors who publish research on sustainability, which were indexed in Scopus and Web of Science (see Table 1.2).

Table 1.3 shows the list of institutions that have published scientific articles on sustainability. It can be seen that universities of North America, Europe, and Australia are dominant. Only one Latin American university is listed.

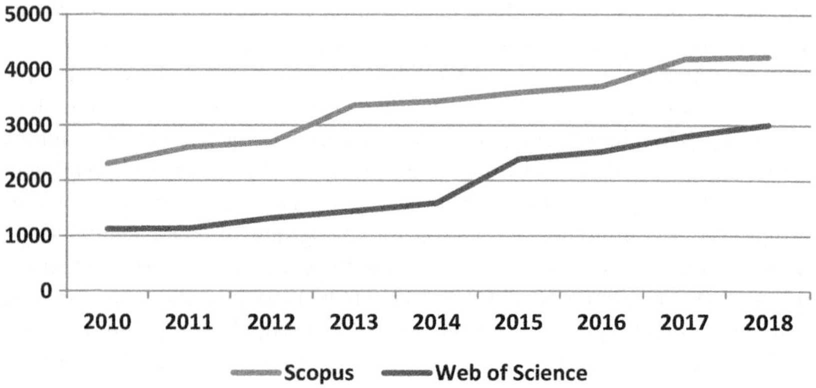

Fig. 1.1 Publications on sustainability indexed in Scopus and Web of Science. (Source: Scopus and Web of Science, 21/12/2018)

Table 1.1 Publications on sustainability most cited in Google Scholar

No	Title	Author / Year	Cites
1.	Agricultural sustainability and intensive production practices	Tilman et al. (2002)	5229
2.	A general framework for analyzing sustainability of social-ecological systems	Ostrom (2009)	4109
3.	A framework for vulnerability analysis in sustainability science	Turner et al. (2003)	3262
4.	Sustainability science	Kates et al. (2001)	3037
5.	Beyond the business case for corporate sustainability	Dyllick & Hockerts (2002)	3027
6.	Towards sustainability in world fisheries	Pauly et al. (2002)	2966
7.	Concepts and definitions of CSR and corporate sustainability: Between agency and communion	Van Marrewijk (2003)	2481
8.	Eight types of product–service system: eight ways to sustainability? Experiences from SusProNet	Tukker (2004)	1613
9.	The emergence of land change science for global environmental change and sustainability	Turner et al. (2007)	1479
10.	Sustainability transitions: An emerging field of research and its prospects	Markard et al. (2012)	1324
11	Urbanization in developing countries: Current trends, future projections, and key challenges for sustainability	Cohen (2006)	1233
12	The multi-level perspective on sustainability transitions: Responses to seven criticisms	Geels (2011)	1227
13	Soil health and sustainability: managing the biotic component of soil quality	Doran & Zeiss (2000)	1223
14	The sustainability balanced scorecard–linking sustainability management to business strategy	Figge et al. (2002)	1176

Table 1.1 (continued)

No	Title	Author / Year	Cites
15	Sustainability science: the emerging research program	Clark & Dickson (2003)	1173
16	An overview of sustainability assessment methodologies	Singh et al. (2009)	1169
17	Categorizing tools for sustainability assessment	Ness et al. (2007)	1152
18	Innovation studies and sustainability transitions: The allure of the multi-level perspective and its challenges	Smith et al. (2010)	1147
19	Agricultural sustainability: concepts, principles and evidence	Pretty (2008)	1122
20	Transdisciplinary research in sustainability science: practice, principles, and challenges	Lang et al. (2012)	1056

Source: Google Scholar, 12/21/2018
Search command: "sustainability"

Table 1.2 Academic journals that publish many studies on sustainability

N°	Journal	Scopus	WoS
1.	Sustainability (MDPI, Switzerland) (ISSN 2071–1050)	1267	1235
2.	Journal of Cleaner Production	1005	982
3.	International Journal of Sustainability in Higher Education	350	235
4.	Ecological Economics	311	338
5.	Abstracts of Papers of the American Chemical Society		306
6.	Ecological Indicators	211	178
7.	Sustainability Science	205	185
8.	ASEE Annual Conference and Exposition Conference Proceedings	203	
9.	WIT Transactions on Ecology and the Environment	196	
10.	Journal of Business Ethics	186	155
11.	Renewable and Sustainable Energy Reviews	177	176
12.	Business Strategy and The Environment	169	102
13.	Sustainable Development	154	133
14.	Environment Development and Sustainability	151	53
15.	Local Environment	150	
16.	International Journal of Sustainable Development and World Ecology	144	143
17.	Journal of Environmental Management	141	134
18.	Current Opinion in Environmental Sustainability	126	128
19.	Sustainability (ISSN 1937 0695)	119	41
20.	Journal of Sustainable Tourism	116	81

Source: Scopus and Web of Science, 12/21/2018. Search command in Scopus: TITLE ("sustainability") Web of Science search command: TI = ("sustainability")
Own elaboration

Table 1.3 Institutions that publish many studies on sustainability

N°	Institution	Scopus	WoS
1.	Wageningen University and Research Centre	362	250
2.	Arizona State University	322	235
3.	University of British Columbia	260	195
4.	Chinese Academy of Sciences	235	182
5.	Universidade de Sao Paulo – USP	217	125
6.	Delft University of Technology	212	113
7.	University of Queensland	206	142
8.	University of Manchester	202	124
9.	Leuphana Universitat Luneburg	180	133
10.	University of Cambridge	179	115
11.	RMIT University	172	79
12.	Virginia Polytechnic Institute and State University	171	73
13.	ETH Zurich	165	107
14.	United States Environmental Protection Agency	159	105
15.	University of Melbourne	159	103
16.	UCL	155	97
17.	INRA Institut National de La Recherche Agronomique	154	158
18.	Lunds Universitet	153	112
19.	University of California, Berkeley	150	115
20.	Pennsylvania State University	147	83

Source: Scopus and Web of Science, 12/21/2018. Search command in Scopus: TITLE ("sustainability") Web of Science search command: TI = ("sustainability")
Own elaboration

This bibliometric information indicates who is developing new research in the field of sustainability and in what direction. The detailed review of this information serves as a starting point for teaching and research activities, since there are mandatory readings that we must do to know the origin and evolution of sustainability in the academic field.

References

Alvarez-Risco, A., López-Odar, D., Chafloque-Céspedes, R., Vílchez-Roman, C., Alemán, L., Asencios Gonzalez, Z., et al. (2018). *Emprendimiento social ¿Las universidades peruanas están cumpliendo su rol promotor?* Lima: Fondo Editorial USMP.

Alves da Costa, F. A., van Mil, J. F., & Alvarez-Risco, A. (Eds.). (2019). *The pharmacist guide to implementing pharmaceutical care*. Switzerland: Springer Nature.

American Anthropological Association. (2014). Strengthening West African health care systems to stop Ebola: Anthropologists offer insights. Retrieved February 5, 2020, from https://pure.mpg.de/rest/items/item_2096578/component/file_2103624/content

CBD (2011). Aichi biodiversity targets. Retrieved February 5, 2020, from https://www.cbd.int/sp/targets

CDC (2011). Burden of foodborne illness: Findings. Retrieved February 5, 2020, from https://www.cdc.gov/foodborneburden/2011-foodborne-estimates.html

Christie, B., & Miller, K. (2015). Academics' opinions and practices of education for sustainable development: Reflections on a nation-wide, mixed-methods, multidisciplinary study. In *Routledge handbook of higher education for sustainable development* (pp. 420–434). Routledge.

Clark, W. C., & Dickson, N. M. (2003). Sustainability science: The emerging research program. *Proceedings of the National Academy of Sciences, 100*(14), 8059–8061.

Cohen, B. (2006). Urbanization in developing countries: Current trends, future projections, and key challenges for sustainability. *Technology in Society, 28*(1–2), 63–80.

Doran, J. W., & Zeiss, M. R. (2000). Soil health and sustainability: Managing the biotic component of soil quality. *Applied Soil Ecology, 15*(1), 3–11.

Dyllick, T., & Hockerts, K. (2002). Beyond the business case for corporate sustainability. *Business Strategy and the Environment, 11*(2), 130–141.

Figge, F., Hahn, T., Schaltegger, S., & Wagner, M. (2002). The sustainability balanced scorecard–linking sustainability management to business strategy. *Business Strategy and the Environment, 11*(5), 269–284.

Forbes. (2018). The world's most sustainable companies 2017. Retrieved February 5, 2020, from https://www.forbes.com/sites/jeffkauflin/2017/01/17/the-worlds-most-sustainable-companies-2017/#483f61f94e9d

Geels, F. W. (2011). The multi-level perspective on sustainability transitions: Responses to seven criticisms. *Environmental Innovation and Societal Transitions, 1*(1), 24–40.

HEI. (2018). State of global air. Retrieved February 5, 2020, from https://www.stateofglobalair.org/sites/default/files/soga-2018-report.pdf

Kates, R. W., Clark, W. C., Corell, R., Hall, J. M., Jaeger, C. C., Lowe, I., et al. (2001). Sustainability science. *Science, 292*(5517), 641–642.

Lang, D. J., Wiek, A., Bergmann, M., Stauffacher, M., Martens, P., Moll, P., et al. (2012). Transdisciplinary research in sustainability science: Practice, principles, and challenges. *Sustainability Science, 7*(1), 25–43.

Leal Filho, W., Azeiteiro, U., Alves, F., Pace, P., Mifsud, M., Brandli, L., et al. (2018). Reinvigorating the sustainable development research agenda: The role of the sustainable development goals (SDG). *International Journal of Sustainable Development & World Ecology, 25*(2), 131–142.

Mani, V., Gunasekaran, A., & Delgado, C. (2018). Enhancing supply chain performance through supplier social sustainability: An emerging economy perspective. *International Journal of Production Economics, 195*, 259–272.

Markard, J., Raven, R., & Truffer, B. (2012). Sustainability transitions: An emerging field of research and its prospects. *Research Policy, 41*(6), 955–967.

Ness, B., Urbel-Piirsalu, E., Anderberg, S., & Olsson, L. (2007). Categorising tools for sustainability assessment. *Ecological Economics, 60*(3), 498–508.

Ostrom, E. (2009). A general framework for analyzing sustainability of social-ecological systems. *Science, 325*(5939), 419–422.

Pauly, D., Christensen, V., Guénette, S., Pitcher, T. J., Sumaila, U. R., Walters, C. J., et al. (2002). Towards sustainability in world fisheries. *Nature, 418*(6898), 689.

Pretty, J. (2008). Agricultural sustainability: Concepts, principles and evidence. *Philosophical Transactions of the Royal Society B: Biological Sciences, 363*(1491), 447–465.

Pullen, S., & Brinkert, K. (2014). SolEn for a sustainable future: Developing and teaching a multidisciplinary course on solar energy to further sustainable education in chemistry. *Journal of Chemical Education, 91*(10), 1569–1573.

Rogers, M., Pfaff, T., Hamilton, J., & Erkan, A. (2015). Using sustainability themes and multidisciplinary approaches to enhance STEM education. *International Journal of Sustainability in Higher Education, 16*(4), 523–536.

Singh, R. K., Murty, H. R., Gupta, S. K., & Dikshit, A. K. (2009). An overview of sustainability assessment methodologies. *Ecological Indicators, 9*(2), 189–212.
Smith, A., Voß, J. P., & Grin, J. (2010). Innovation studies and sustainability transitions: The allure of the multi-level perspective and its challenges. *Research Policy, 39*(4), 435–448.
Tilman, D., Cassman, K. G., Matson, P. A., Naylor, R., & Polasky, S. (2002). Agricultural sustainability and intensive production practices. *Nature, 418*(6898), 671.
Tukker, A. (2004). Eight types of product–service system: Eight ways to sustainability? Experiences from SusProNet. *Business Strategy and the Environment, 13*(4), 246–260.
Turner, B. L., Kasperson, R. E., Matson, P. A., McCarthy, J. J., Corell, R. W., Christensen, L., et al. (2003). A framework for vulnerability analysis in sustainability science. *Proceedings of the National Academy of Sciences, 100*(14), 8074–8079.
Turner, B. L., Lambin, E. F., & Reenberg, A. (2007). The emergence of land change science for global environmental change and sustainability. *Proceedings of the National Academy of Sciences, 104*(52), 20666–20671.
UN. (2012). Rio + 20. Retrieved February 5, 2020, from https://rio20.un.org/sites/rio20.un.org/files/a-conf.216l-1_english.pdf.pdf
UN. (2015). Sustainable development goals. Retrieved February 5, 2020, from https://sustainabledevelopment.un.org/?menu=1300
UNESCO. (1992). The Rio Declaration on environment and development. Retrieved February 5, 2020, from http://www.unesco.org/education/pdf/RIO_E.PDF
UNESCO. (2002). Education for sustainability: from Rio to Johannesburg, lessons learnt from a decade of commitment. Retrieved February 5, 2020, from https://unesdoc.unesco.org/ark:/48223/pf0000127100
UNFCCC. (2018). The Paris Agreement. Retrieved February 5, 2020, from https://unfccc.int/process-and-meetings/the-paris-agreement/the-paris-agreement
Van Marrewijk, M. (2003). Concepts and definitions of CSR and corporate sustainability: Between agency and communion. *Journal of Business Ethics, 44*(2–3), 95–105.
WHO. (2018a). Drinking-water. Retrieved February 5, 2020, from https://www.who.int/news-room/fact-sheets/detail/drinking-water
WHO. (2018b). Noncommunicable diseases. Retrieved February 5, 2020, from https://www.who.int/news-room/fact-sheets/detail/noncommunicable-diseases
Wolsko, C., Marino, E., Doherty, T. J., Fisher, S., Green, A. S., Goodwin, B., et al. (2016). Systems of access: A multidisciplinary strategy for assessing the social dimensions of sustainability. *Sustainability: Science, Practice and Policy, 12*(1), 88–100.

Chapter 2
Sustainability: Concepts, Definitions, and Applications

Marc A. Rosen

Abstract People and societies face important sustainability challenges, which are expected to become more significant in the future. Making societies and their development more sustainable requires consideration of economic, social, and environmental factors among others. Assessment tools are needed to evaluate sustainability and how it is affected by modifications. To account for all relevant factors, a comprehensive set of indicators is required, including practical quantitative indicators and qualitative indicators where necessary. In this chapter, sustainability definitions are reviewed and the historical context for sustainability is briefly described. Then sustainability is discussed, focusing on its economic, environmental, and social dimensions, and the related concept of sustainable development is examined. Finally, assessment measures for sustainability are examined, and several applications are presented.

Keywords Sustainability · Sustainable development · Environment · Economy · Society · Climate change · Equity · Resources · Carrying capacity

2.1 Introduction

Achieving sustainability, which fundamentally relates to the ability to sustain humanity, civilizations, and ecosystems on Earth, is one of the most important objectives of a society and its people. Yet people and societies today face challenges to sustainability. These challenges are expected to become more significant in the future.

M. A. Rosen (✉)
Faculty of Engineering and Applied Science, University of Ontario Institute of Technology, Oshawa, ON, Canada
e-mail: marc.rosen@uoit.ca

A. Alvarez-Risco et al. (eds.), *Building Sustainable Cities*,
https://doi.org/10.1007/978-3-030-45533-0_2

Sustainability concerns are wide ranging, covering many diverse issues: industrial activities including patterns of consumption and production, resource supply (water, energy, mineral, and food), climate change and pollution, sanitation and management of wastes, use and exploitation of land and water including loss of freshwater quality and quantity and corresponding droughts and desertification, degradation of ecological populations and ecosystems including biodiversity reduction and extinction of species, growth in population as well as requirements and desires of populations, continued trends towards increased urbanization and globalization, social development and stability, including protection of cultures, and impacts of major accidents and disasters (natural and anthropogenic). The role of governments and their actions and policies affect all of the above issues and thus play an important role in shifting towards sustainability.

The wide range of issues affecting sustainability indicates that there is a need for an approach to them that is both comprehensive and holistic. Sustainability is being reflected increasingly in the strategies of businesses and governments, and related activities are increasingly appearing in their plans and operations. For such activities to contribute to attaining sustainability objectives, they must impact the pillars of sustainability. Achieving sustainability in societies is a challenge not limited to the present day, even though the global nature of the problem has come to the fore in recent decades. Sustainability was an objective, at least implicitly, for early civilizations. Some examples that illustrate the historical context of this chapter follow:

- Since humans transitioned from mobile hunter-gatherers to agriculture and settlements during the agricultural revolution approximately 12,000 years ago, the sustainability of the lifestyle was crucial in determining if society would thrive or collapse.
- The decline of the Western Roman Empire illustrates a society that failed to sustain itself and ultimately collapsed, in part due to waning returns from natural resource production as well as social factors contributed to the decline (Tainter, 1998).
- The Polynesians on Easter Island exhausted the resources of their remote habitat to such an extent that they could no longer feed themselves or even build canoes to escape, leading to ecological damage to Easter Island that led to a total societal collapse, decimating its population compared to its peak in the 1600s (Tainter, 1998).

2.2 Sustainability Concepts and Definitions

There are many definitions of sustainability, but none are universally accepted and applicable in all situations. Transforming sustainability from the theoretical to the practical and operational is undoubtedly a challenge, one which is made even more complicated when the many sustainability definitions are taken into account.

The theoretical concept of sustainability can be taken to mean "lasting" or "enduring in perpetuity." For example, Ehrenfeld proposed a simple definition of sustainability as "the possibility that human and other forms of life will flourish on the planet forever." However, such an interpretation is not helpful to practitioners and lacks practicality. Certainly, many items or activities are sustainable over very short timescales (several years, say), but very little—if anything—can be sustained indefinitely. A period reflecting perhaps two to four generations (e.g., 50–100 years) is often viewed as a practical trade-off for sustainability considerations. Such a timescale certainly is more tractable and practical than that put forth by Ehrenfeld.

From a technological viewpoint, sustainability can be considered from the perspective of carrying capacity, i.e., the maximum number of people supportable in a given area, accounting for resource availability and the ability of the environment to accept waste emissions. The carrying capacity of a region is dependent on the demand and supply of natural resources. But sustainability cannot be defined only from an environmental perspective and involves more than technical factors.

Usually, sustainability is conceptualized in terms of three dimensions or pillars: economic, social, and environmental. A significant facet of this conceptualization is its broadening or extending of sustainability far beyond carrying capacity, by incorporating social and economic factors in addition to or in concert with the more traditional environmental ones. A perspective on sustainability that is multidimensional is compatible with an appreciation of the fact that many factors affect if and how societies flourish or decline. Achieving a sustainable balance is challenging as these three dimensions are often in tension (e.g., environmental and social sustainability may be achieved to the detriment of economic sustainability). Figure 2.1 shows the main dimensions of sustainability. Also, Fig. 2.2 shows the intersections of dimensions of sustainability. Figure 2.3 shows the relations between sustainability dimensions. Figure 2.4 shows the hierarchy of its main dimensions.

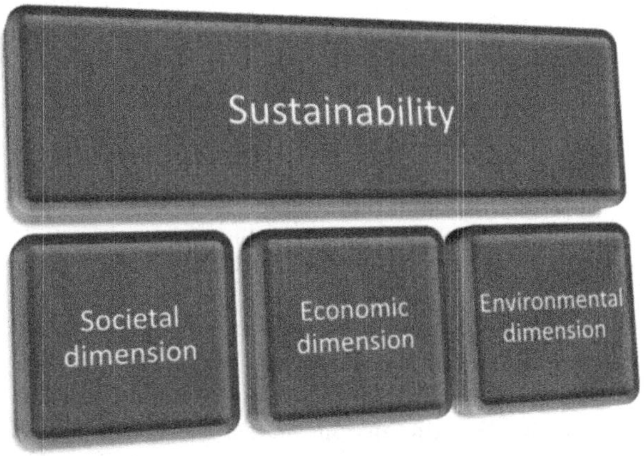

Fig. 2.1 Sustainability illustrated as supported by pillars representing its main dimensions

Fig. 2.2 Sustainability illustrated as the common intersection of the three spheres of environmental, economic, and societal sustainability

Fig. 2.3 Relations between sustainability dimensions. Achieving environmental sustainability requires societal sustainability since the environment includes society. Achieving societal sustainability requires economic sustainability since society includes economics and industry

Sustainability and its three main dimensions (which are examined individually subsequently) can be conceptually illustrated in various ways, demonstrating their interconnected nature:

- Sustainability can be viewed as supported by three legs or pillars, representing environmental, economic, and social sustainability (see Fig. 2.1). This system can only remain standing and in balance if environmental, economic, and social sustainability dimensions are satisfied, even though they are often in conflict.

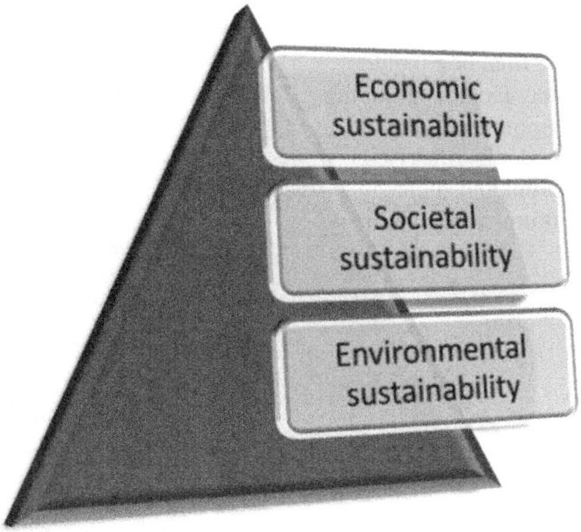

Fig. 2.4 Sustainability illustrated as a hierarchy of its main dimensions

- Sustainability can be viewed as the common overlap of three intersecting circles, representing the environmental, economic, and social dimensions of sustainability (see Fig. 2.2).
- Sustainability can be viewed with using concentric circles representing its main dimensions (see Fig. 2.3). Economic sustainability, being an activity of people and their communities, is a subset of social sustainability, which involves many social and cultural dimensions as well as economic factors. Social sustainability is a subset of environmental sustainability, since activities of people and societies are carried out within the environment.
- The three dimensions of sustainability can also be viewed in a hierarchal structure, where the economy is a subsystem of society, and which itself is embedded in the environment (see Fig. 2.4).

The above points also demonstrate the multidisciplinary nature of sustainability, and its linkages to diverse fields such as science, engineering, environment, ecology, economics, business, sociology, and philosophy. Addressing sustainability requires consideration of topics such as resource use (energy and material), economic development, social development, health, environmental stewardship, engineering, design, architecture, and how different disciplines interact.

Sustainability concepts can be focused around a discipline when needed or useful. For instance, many factors that need to be addressed in moving towards engineering sustainability are shown in Fig. 2.5, including appropriate selection of resources, use of sustainable engineering processes, enhancement of efficiency of engineering processes and resource use, holistic adoption of environmental stewardship in engineering activities, and satisfying other key sustainability facets (e.g., economics, equity, land use, lifestyle, sociopolitical factors, and population).

Fig. 2.5 Engineering sustainability requirements. The final box includes many factors, including economics, living standards, lifestyles, health, and social and cultural acceptability

2.3 Environmental Sustainability

For virtually all energy and material interactions on our planet, the Earth's environment is both the sink and the source. Sustaining civilization and humanity requires that the Earth maintain an ability to support the activities of societies and people. The economies of societies and human populations have grown so greatly in scale that presently anthropogenic activities lead to wide and varied impacts and consequences, ranging from local to global and short- to long-term. Many environmental issues affect sustainability, e.g., loss of biodiversity, deforestation, and destruction of natural habitats from development, often lead to animal migrations, and the extraction or use of limited resources and the release into the environment of emissions and wastes (Aghbashlo & Rosen, 2018). The health of living species including people as well as air quality can be adversely affected by pollutants released to the air. The sources of such emissions are widespread, and particularly linked to combustion in industrial processing, electricity generation, transportation, and other operations. Emissions of liquid wastes like agricultural runoffs and wastewater releases also can lead to environmental problems like eutrophication of waters, acidification, and bioaccumulation of toxic substances in aquatic species.

Two challenges relating to environmental sustainability are particularly important:

- **Climate change**. Mitigating the negative effects of climate change, by avoiding increases or even achieving reductions in atmospheric greenhouse gas (GHG)

concentrations, is one of the most significant challenges facing humanity currently, according to most studies. The primary GHG is carbon dioxide (CO_2). The main anthropogenic sources of GHG emissions include fossil fuel combustion, agricultural nitrogen utilization, and enteric fermentation in ruminant animals. The fifth assessment report of the Intergovernmental Panel on Climate Change (IPCC), finalized in 2014, reports warming trends due to anthropogenic activities as "very likely."
- **Stratospheric ozone depletion**. The layer of ozone (O_3) in the stratosphere absorbs ultraviolet radiation from the sun, particularly UV-B, which can harm animals and plants. The main ozone depleting substances are chlorofluorocarbons (CFCs), used for decades as refrigerants. A phase out of CFCs was agreed to in 1989 via the Montreal Protocol, but they are still used in low-income countries.

2.4 Economic Sustainability

Sustainability in society requires an economy that is capable of providing the services needed by people, acceptable lifestyles and living standards, and careers and jobs, among other effects. To be sustainable, a society requires over the long term not only economic growth (typically quantified as increase in gross domestic product), but predominantly economic development. Nevertheless, the focus currently is normally on economic growth, which is the basis of most capitalist economies and the wealth generation and jobs creation they provide. But, as the economy exists on a planet where most resources are finite (Aghbashlo & Rosen, 2018), an ever-growing economy is not necessarily sustainable over the long term. The most advantageous options for economic sustainability likely vary by country. For instance, poor nations likely derive greater benefits from economic growth, whereas wealthy ones may achieve greater benefits by developing—rather than growing—their economies so as to preserve resources and capacities for waste disposal.

Economic sustainability concepts can be separated into weak and strong sustainability perspectives, as follows:

- **Weak sustainability:** This is a perspective premised on the idea of a fixed overall capital stock (i.e., natural and human capital). Natural capital includes the Earth's stock of natural resources, and encompasses all living organisms as well as the ground (e.g., soils, rock), water, and air. Human capital comprises the stock of labor, knowledge, creativity, and human attributes. A key feature of weak sustainability is that it permits natural capital to substitute for human capital, and vice versa. Thus, an increase in human capital, such as knowledge and skill, can mitigate or offset a reduction in natural resources, from this perspective of sustainability. For these reasons, some economists prefer this perspective.
- **Strong sustainability:** This is a differing perspective of sustainability, premised on natural and human capital not being substitutable, during the provision of

goods and services. This perspective of non-interchangeability of natural and human capital does permit them to be complementary. Environmentalists favor this sustainability perspective frequently.

2.5 Societal Sustainability

Societal sustainability is broad in concept covering health, equity, cultural development, and other factors and not easily defined. What constitutes societal sustainability may change temporally, e.g., people may inhabit smaller houses, live in higher density neighborhoods, possess less material goods, and travel less in the future. This may lead to a higher quality of life even if it entails a lower gross domestic product per capita. Understanding the psychological relating to sustainability and sustainable development is also an important facet (Di Fabio & Rosen, 2018).

Two major factors in societal sustainability follow:

- **Equity**. The concept of societal sustainability includes equity within and between generations. Intragenerational equity requires more balanced distribution of wealth among wealthy and poor nations. Intergenerational requires future generations to be able to attain a reasonably good quality of life compared to preceding generations. Intergenerational equity spanning two to four generations into the future, roughly equivalent to 50 to 100 years, is often envisioned for societal sustainability.
- **Health**. Human health and well-being, including life expectancy and infant mortality, is an important factor social sustainability. Many factors contribute to human health, including access to healthy and clean food and drinking water, safe waste disposal, and an environment without harmful substances that can lead to diseases.

2.6 Sustainable Development

Sustainable development can be viewed as development that can be sustained into the future for some lengthy period of time, and continuously improved (Rosen, 2017a). Development implies a qualitative improvement and differs from simple growth, which is quantitative. The term sustainable development was coined by the World Commission on Environment and Development (1987) of the United Nations in its 1987 report "Our Common Future," as "development that meets the needs of the present without compromising the ability of future generations to meet their own needs." Note that the term sustainable development differs somewhat from sustainability, with sustainable development implying a course of action that improves the quality of life of people and societies, and that can endure into the future, and sustainability implying a state that can be maintained into the future.

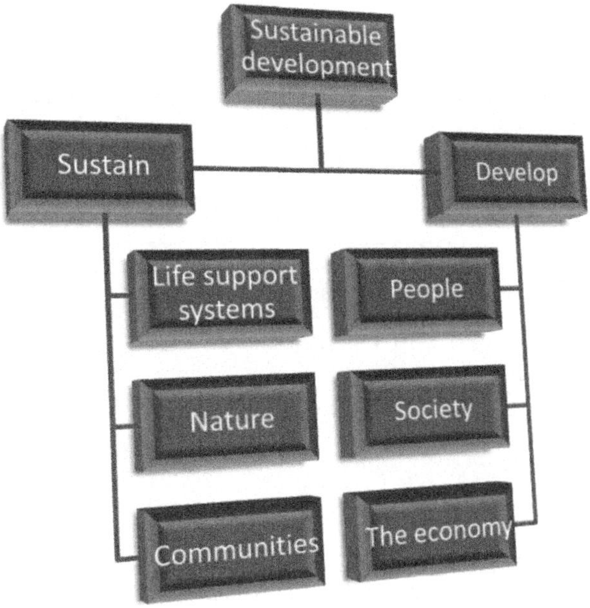

Fig. 2.6 Interpretation of sustainable development based on both what is to be sustained and what is to be developed

Sustainable development can be interpreted based on what is to be sustained and what is to be developed. Figure 2.6 provides the following interpretation of sustainable development (U.S. National Research Council Board on Sustainable Development, 1999):

- **To be developed**. People (e.g., life expectancy, quality of life, education, and equity), society (e.g., security, institutions) and the economy (e.g., wealth creation, and jobs) are to be developed. Social and economic systems of people can be enhanced through development, e.g., quality of life of people in developing countries can be enhanced via more equitable wealth distribution, better education, and reduced reliance on non-renewable resources.
- **To be sustained**. Life support systems (e.g., resources, land, soil, water, and air), nature (e.g., ecosystems, habitats, and biodiversity), and communities (e.g., cultures) are to be sustained. Earth and its ecosystems provide ecosystem services (e.g., climate stability, water, and air purification) and natural resources, which permit communities to endure and flourish.

Ensuring that sustainable development is comprehensively and pervasively incorporated into industry, government, planning, and people's lives is important. Sound, coordinated, meaningful, realistic, and achievable policies and strategies for sustainable development are required. Significant political efforts have recently been taken in this regard (Rosen, 2017b) with the adoption of the UN Sustainable Development Goals for 2015–2030 at the 70th Session of the United Nations

Fig. 2.7 Sustainable Development Goals adopted in 2015 by the United Nations (public domain material provided by United Nations at http://www.un.org/sustainabledevelopment/news/communications-material/)

General Assembly in 2015, as part of the 2030 Agenda for Sustainable Development (United Nations, 2015). The UN Sustainable Development Goals encompass the 17 broad goals illustrated in Fig. 2.7.

2.7 Assessing Sustainability

Sustainability assessment is important for "operationalizing" sustainability and measuring and monitoring progress towards it, but it is challenging and methods are not universally accepted. Nonetheless, various assessment methods have been developed:

- Some assessment methods utilize sustainability indicators, which are typically simple quantitative proxies that measure economic, social, and environmental factors.
- Some sustainability indexes are based on an aggregate or composite of selected sustainability indicators. Such single-value measures, although beneficial for understanding and communication, can mask details associated with the multidimensional nature of sustainability.

Sustainability assessment methods usually focus on only one dimension of sustainability rather than considering the environment, the economy, and society together, and they often lack of a systems approach that regards the system being considered as a whole and accounts for the interactions among its environmental, economic, and societal dimensions. Coupled human–environment systems have interactions among different systems that lead to trade-offs, e.g., reducing costs may cause a process to have higher emissions or lower efficiency. Life cycle analysis is usually part of a systems approach.

2.8 Applications

Sustainability practices and principles have been applied in recent years in a variety of areas. Some examples:

- The sustainability of energy, water, and environment systems has been investigated (Krajacic et al., 2018), and energy has received much attention, e.g., Gomez-Echeverri, Johansson, Nakicenovic, and Patwardhan (2012) led a global energy assessment to identify sustainable routes.
- Infrastructure and building sustainability has been examined, e.g., Bingham, Agelin-Chaab, and Rosen (2019) assessed a residential home with PV and battery storage in The Bahamas, while Russell-Smith, Lepech, Fruchter, and Meyer (2015) improved buildings with sustainable target value design.
- The sustainability of manufacturing has been examined (Nazzal, Abuamarah, Kishawy, & Rosen, 2013).
- Regional sustainability has been investigated, e.g., Ali Mansoori, Enayati, and Barnie Agyarko (2016) examined sustainable development for a state, while Gomez-Echeverri et al. (2012) examined the world.
- Some grand sustainability initiatives have been undertaken, like the Red Sea–Dead Sea canal project (Rosen & Abu Rukah, 2011). The Dead Sea and its unique environment are degrading due to anthropogenic activity affecting its water balance. Israel and Jordan announced jointly in 2002 a desire to halt the water quality deterioration and water level decline of the Dead Sea. To do this, they proposed the construction of a canal to transport water from the Red Sea to the Dead Sea. Beyond avoiding the loss of the Dead Sea, the project would also generate electricity and provide freshwater by seawater desalination for Jordan, Israel, and the Palestinian National Authority. The project, with continual modifications of details, now seems ready to proceed, with Israel and Jordan advancing $800 million US for the project recently.

Closing Remarks

Sustainability is multidimensional, encompassing economic, social, environmental, and other factors. Although it can be vague and often complex, sustainability is an important aim for humanity and societies. Tools for sustainability assessment are needed to assess processes/systems and how modifications affect their sustainability. This often requires a comprehensive set of indicators regarding environmental, economy, and societal sustainability as well as related topics like technological and institutional sustainability. This chapter, including its treatment of sustainability definitions and historical contexts, its economic, environmental, and social dimensions, sustainable development, sustainability assessment measures, and relevant applications, is intended aid efforts to address sustainability challenges of today and tomorrow.

References

Aghbashlo, M., & Rosen, M. A. (2018). Exergoeconoenvironmental analysis as a new concept for developing thermodynamically, economically, and environmentally sound energy conversion systems. *Journal of Cleaner Production, 187*, 190–204.

Ali Mansoori, G., Enayati, N., & Barnie Agyarko, L. (2016). *Energy: Sources, utilization, legislation, sustainability, Illinois as model state*. Singapore: World Scientific.

Bingham, R., Agelin-Chaab, M., & Rosen, M. A. (2019). Whole building optimization of a residential home with PV and battery storage in the Bahamas. *Renewable Energy, 132*, 1088–1103.

Board on Sustainable Development. (1999). *Our common journey: A transition toward sustainability*. Washington, DC: National Academy Press.

Di Fabio, A., & Rosen, M. A. (2018). Opening the black box of psychological processes in the science of sustainable development: A new frontier. *European Journal of Sustainable Development Research, 2*(4), 47.

Gomez-Echeverri, L., Johansson, T. B., Nakicenovic, N., & Patwardhan, A. (Eds.). (2012). *Global energy assessment: Toward a sustainable future*. Vienna and Cambridge: International Institute for Applied Systems Analysis and Cambridge University Press.

Krajacic, G., Vujanovic, M., Duic, N., Kilkis, S., Rosen, M. A., & Al-Nimr, M. A. (2018). Integrated approach for sustainable development of energy, water and environment systems. *Energy Conversion and Management, 159*, 398–412.

Nazzal, Y., Abuamarah, B. A., Kishawy, H. A., & Rosen, M. A. (2013). Considering environmental sustainability as a tool for manufacturing decision making and future development. *Research Journal of Environmental and Earth Sciences, 5*(4), 193–200.

Rosen, M. A. (2017a). Sustainable development: A vital quest. *European Journal of Sustainable Development Research, 1*(1), 2.

Rosen, M. A. (2017b). How can we achieve the UN sustainable development goals? *European Journal of Sustainable Development Research, 1*(2), 06.

Rosen, M. A., & Abu Rukah, Y. (2011). A pragmatic approach for sustainable development of the red-Mediterranean-dead seas canal project: A case study. *International Journal of Ecology and Development, 19*(S11), 63–75.

Russell-Smith, S. V., Lepech, M. D., Fruchter, R., & Meyer, Y. B. (2015). Sustainable target value design: Integrating life cycle assessment and target value design to improve building energy and environmental performance. *Journal of Cleaner Production, 88*, 43–51.

Tainter, J. A. (1998). *The collapse of complex societies*. Cambridge, UK: Cambridge University Press.

United Nations. (2015). *Resolution adopted by the General Assembly on 25* September 2015. A/RES/70/1. Seventieth session. United Nations.

World Commission on Environment and Development. (1987). *Our common future*. New York: Oxford University Press.

Chapter 3
Energy Sustainability

Marc A. Rosen

Abstract Energy sustainability is an important component of sustainability, especially for urban regions due to their relatively large energy use. The importance of energy sustainability is due to the notable environmental impacts of energy processes and the importance of energy in economic development, lifestyles, and living standards, in combination with the pervasiveness of energy use. Many factors need to be addressed appropriately for energy sustainability, including appropriate energy sources, energy carriers, efficiency enhancement, holistic environmental stewardship, and satisfying various other sustainability factors like economics, equity, land use, lifestyle, socio-political actions, and population. In this chapter, energy sustainability definitions are reviewed and the requirements of energy sustainability are discussed, focusing on its technical, environmental, economic, social, and other dimensions. Then energy sustainability is examined, along with means for enhancing it, and energy sustainability evaluation is discussed. An illustrative urban energy system is presented.

Keywords Energy · Energy use · Sustainability · Environment · Economy · Society · Equity · Resources · Renewable energy · Fossil fuels

3.1 Introduction

Energy sustainability is more holistic than simply sustainable energy sources, involving the sustainable use of energy in overall energy systems. Such systems include the harvesting of energy sources, their conversion to useful energy forms, energy transport and storage, and the utilization of energy to provide energy

M. A. Rosen (✉)
Faculty of Engineering and Applied Science, University of Ontario Institute of Technology, Oshawa, ON, Canada
e-mail: marc.rosen@uoit.ca

A. Alvarez-Risco et al. (eds.), *Building Sustainable Cities*,
https://doi.org/10.1007/978-3-030-45533-0_3

services. Sustainability is often considered to have environmental, economic, and social dimensions, and energy is indirectly linked to each. Energy resources are obtained from the environment, and wastes from energy processes are emitted to the environment. Energy resources drive much of the world's economic activity. Energy services support good standards of living, social and cultural development, and social stability.

Significant social and environmental challenges are linked to development, including high energy consumption, high greenhouse gas emissions, a large ecological footprint, and rapid resource consumption, including food and water. People are continually moving from rural areas to cities due to study and work opportunities, increasing the energy needs of citizens, including for communications. From the United Nations Climate Change Conferences, messages are conveyed to populations to reduce energy use and environmental impacts through actions such as reducing unnecessary use of lights in offices and homes. Most countries are increasingly aspiring to achieve sustainability and sustainable development.

In most countries, urban energy use is significant and often accounts for the majority of energy use. Given that the world's population and economy are becoming increasingly urbanized, the sustainability of urban energy use is a key factor in the quest for energy sustainability.

In this chapter, energy is described, and energy sustainability definitions are examined. The requirements of energy sustainability are discussed, focusing on technical, environmental, economic, social, and other factors. Then energy sustainability is examined, and means for enhancing energy sustainability are discussed. An interesting urban energy system is presented.

3.2 Energy

Energy exists in various forms, e.g., fossil fuels and fossil fuel-based products like gasoline, uranium, electricity, work, heat, and electromagnetic radiation (including light). Energy can be converted from form to form with energy-conversion technologies. We use energy carriers (often simply referred to as energy), produced from energy sources, in all aspects of living. Energy sources, or primary energy forms, are found in the natural environment. Many are listed in Table 3.1, where it is seen that some like fossil fuels are finite (non-renewable) while others like solar energy are renewable. Fossil fuels are the basis for most industrialized countries. Energy carriers are the energy forms that we transport and use (see Table 3.2), e.g., secondary chemical fuels and non-conventional chemical fuels like hydrogen and methanol. Many energy carriers do not exist naturally, including such energy forms as work, electricity, and non-ambient thermal energy.

Table 3.1 Energy sources, broken down by non-renewable and renewable types

Non-renewable	Renewable
Fossil fuels (mainly oil, natural gas, coal)	Solar radiation
Non-fossil fuels	Solar-related
Biomass (use rate > replenishment rate)	Hydraulic (very large to micro)
Uranium	Wind
Fusion material (e.g., deuterium)	Wave
Wastes (direct use or indirect via conversion)	Ocean thermal (temperature difference of deep and surface water)
	Biomass (use rate < replenishment rate)
	Non-solar-related
	Geothermal (internal earth heat and ground-source)
	Tidal (moon/sun gravitation and earth rotation)

Table 3.2 Energy carriers, broken down by chemical and non-chemical types

Chemical	Non-chemical
Fossil fuels	Electricity
Secondary chemical fuels	Work
Oil products (e.g., gasoline, diesel fuel, naphtha)	Thermal energy
Synthetic gaseous fuels (e.g., from coal gasification)	Heat (or heated medium)
Coal products (e.g., coke)	Cold (or cooled medium)
Non-conventional (hydrogen, methanol, ammonia)	

3.3 Energy Sustainability Definitions

Energy sustainability can be developed by extending the general definitions of sustainability to energy, but it is more complex. Energy is the basis of ecosystems, civilizations, and life, and is used in almost all facets of living. The standards of living achieved in countries are often a function of energy-related factors.

The present author defines energy sustainability as the provision of energy services, for all people now and in the future, in a sustainable manner. Providing energy services in a sustainable manner implies they are sufficient to provide basic necessities, affordable, not detrimental to the environment, and acceptable to communities and the people living within them. But universal agreement on a definition of energy sustainability has not yet been achieved, and other definitions and descriptions have been presented (Rosen, 2017a).

3.4 Energy Sustainability Requirements

There are several factors affecting how energy resources can be used sustainably in society, defining key requirements for energy sustainability, which are described in this section.

3.4.1 Capture/Production of Sustainable Energy Sources (Requirement 1)

Non-fossil fuel energy options reduce or eliminate greenhouse gas emissions, and are renewable, and thus often facilitate sustainable energy options. The main types of renewable energy include solar energy, wind energy, hydraulic energy, geothermal energy (both the internal heat of the earth and the natural temperatures of the ground), biomass energy (provided its rate of use does not exceed its rate of replenishment), wave energy, tidal energy, and ocean thermal energy. Waste materials and energy can be used in waste-to-energy incinerators to generate electricity are also sometimes considered renewable energy. Nuclear energy is not renewable, but it avoids greenhouse gas emissions and thus contributes to avoiding climate change.

3.4.2 Conversion of Sustainable Energy Sources to Suitable Energy Carriers (Requirement 2)

Energy carriers are an important consideration in energy sustainability because conventional and non-fossil fuel energy options are not necessarily readily utilizable in their natural forms. For instance, conversion systems are often needed to make non-fossil energy more conveniently utilizable. Energy carriers include electricity, work, thermal energy, and secondary chemical fuels. An example of a non-conventional energy carrier is hydrogen, which may become an important energy carrier in the future hydrogen economy (Gnanapragasam & Rosen, 2017; Rosen, 2017b; Scott, 2007).

3.4.3 Increased Efficiency in Provision of Energy Services (Requirement 3)

High efficiency allows the greatest benefits to be attained from energy options, including renewable ones, and thus supports efforts to achieve energy sustainability. Increased efficiency elongates the lives of finite energy resources, and reduces the capacities required of energy devices. In a holistic sense, efficiency enhancement measures include technology efficiency improvements, improved energy management, energy conservation, fuel substitution, more efficient utilization of both energy quantity and quality, and better matching of energy supply and demand. Efficiency improvement efforts are often best assessed and developed with exergy analysis, an alternative analysis method often reveals insights not identified with conventional energy analysis (Dincer & Rosen, 2013; Rosen, 2012).

3.4.4 Reduced Environmental Impact Over Life Cycle of Energy Processes (Requirement 4)

A wide range of environmental impacts are associated with energy processes. Some notable examples follow:

- Climate change due to emissions of greenhouse gases, particularly carbon dioxide.
- Stratospheric ozone depletion permits the increased ultraviolet radiation to reaching the earth.
- Acidification of water and soil.
- Abiotic depletion of non-renewable raw materials due to their extraction.
- Ecotoxicity exposure and the ensuing health problems.
- Radiological releases that increase rates radiogenic cancer mortality and morbidity.

These energy-related environmental impacts need to be addressed for energy sustainability. For instance, non-fossil fuel energy options are needed to help humanity mitigate climate change (Rosen et al., 2012). To comprehensively assess the environmental impact of an energy activity, its entire life cycle must be considered, from acquisition of energy sources and other resources to their utilization and ultimate disposal. Life cycle assessment (LCA) is such a technique and it creates an inventory of environmental effects like resource depletion, waste generation and energy consumption, and their environmental impacts (Graedel & Allenby, 2010; Jianu, Pandya, Rosen, & Naterer, 2016). Guidelines for LCA have been developed by the International Organization for Standardization.

3.4.5 Consideration of Other Facets of Sustainability (Requirement 5)

Many other sustainability issues are related to energy sustainability, and some are now described.

3.4.5.1 Economic Affordability

For sustainability, the energy services required for basic needs must be economically affordable by all societies and people (Rosen, 2011). Sometimes efficiency improvement and environmental mitigation measures can over time save money or be revenue neutral, aiding affordability.

3.4.5.2 Community Involvement and Social Acceptability

For energy sustainability, people and communities must be involved in and supportive of energy-related decisions. Communities support is critical, and almost always requires consultation.

3.4.5.3 Lifestyles

Modifying lifestyles and moderating energy-related desires can support energy sustainability, although this is often challenging because people's aspirations tend to increase. Translating future energy-related threats into immediate priorities will likely remain a difficult challenge for decision and policy makers.

3.4.5.4 Appropriate Land Use

Land uses for energy need to be balanced with uses for such other purposes as agriculture and recreation, e.g., land use for the growth of energy plants needs to be balanced with agricultural needs while land requirements for electricity transmission need to be balanced with the needs of the ecosystems they traverse sensitive areas.

3.4.5.5 Equity

All societies must have access to energy resources, regardless of geographic location, and future generations must also be able to access energy resources. In addition, equity among developed and developing countries must be achieved in terms of energy opportunities.

3.4.5.6 Meeting Increasing Energy Demands

The increasing use of energy resources, especially as developing countries become more urbanized, industrialized, and attain higher living standards, must be satisfied. This issue is exacerbated by populations rise.

3.4.5.7 Aesthetics

The cleanliness of the environment affects the well-being of people and thus pertains to energy sustainability. Even renewable energy technologies can mar the landscape, e.g., large wind turbines farms and rooftop or ground-mount photovoltaic panels.

3.5 Urban Energy Sustainability

The energy needs of cities are large and increase with urban growth. In general, urbanization causes major shifts in land-use patterns, and changes the ways societies use energy. Urbanization is accelerating in many developing countries. Increasing urban energy efficiencies can often improve living standards, quality of life, and satisfaction of people. For example, Sweden has a per capita gross domestic product near to that of the USA but uses 40% less energy per capita and outranks the USA on most social indicators. Some contributing factors are lower energy wastes in residential and commercial buildings, and transportation advantages such as better public transit, smaller automobiles, higher gasoline taxes, and compactness of geography.

Energy measures can sometimes be introduced voluntarily and be successful, while at other times governments need to use incentives and enforcement measures. In urban communities, for example, the combination of inexpensive energy and moderate environmental constraints in North America has led to a preference for urban travel by automobile rather than public transportation. Options related to energy issues to improve living standards can be technical and nontechnical (e.g., changing lifestyles and increasing awareness). China and its cities provide a notable recent example of the challenges of urban energy sustainability. Energy use in China rose notably from 2000 to 2005, especially in urban areas, increasing wealth and living standards and fostering trends of greater resource consumption and purchase of consumer goods. This change in behavior can further increase demands for resources, including energy. A reinforcing effect can thus develop, where increased energy use raises living standards, which in turn drives further increases in energy use (Zhang & Wang, 2013).

3.6 Improving Energy Sustainability

Numerous methods and technologies can be used to enhance energy sustainability, and often apply strongly in urban contexts. Some of these are described below.

High-Efficiency Devices and Lighting

Energy sustainability can be supported by the use of high-efficiency devices, e.g., high-efficiency home appliances, furnaces and air conditioners, and motors and fans. New lighting systems have significantly higher efficiencies and longer lives than older equipment.

Energy Storage
Energy storage can be used to improve system efficiency by storing energy between times when it is available and when it is needed (Dincer & Rosen, 2011), e.g., solar thermal energy collected in the day for space heating at night.

Energy Loss Prevention and Waste Recovery
Efficiency can be improved by preventing energy losses and recovering energy wastes, by inspecting periodically to detect and mitigate losses, and using technologies and processes to avoid losses (e.g., insulating).

Maintenance, Monitoring, and Control
The efficiency and life span of equipment can be enhanced via maintenance, e.g., regular cleaning of equipment, replacement of consumable items, and lubrication of moving parts, periodic equipment overhauls, and periodic calibration, tuning and testing of equipment.

Improved Matching of Energy Supplies and Demands
Instead of supplying energy of a much higher quality than required for a demand, it can be more efficient to supply an energy form of a better matched quality, e.g., supplying furnace combustion gases at 600 °C to space heat a building to 22 °C is a poorer match than heat at 45 °C.

Building Envelopes
The energy efficiency of a building can be improved via insulation to reduce heat gains in summer and losses in winter, weather stripping and caulking, to reduce air leakages, high-efficiency windows with multiple glazing and low-emissivity window coatings, and window shades or blinds equipped with sensors that adjust to keep out excessive sunlight.

Passive Methods and Technologies
Passive, as opposed to active, methods can be used to enhance energy sustainability, e.g., utilizing daylight harvesting to offset artificial lighting and using solar energy to heat buildings, as well as placing trees, windows, and window shades so as to keep buildings cool during summers.

District Energy and Integrated Energy Systems
Thermal energy can be a particularly useful energy carrier for urban energy use. It can be produced in centralized heating or cooling facilities and transported to users over long distances in district heating and/or cooling systems, which are used in many cities and industrial parks (Rosen & Koohi-Fayegh, 2016). Efficiency can also be increased by linking separate systems advantageously to create integrated energy systems, e.g., cogeneration, trigeneration, and multigeneration (Rosen & Koohi-Fayegh, 2016).

Use of Exergy Analysis and Other Tools

It has been suggested that thermodynamic performance is best evaluated and improved using exergy analysis in addition to or in place of energy analysis, where exergy is a measure of the usefulness or quality or value of energy (Dincer & Rosen, 2013). Exergy analysis provides more meaningful efficiencies energy analysis, and specifies process inefficiencies (types, causes, and locations) better. Many applications of exergy analysis have been reported, including advanced optimization methods (Dincer, Rosen, & Ahmadi, 2017).

3.7 Illustrative Example: Net-Zero Energy Buildings and Communities

Buildings are responsible for a significant portion of the energy use and environmental impacts in many countries. The design and operation of buildings and communities can be transformed to reduce energy use and emissions, by allowing buildings to act as a net energy generator. Thus, net-zero energy buildings can contribute to energy sustainability, especially in urban areas (Rosen, 2015). A net-zero energy building is defined as one that, in an average year, produces as much electrical plus thermal energy from renewable energy sources as it consumes. A net-zero energy community is similar, but applicable to communities. Smart net-zero energy buildings and communities can reduce environmental impacts, localized generation strategies, manage loads, reduce utility electrical demands, and transportation energy savings via electric or hybrid cars that use electricity from renewable building-integrated energy systems (Garmsiri, Kouhi-Fayegh, Rosen, & Smith, 2016).

Work on net-zero energy buildings has been reported. The Smart Net-zero Energy Buildings Strategic Research Network was launched in Canada in 2011 (http://www.solarbuildings.ca), and the International Energy Agency has an annex entitled "Towards Net-zero Energy Solar Building." Design, optimization, and modeling issues have been investigated (Berggren, Hall, & Wall, 2013; Bucking, Athienitis, & Zmeureanu, 2013; Cellura, Guarino, Longo, & Mistretta, 2014; Mohamed, Hasan, & Sirén, 2014), as has the relation between net energy use and the urban density of solar buildings. The integration of net-zero energy buildings into the electrical grid has been examined (Gaiser & Stroeve, 2014). Work on net-zero energy buildings was extended in recent years to net-zero energy communities, e.g., the design of solar-optimized neighborhoods, infrastructure interactions in the design of sustainable neighborhoods, and benefits of seasonal storage of solar energy for space heating in the Drake Landing Solar Community in Canada (Rad, Fung, & Rosen, 2017).

Net-zero energy buildings and communities can make significant contributions to energy sustainability:

- **Sustainable energy sources**. The main benefit of net-zero energy buildings and communities, averaged over the year, is that they typically utilize renewable energy sources like solar and geothermal (Rosen & Koohi-Fayegh, 2017).
- **Sustainable energy carriers**. Thermal energy (heat) is used, which facilitates the use of renewable energy sources like solar and geothermal.
- **Increased efficiency**. The efficiency of net-zero energy buildings and communities is typically high.
- **Reduced environmental impact**. Net-zero energy buildings and communities have little environmental emissions associated with their operation, even after accounting for the full life cycles of the buildings and communities.
- **Fulfillment of other aspects of sustainability**. Net-zero energy buildings and communities are anticipated to contribute to economic affordability of energy, alleviate resource demands on societies, and be acceptable, noting that net-zero energy buildings and communities are likely to be implemented only where they are viewed as socially acceptable.

Closing Remarks

Energy sustainability is important for overall sustainability. Various factors need to be appropriately addressed to achieve energy sustainability, including appropriate selection of energy resources, proper energy carriers to facilitate sustainable energy resources, enhancement of the efficiency of energy activities, environmental stewardship in energy processes, and consideration of other key sustainability measures, such as economics, lifestyles, living standards, and equity. The material provided in this chapter, which includes energy types, energy sustainability definitions and requirements, energy sustainability, and means for enhancing it, is intended to help in achieving, or at least shifting closer to, energy sustainability by doing so, it is hoped that sustainability can be given a greater focus.

References

Berggren, B., Hall, M., & Wall, M. (2013). LCE analysis of buildings: Taking the step towards net zero energy buildings. *Energy and Buildings, 62*, 381–391.

Bucking, S., Athienitis, A., & Zmeureanu, R. (2013). An information driven hybrid evolutionary algorithm for optimal design of a net zero energy house. *Solar Energy, 96*, 128–139.

Cellura, M., Guarino, F., Longo, S., & Mistretta, M. (2014). Energy life-cycle approach in net zero energy buildings balance: Operation and embodied energy of an Italian case study. *Energy and Buildings, 72*, 371–381.

Dincer, I., & Rosen, M. A. (2011). *Thermal energy storage: Systems and applications* (2nd ed.). London: Wiley.

Dincer, I., & Rosen, M. A. (2013). *Exergy: Energy, environment and sustainable development* (2d ed.). Oxford: Elsevier.

Dincer, I., Rosen, M. A., & Ahmadi, P. (2017). *Optimization of energy systems*. London: Wiley.

Gaiser, K., & Stroeve, P. (2014). The impact of scheduling appliances and rate structure on bill savings for net-zero energy communities: Application to West Village. *Applied Energy, 113*, 1586–1595.

Garmsiri, S., Kouhi-Fayegh, S., Rosen, M. A., & Smith, G. R. (2016). Integration of transportation energy processes with a net zero energy community using captured waste hydrogen from electrochemical plants. *International Journal of Hydrogen Energy, 41*(19), 8337–8346.

Gnanapragasam, N. V., & Rosen, M. A. (2017). A review of hydrogen production using coal, biomass and other solid fuels. *Biofuels, 8*(6), 725–745.

Graedel, T. E., & Allenby, B. R. (2010). *Industrial ecology and sustainable engineering*. Upper Saddle River, NJ: Prentice Hall.

Jianu, O., Pandya, D., Rosen, M. A., & Naterer, G. (2016). Life cycle assessment for Thermolysis and electrolysis integration in the copper-chlorine cycle of hydrogen production. *WSEAS Transactions on Environment and Development, 12*(Art. #26), 261–267.

Mohamed, A., Hasan, A., & Sirén, K. (2014). Fulfillment of net-zero energy building (NZEB) with four metrics in a single family house with different heating alternatives. *Applied Energy, 114*, 385–399.

Rad, F. M., Fung, A. S., & Rosen, M. A. (2017). An integrated model for designing a solar community heating system with borehole thermal storage. *Energy for Sustainable Development, 36C*, 6–15.

Rosen, M. A. (2011). *Economics and exergy: An enhanced approach to energy economics*. Hauppauge, NY: Nova Science Publishers.

Rosen, M. A. (2012). *Environment, ecology and exergy: Enhanced approaches to environmental and ecological management*. Hauppauge, NY: Nova Science Publishers.

Rosen, M. A. (2015). Net-zero energy buildings and communities: Potential and the role of energy storage. *Journal of Power and Energy Engineering, 3*(4), 470–474.

Rosen, M. A. (2017a). Sustainable development: A vital quest. *European Journal of Sustainable Development Research, 1*(1), 2.

Rosen, M. A. (2017b). Enhancing renewable energy prospects via hydrogen energy systems. *Journal of Fundamentals of Renewable Energy and Applications, 7*(5), 22.

Rosen, M. A., & Koohi-Fayegh, S. (2016). *Cogeneration and district energy systems: Modelling, analysis and optimization*. London: Institution of Engineering and Technology.

Rosen, M. A., & Koohi-Fayegh, S. (2017). *Geothermal energy: Sustainable heating and cooling using the ground*. London: Wiley.

Scott, D. S. (2007). *Smelling land: The hydrogen defense against climate Catastrophe*. Ottawa: Canadian Hydrogen Association.

Zhang, Y., & Wang, Y. (2013). Barriers' and policies' analysis of China's building energy efficiency. *Energy Policy, 62*, 768–773.

Chapter 4
Cities

Shyla Del-Aguila-Arcentales, Aldo Alvarez-Risco, M. Chandra Sekar, and Dora Marinova

Abstract When cities are presented as centers of human interaction, it is necessary to know the evolution that cities have had since their formation, how they are currently formed and how their transformation has been taking place since the arrival of the world interconnected with the Internet. It is necessary to emphasize how valuable cities have been from their history and culture to their potential to host families and to provide everything necessary for their harmonious development. Knowing the cities, you can have more clarity to understand the urgent need to build sustainable cities, those same cities whose families hope to have a healthy and safe environment.

Keywords Cities · Population · Urbanization · Education · Citizens · People

4.1 Cities

Usually when you visualize the term "city," you imagine its size, organization, cultural component, and the complex dynamics underlying its inhabitants. What you know for real is the growing urbanization of the world. While around 5500 years ago people began to live in cities, in the last 100 years, the population density in

S. Del-Aguila-Arcentales
Escuela Nacional de Marina Mercante "Almirante Miguel Grau", Callao, Peru

Universidad Nacional de la Amazonia Peruana, Iquitos, Peru
e-mail: sdelaguila@enamm.edu.pe

A. Alvarez-Risco (✉)
Facultad de Ciencias Administrativas y Recursos Humanos, Universidad de San Martín de Porres, Lima, Peru
e-mail: aldo.alvarez@redsaf.org

M. C. Sekar
University of Findlay, Findlay, OH, USA
e-mail: sekar@findlay.edu

D. Marinova
Curtin University Sustainability Policy (CUSP) Institute,
Curtin University, Perth, WA, Australia
e-mail: d.marinova@curtin.edu.au

A. Alvarez-Risco et al. (eds.), *Building Sustainable Cities*,
https://doi.org/10.1007/978-3-030-45533-0_4

cities began to grow rapidly. This social dynamic leads to generating the question: What are the causes of the origin of cities?

Present evidence demonstrates how groups are formed, efficiently using their energy in acquiring food and thereby ensuring personal and collective survival. This scenario for a given target population does not require significant specialization in work, thereby resulting in minimal social tension among the members. While a few human groups are able to maintain this dynamic, over time the relationship between members becomes more complicated due to the advancement in technology and organizational structures. This can be evidenced by changes in cities since the industrial revolution, as well as with the arrival of mass communication devices such as radio. The dissemination of news generates a change in the behavior of the population further evidenced by the arrival of television transmission which is often accused for a break in the family union. More recently, the Internet further transformed cities, making them more participatory but at the same time less cohesive.

The extent and type of activities taking place in the city are a crucial element that contributed to the evolution of cities, considering that both, the number and type of participants have transformed over the years. At the beginning, the main participants involved in the "life" of the city were limited to an elite group, those belonging to the officials of the governments or the church. Availability of cheap energy contributed to the growth of the cities. Originally, most productive activities in the city required the use of muscle force, but with the invention of electricity and its ready availability decreased the dependency on muscular power making participation in city life less laborious thereby increasing the number of participants.

4.2 How Are Cities Regulated?

At the core of any city, there is an elite group of people that establishes social and commercial rules for its functioning. For example, the surplus of agricultural production may be stored for a later sale, following the guidelines established by the city's governing group. Faced with the need to manage the city efficiently, diversification in the people's professional work occurs, where one group takes charge of maintaining the health of people, another group looks after the city's services, etc. Historically, the clergy always had a transcendental role in the cities. Its influences transcended the city's borders, since the religious belief generates an alignment of thought and behavior of people who profess the same creed.

4.3 Urbanization

Surplus food is a common element that led to emergence of cities. Availability of surplus food enables humans to engage in other activities, as their survival is assured by the availability of the main necessity (Jacobs, 1969). That food surplus enables the governing groups to maintain their hold on the commercial, political, and religious aspects of the city. This conglomerate of people will eventually occupy together the same geographic space to form urban or proto-urban places (Braudel, 1995 [1949]; Childe, 1950).

Such conglomeration is based on the interconnection of interrelated productive activities so that it becomes more efficient to live closely and to be able to easily avail the services of all participants. For example, professionals living in the urban area, such as a baker, priest or banker would have "captive clients" who need their services enabling them to earn their living. These activities formed the nucleus of the urban process (Wheatley, 1971).

Archeological evidence indicates the first urban center appeared around the year 7500 BC in the Middle East and was followed by establishment of other urban centers in Mesopotamia, Egypt, and India by 3500 BC. Records indicate the emergence of Greek cities around 2000 BC, and the Chinese and cities of ancient Rome in the range 2000–1500 BC. Technological advances transform the movement of people in the cities. Most of us can easily relate to this transformation, as 40 years ago it was both difficult and expensive to be transported across town in a taxi, while now with a simple app, the same trip is much easier and cheaper to complete. In the old cities of bygone days, the shortening of distances brought about by technological innovation increased trade, but also secured power for the rulers. Stability led to increased consumption of goods and intensified economic specialization in the cities, which became famous and well-known for specific products across a larger geographical area (McCormick, 2001).

A greater supply of products and increased trade between cities contributed to decrease in transportation costs that in turn stimulated greater interconnection between urban centers, making it possible to demonstrate for the first time the concept of commercial networks, which are maintained until today, supported by ever-increasing interconnectivity. Undoubtedly, it was the industrial revolution that generated the economic leap of the cities (Bourguignon & Morrisson, 2002). The industrial revolution, while it brought with it various production changes, optimizing of the processes, standardization and decreases in cost per unit item, it also generated simultaneously a social change that continues until now—the factory model. Change in the educational training of the people needs to occur, so that they have the necessary skills to be hired at the new production factories. This in turn changed the attitude of the student toward education and training, who now wants to learn the essential to be a useful worker in the system. One result was companies with human power who would have people with basic education, without creativity or vision, and in many cases without a desire or ability to know and demand their labor rights.

As the above paragraphs indicate, education plays a transforming role in the cities as it changes its objective to the needs of the market. The latter leads us to ask ourselves: How then did education evolve in the cities?

Considering human civilization as a whole, schools are relatively new institutions. As hunter gatherers, for hundreds of thousands of years, adults were in charge of survival tasks, leaving children and adolescents alone, whom they educated themselves through exploration marked by trial and error. Leisure time and opportunity for formal education became available with the advent of agriculture and later with the industrial revolution. However, these resulted in children leaving the natural discovery process as a tool for learning but instead being trained as laborers for specific tasks. While this helped children learn specific activities for future survival, it destroyed the child's creative and innovative spirit. The invention of agriculture, which began 10,000 years ago, generated great social changes, specifically in people's ways of life. Prior knowledge requirement focused on plants and animals on which they depended, and on the mastery of topography of the surroundings as well as skills needed for hunting and gathering. Agriculture gradually changed the lifestyle of the nomadic tribes. For example, with agriculture, people could produce more food, increased their stability by being able to live near their land and allowed them to feed more children. Agriculture required long hours of unskilled and repetitive work, which could mostly be done by children, so that families with more children had more labor to spare.

The ownership of the land generated differences in social classes, those who had land and those who worked in those lands for a wage. This social-economic dependence decreased over time but still persists in many sectors of the society even today. With the industrial revolution, a new class of bourgeoisie was created, causing feudalism to diminish although the social impact on the population that was held in servitude for so many years was not immediately evident. In England, for example, the industrial revolution resulted in the need for children along with their parents at the factory floor, and in changes in social adoption laws to protect against child labor. A profound change in education came from the church, which gradually spread the proposal of universal and compulsory education. This resulted in the emergence of the social consensus that childhood should be a time for learning, and schools were considered as places where learning takes place. The idea and practice of universal and compulsory public education developed gradually in Europe, from the early sixteenth century to the nineteenth century. While universal schooling for children had many supporters, everyone had their own ideas about the content students should learn. The formal education taught in class is quite similar in many countries and essentially the same over thousands of years, with the teacher standing in front of the class and teaching its students as the principal teaching methodology. It is specifically non-formal education that has undergone major changes. Let us give an example. Today, a person may not know which is the capital of Hungary or the currency of Kenya. Before the emergence of the Internet, this knowledge would have been delivered to him or her in a didactic fashion. We would have been expected to learn 190 capitals of countries, their currencies, language, etc. It is possible that students learned only 30 of each and the rest were listed in the books, including the famous World Almanac (The World Almanac, 2018). Currently with the Internet, when we want more information on a topic, we use our computers, laptops, and smartphones to clarify any doubts. The most used search engine is Google (87.66%), followed by Bing (4.98%), Yahoo (2.72%), Yandex Ru (0.82%), and Baidu (0.74%) (Statcounter, 2019) and all these options provide immediate information.

4.4 Have Cities Changed with the Evolution of Technology and Education?

Technology has always had a relevant role in education. In the nineteenth century, the slate changed the way teachers interacted in the classroom (Concordia University, 2012). In the twentieth century, the arrival of record players, radios, televisions and, finally, computers, created distance delivery and increased student access to education. This personalized and depersonalized education simultaneously, since it allowed each participant to have a unique personalized experience and to learn at their own speed, according to their preferences. At the same time, in spite of the discussion forums, there is clear absence of face-to-face interaction among the students in the same class. Is it possible that distance education can continue to implement features that allow a closer experience among the participants? We will see what the following years will bring.

Big Data emerged toward 1997 as the main component of the Internet, a year in which it was calculated that the Internet was made up of 12,000 petabytes (10^{15} bytes) and that these would increase by ten times each year. Since 1999, Big Data has been impacting various activities in the world such as health (Wang, Kung, Wang, & Cegielski, 2018), travel (Chen, Ma, Susilo, Liu, & Wang, 2016), accounting (Janvrin & Watson, 2017), energy management (Zhou, Fu, & Yang, 2016), telecommunication (Shrivastava, Sahoo, & Pandey, 2018), sports (Baerg, 2017) among others. Another aspect that has also changed education and therefore the dynamics in cities is the educational repositories, such as Openstax. Already in this century, different tools have appeared that have transformed the way people live in the cities: Moodle to expand more virtual classes, Facebook allowing for 24/7 connection, YouTube to allow participants to upload and watch varied types of videos in the world Academic YouTube Edu. Likewise, Twitter arrived for communication with a few words, Kindle for readings of electronic books, among many other tools.

What aspect will alter these dynamics in the cities in the next 10 years? We witness that the Internet of Things is already transforming the way citizens and cities will behave. For example, there are energy management reports (Hsieh, Chen, & Lee, 2019), disaster management approaches (Al-Turjman, 2019), forensic investigation tools (Yaqoob, Hashem, Ahmed, Kazmi, & Hong, 2019), transportation information (Babar & Arif, 2018), and many others in development.

Armed conflicts in the world have had a great impact on global migration, generating social and political transformations (Bauböck, 2018). The cities now are more than ever in a process of transculturation (Rabiau, 2019). For example, the great Syrian migration has strongly impacted the 3 neighboring countries of Turkey, Lebanon, and Jordan which were forced to receive almost 5.5 million people (ACNUR, 2018a). While Turkey with a population of 80 million inhabitants received 3.6 million Syrian refugees (representing 4.5% of the total population), Lebanon with a total population of about 4 million inhabitants ended up receiving 950 thousand refugees representing a 20% increase in its number of inhabitants. This clearly will have an impact on the cultural, social, economic, and historical dynamics on those societies. More recently, the same phenomenon was seen in Latin America, when almost 3 million people from Venezuela along with 2.4 million people living in other countries in Latin America and the Caribbean were economically dislocated. Peru ended up receiving over half-million Venezuelan refugees (ACNUR, 2018b). The above example illustrates that cities will change with the continued social and demographic flux. Many cities over the last two decades have changed dramatically with the acceleration of transformation becoming faster, so city planning and development need to take into account many of these aspects, especially the technological ones and at the same time, evaluate the resources that are available for commercial and productive activities.

4.5 What Initiatives in Relation to Cities Have Been Created?

The United Nations Educational, Scientific and Cultural Organization (UNESCO) seeks to build peace through international cooperation in education, the sciences and culture. Its programs contribute to the achievement of the Sustainable Development Goals defined in Agenda 2030, adopted by the UN General Assembly in 2015 (UNESCO, 2018a). In 2004, the UNESCO Creative Cities Network (UCCN) was created with the aim to promote cooperation with and among cities which have identified creativity as a strategic factor for sustainable urban development. The 180 cities that currently make up this network work together toward a common objective: placing creativity and cultural industries at the heart of their development plans at the local level and cooperating actively at the international level (UNESCO, 2018b).

Thus, they stand out in different areas:

Music
Salvador (Brazil)
Pesaro (Italy)
Chennai (India)
Kansas (USA)
Kinshasa (Democratic Republic of the Congo)
Hamamatsu (Japan).

Literature
Prague (Czech Republic)
Quebec (Canada)
Bucheon (South Korea)
Ulyanovsk (Russia)
Montevideo (Uruguay)
Krakow (Poland).

Design
Buenos Aires (Argentina)
Istanbul (Turkey)
Graz (Austria)
Berlin (Germany)
Shanghai (China)
Dubai (United Arab Emirates).

Crafts and folk art
Pekalongan (Indonesia)
Isfahan (Iran)
Bamiyan (Afghanistan)
Chordeleg (Ecuador)
Nassau (Bahamas)
Sokode (Togo).

Film
Galway (Ireland)
Busan (South Korea)
Sydney (Australia)
Bristol (England)
Rome (Italy)
Yamagata (Japan).

Media arts
Enghien-les-Bains (France)
Kosice (Slovakia)
Dakar (Senegal)
Tel Aviv (Israel)
Guadalajara (Mexico)
Braga (Portugal).

Gastronomy
Tsuruoka (Japan)
Belem (Brazil)
Parma (Italy)
Rasht (Iran)
Hatay (Turkey)
Macao (China).

Likewise, UNESCO has declared certain cities as World Heritage Sites (UNESCO, 2018c)

World Heritage Convention.

Available in https://whc.unesco.org/en/list/

The search of the World Heritage Convention website, placing the following places as the search criteria (with the year in which they were declared World

Heritage Sites), generates many locations which require the maximum national and international effort for their preservation:

Afghanistan
Minaret and archaeological remains of Jam (year 2002)
Cultural landscape and archaeological remains of the Bamiyan Valley (year 2003)
Austria
Historical center of Vienna (year 2001)
Bolivia (Plurinational State of)
City of Potosí (year 1987)
Central African Republic
National Park of the Manovo-Gounda St. Floris (year 1988)
Chile
Saltpeter offices of Humberstone and Santa Laura (year 2005)
Côte d'Ivoire
Comprehensive natural reserve of Mount Nimba (year 1981, 1982)
Democratic Republic of Congo
National Park of Garamba (year 1980)
National Park of Kahuzi-Biega (year 1980)
Virunga National Park (year 1979)
Salonga National Park (year 1984)
Wildlife Reserve of Okapis (year 1996)
Egypt
Abu Mena (year 1979)
Guinea
Comprehensive natural reserve of Mount Nimba (year 1981, 1982)
Honduras
Biosphere Reserve of Río Plátano (year 1982)
Indonesia
Heritage of the tropical rainforests of Sumatra (year 2004)
Iraq
Asur (Qal'at Sherqat) (year 2003)
Archaeological city of Samarra (year 2007)
Hatra (year 1985)
Israel
Old city of Jerusalem and its walls (year 1981)
Kenya
National Parks of Lake Turkana (year 1997, 2001)
Libya
Old City of Ghadames (year 1986)
Archaeological site of Cyrene (year 1982)
Archaeological site of Leptis Magna (year 1982)
Archaeological site of Sabratha (year 1982)
Rock site of Tadrart Acacus (year 1985)
Madagascar
Rainforests of Atsinanana (year 2007)
Mali

Ancient cities of Djenné (year 1988)
Timbuktu (year 1988)
Tomb of the Askia (year 2004)
Micronesia (Federated States of)
Nan Madol: Ceremonial Center of Eastern Micronesia (year 2016)
Niger
Natural reserves of the Air and the Ténéré (year 1991)
Palestine
Old City of Hebron/Al-Khalil (year 2017)
The Birthplace of Jesus: Church of the Nativity and pilgrimage route in Bethlehem (year 2012)
Palestine: land of olive groves and vineyards—Cultural landscape of southern Jerusalem, Battir (Palestine) (year 2014)
Panama
Fortifications of the Caribbean coast of Panama: Portobelo and San Lorenzo (year 1980)
Peru
Archaeological site of Chan Chan (year 1986)
Senegal
National Park of Niokolo-Koba (year 1981)
Serbia
Medieval monuments of Kosovo (year 2004, 2006)
Solomon Islands
Rennell East (year 1998)
Syrian Arab Republic
Ancient Villages of Northern Syria (year 2011)
Old City of Aleppo (year 1986)
Old City of Bosra (year 1980)
Old City of Damascus (year 1979)
Crac of the Knights and Qal'at Salah Al Din (year 2006)
Site of Palmira (year 1980)
Uganda
Tombs of the Kings of Buganda in Kasubi (year 2001)
United Kingdom of Great Britain and Northern Ireland
Liverpool—Mercantile maritime port (year 2004)
United Republic of Tanzania
Reserve of hunting of Selous (year 1982)
United States of America
Everglades National Park (year 1979)
Uzbekistan
Historical center of Shakhrisyabz (year 2000)
Venezuela (Bolivarian Republic of)
Choir and its port (year 1993)
Yemen
Historical city of Zabid (year 1993)
Old walled city of Shibam (year 1982)
Old City of Sana'a (year 1986)

4.6 Efforts in Urbanization

To promote a human face of urbanization on World Cities Day: October 31, UNESCO advocates the humanization of urbanization, an increasingly pressing need as cities continue to grow. Cities are now home to 3.9 billion people, half the world's population, and their number is expected to reach 5 billion by 2030 (UNESCO, 2018d). Today, more than 4 billion people around the world—more than 50% of the world's population—live in cities. For example, only East Asia and the Pacific cities are home to 1200 million people, almost the same as the population of India. This number continues to grow rapidly, especially in Asia and Africa, as people and families increasingly migrate to urban areas in search of better livelihoods. By 2050, with an urban population that will double its current size, almost 70 out of every 100 people in the world will live in cities (World Bank, 2018). In October 2016, at the Habitat III conference, countries around the world approved the historic New Urban Agenda that sets a new global standard for sustainable urban development and guides global efforts to achieve the Sustainable Development Goals in the era of climate change. At the beginning of February 2018, national and municipal leaders met again at the ninth session of the World Urban Forum (WUF9) in Kuala Lumpur, Malaysia, to continue with more in-depth discussions on the theme: Cities 2030, Cities for all: Implement the New Urban Agenda. Cities around the world will continue to evolve as they respond to the global challenges.

Closing Remarks
Cities are the main habitat for human population nowadays and into the future. Efforts at the city level are increasing to improve people's livelihoods and opportunities and are carried out in an institutional and planned way, with support networks. The 2030 horizon for the Sustainable Development Goals is driving efforts from cities and citizens across the globe. Many authorities and citizens are linked in the agenda to improve the cities and create programs to promote a sustainable way of living that provides opportunities for all and restores the urban environment.

References

ACNUR. (2018a). Total persons of concern by country of Asylum. Retrieved February 5, 2020, from https://data2.unhcr.org/en/situations/syria#_ga=2.176887783.37113213.1546572169-339131225.1546572169

ACNUR. (2018b). La cifra de personas refugiadas e inmigrantes venezolanas alcanza los 3 millones. Retrieved February 5, 2020, from https://www.acnur.org/noticias/press/2018/11/5be443b54/la-cifra-de-personas-refugiadas-e-inmigrantes-venezolanas-alcanza-los-3.html

Al-Turjman, F. (2019). Cognitive routing protocol for disaster-inspired internet of things. *Future Generation Computer Systems, 92*, 1103–1115.

Babar, M., & Arif, F. (2018). Real-time data processing scheme using big data analytics in internet of things based smart transportation environment. *Journal of Ambient Intelligence and Humanized Computing, 10*, 4167–4177.

Baerg, A. (2017). Big data, sport, and the digital divide: theorizing how athletes might respond to big data monitoring. *Journal of Sport and Social Issues, 41*(1), 3–20.
Bauböck, R. (2018). The crossing and blurring of boundaries in international migration. Challenges for social and political theory. In *Blurred boundaries* (pp. 17–52). Routledge.
Bourguignon, F., & Morrisson, C. (2002). Inequality among world citizens: 1820–1992. *American Economic Review, 92*(4), 727–744.
Braudel, F. (1995 [1949]). *The Mediterranean and the Mediterranean world in the age of Philip II*. Berkeley and Los Angeles: University of California Press.
Chen, C., Ma, J., Susilo, Y., Liu, Y., & Wang, M. (2016). The promises of big data and small data for travel behavior (aka human mobility) analysis. *Transportation Research Part C: Emerging Technologies, 68*, 285–299.
Childe, V. G. (1950). The urban revolution. *Town Planning Review, 21*(1), 3–17.
Concordia University. (2012). The history of the classroom blackboard. Retrieved February 5, 2020, from http://education.cu-portland.edu/blog/reference-material/the-history-of-the-classroom-blackboard
Hsieh, J. Y., Chen, W. T., & Lee, J. T. (2019). An intelligent power manager with energy harvesting for internet of things applications. *Microwave and Optical Technology Letters, 61*(1), 271–274.
Jacobs, J. (1969). *The economy of cities*. New York: Vintage.
Janvrin, D. J., & Watson, M. W. (2017). "Big Data": A new twist to accounting. *Journal of Accounting Education, 38*, 3–8.
McCormick, M. (2001). *Origins of the European economy: communication and commerce AD 300–900*. New York: Cambridge University Press.
Rabiau, M. A. (2019). Culture, migration, and identity formation in adolescent refugees: a family perspective. *Journal of Family Social Work, 22*(1), 83–100.
Shrivastava, P., Sahoo, L., & Pandey, M. (2018). Recognition of telecom customer's behavior as data product in CRM big data environment. In *Proceedings of First International Conference on Smart System, Innovations and Computing* (pp. 165–173). Singapore: Springer.
Statcounter. (2019). Global market share of search engines 2010–2019. Retrieved February 5, 2020, from https://gs.statcounter.com/search-engine-market-share/desktop/worldwide/#monthly-201907-201907-bar
The World Almanac. (2018). Retrieved February 5, 2020, from http://www.worldalmanac.com
UNESCO. (2018a). UNESCO in brief. Retrieved February 5, 2020, from https://en.unesco.org/about-us/introducing-unesco
UNESCO. (2018b). Creative cities network. Retrieved February 5, 2020, from https://en.unesco.org/creative-cities/creative-cities-map
UNESCO. (2018c). Creative cities network. Retrieved February 5, 2020, from https://whc.unesco.org/en/list
UNESCO. (2018d). UNESCO to promote human face of urbanization on World Cities Day, 31 October. Retrieved February 5, 2020, from https://en.unesco.org/news/unesco-promote-human-face-urbanization-world-cities-day-31-october
Wang, Y., Kung, L., Wang, W. Y. C., & Cegielski, C. G. (2018). An integrated big data analytics-enabled transformation model: Application to health care. *Information & Management, 55*(1), 64–79.
Wheatley, P. (1971). *The pivot of the four quarters: a preliminary enquiry into the origins and character of the ancient Chinese city*. Chicago: Aldine.
World Bank. (2018). 3 Big Ideas to achieve sustainable cities and communities. Retrieved February 5, 2020, from https://www.worldbank.org/news/immersive-story/2018/01/31/3-big-ideas-to-achieve-sustainable-cities-and-communities
Yaqoob, I., Hashem, I. A. T., Ahmed, A., Kazmi, S. A., & Hong, C. S. (2019). Internet of things forensics: Recent advances, taxonomy, requirements, and open challenges. *Future Generation Computer Systems, 92*, 265–275.
Zhou, K., Fu, C., & Yang, S. (2016). Big data driven smart energy management: From big data to big insights. *Renewable and Sustainable Energy Reviews, 56*, 215–225.

Chapter 5
Sustainable Cities

Shyla Del-Aguila-Arcentales, Aldo Alvarez-Risco, and Marc A. Rosen

Abstract When discussing sustainable cities, we imagine cities that achieve sustainable economic development, that provide the basis for the social development of citizens, and that are harmonized with and care for the environment. In line with progress to make cities sustainable, various national and supranational initiatives are being developed, leading to agreements to cooperate and work together to create and maintain sustainable cities. One also hears of smart cities, eco-cities, and green cities and reports about them and their particularities, but these types of cities are in many ways subsets of sustainable cities given the broader criteria implied by sustainability. In this chapter, sustainable cities and the various aspects that they comprise are described, and some of the most relevant initiatives carried out by governments and municipalities to promote and foster their development are discussed.

Keywords Cities · Sustainable cities · Sustainable · Development · Sustainability

5.1 Cities: What Are They Doing to Be Sustainable?

There are numerous initiatives that seek to contribute to cities and make them more sustainable. For example, the World Bank put forth the initiative *Systems of Cities* because successful cities can enhance their ways, improve their finances, attract

S. Del-Aguila-Arcentales
Escuela Nacional de Marina Mercante "Almirante Miguel Grau", Callao, Peru

Universidad Nacional de la Amazonia Peruana, Iquitos, Peru
e-mail: sdelaguila@enamm.edu.pe

A. Alvarez-Risco (✉)
Facultad de Ciencias Administrativas y Recursos Humanos, Universidad de San Martín de Porres, Lima, Peru
e-mail: aldo.alvarez@redsaf.org

M. A. Rosen
Faculty of Engineering and Applied Science, University of Ontario Institute of Technology, Oshawa, ON, Canada
e-mail: marc.rosen@uoit.ca

A. Alvarez-Risco et al. (eds.), *Building Sustainable Cities*,
https://doi.org/10.1007/978-3-030-45533-0_5

private investors, and take care of the poor and disadvantaged. The new Urban and Local Government Strategy promoted by the World Bank is intended to help governments at all levels make cities more equitable, efficient, sustainable, and environmentally friendly. The strategy draws on two principles. The first is that density, agglomeration, and proximity are fundamental to human advancement, economic productivity, and social equity. The second is that cities need to be well managed and sustainable (World Bank, 2017). The new Urban Strategy realigns the World Bank's urban business with five business lines considered critical for cities and local governments in the decade ahead (World Bank, 2017):

- Focusing on the core elements of the city system: City management, finance, and governance
- Making pro-poor policies a city priority: Reducing urban poverty and upgrading slums
- Supporting city economies: Cities and economic growth
- Encouraging progressive urban land and housing markets: Urban land, housing, and planning
- Promoting a safe and sustainable urban environment: Urban environment, climate change, and disaster management.

Another initiative of the World Bank is called Inclusive Cities (World Bank, 2018), which is based on the following interventions:

1. Adopting multi-sector solutions for a multi-dimensional approach
2. Combining "preventive" and "curative" solutions
3. Sequencing, prioritizing, and scaling up investments
4. Harnessing communities' potential as drivers of inclusion
5. Strengthening capacity at the local level
6. Fostering partnerships

Concrete results can be seen as a consequence of various initiatives, including the Vietnam Urban Upgrading Project, the Cameroon Inclusive and Resilient Cities Development Project (PDVIR), and the Metropolitan Buenos Aires Urban Transformation Project (World Bank, 2018). In 2016, during the United Nations Conference on Housing and Sustainable Urban Development, held in Quito, Ecuador, the New Urban Agenda was approved (Habitat III, 2017). That Agenda was subsequently endorsed by the General Assembly of the United Nations. The New Urban Agenda needs to be implemented by encouraging the acceleration of its implementation through the following ten points, grouped by area (WUF9, 2016).

5.2 Frameworks

1. Encourage the formulation of implementation frameworks for the New Urban Agenda at all levels, including monitoring mechanisms, providing a coordinated space for an effective contribution from all stakeholders, aligning to the efforts

and actions of the 2030 Agenda and other international, regional, national, subnational, and local development frameworks.

2. To support for the creation and consolidation of inclusive platforms and agendas for dialogue among all levels of government, decision makers, and stakeholders such as regional, national, and local Urban Forums and Committees that can strengthen policy review and assessment of. These can also foster exchange of experiences and cooperation, as well as scaling up voluntary commitments and actions from all partners.
3. Further develop and advocate for integrated territorial development, which includes integration of sectoral policies, institutions, and investment; integration among the different spheres of government; spatial integration across the urban–rural continuum; improved coordination across actors; and enhanced alignment of national, subnational, and local policies with international agendas.
4. Adapt innovative and robust mechanisms for the diversification and expansion of the means of implementation, to cater for complex and integrated approaches promoted by the New Urban Agenda. Technological innovations and improvements, research, capacity building, technical assistance, and partnership development, among others, may require enhanced resourcing.

5.3 Governance and Partnerships

5. Adopt multiple collaborative governance mechanisms that actively engage national, subnational and local governments, all groups of society, including youth, women, and grassroots organizations, and particularly the excluded, vulnerable, and disadvantaged groups. This work in solidarity is critical to promote more buy-in and co-responsibility in the activities towards sustainable urban development, and to ensure the sustainability of the results.
6. Promote multi-stakeholder constituency-based coalitions to use the implementation of the New Urban Agenda to improve prevention, prepare, and respond to urban crises.

5.4 Innovative Solutions

7. Foster a culture of creativity and innovation to be embedded in the way cities and human settlements operate.
8. Develop monitoring and data collection mechanisms, including community generated data, to enhance availability of information and disaggregated and comparable data at city, functional urban areas, and community levels. This would promote informed and evidence-based decision making and policy formulation, assessing progress and impact at all levels.
9. Create an enabling environment and develop capacities for scaling up of good practices including municipal finance, sustainable private and public invest-

ments in urban development and job creation, and generating value while advancing the public good.
10. Adopt accessibility and universal design as a core to national, subnational, and local action plans for implementing the New Urban Agenda through inclusive, accessible, and participatory processes and consultations.

It is necessary and helpful to review the commitments in the New Urban Agenda, both at governmental and institutional levels (Habitat III, 2017).

In specific fields such as health, there are also intergovernmental agreements, such as the Shanghai Declaration on promoting health in the 2030 Agenda for Sustainable Development (WHO, 2016). This agreement involves the following commitments, grouped by area:

Good governance is important for health, necessitating a commitment to:

- Apply fully the mechanisms available to government to protect health and promote well-being through public policies;
- Strengthen legislation, regulation, and taxation of unhealthy commodities;
- Implement fiscal policies as a powerful tool to enable new investments in health and well-being including strong public health systems;
- Introduce universal health coverage as an efficient way to achieve both health and financial protection;
- Ensure transparency and social accountability and enable the broad engagement of civil society;
- Strengthen global governance to better address cross-border health issues;
- Consider the growing importance and value of traditional medicine, which could contribute to improved health outcomes, including those in the SDGs;
- Introduce universal health coverage as an efficient way to achieve both health and financial protection;
- Ensure transparency and social accountability and enable the broad engagement of civil society;
- Strengthen global governance to better address cross-border health issues;
- Consider the growing importance and value of traditional medicine, which could contribute to improved health outcomes, including those in the SDGs.

Cities and communities are critical settings for health, necessitating a commitment to:

- Prioritize policies that create co-benefits between health and well-being and other city policies, making full use of social innovation and interactive technologies;
- Support cities to promote equity and social inclusion, harnessing the knowledge, skills, and priorities of their diverse populations through strong community engagement;
- Re-orient health and social services to optimize fair access and put people and communities at the center.

Health literacy empowers and drives equity, necessitating a commitment to:

- Recognize health literacy as a critical determinant of health and invest in its development;
- Develop, implement, and monitor intersectoral national and local strategies for strengthening health literacy in all populations and in all educational settings;
- Increase citizens' control of their own health and its determinants, through harnessing the potential of digital technology;
- Ensure that consumer environments support healthy choices through pricing policies, transparent information, and clear labeling.

It can be seen that metropolitan areas around the world are engaged in a variety of initiatives aimed at upgrading urban infrastructure, products, and services, in a joint effort to create better social, environmental, and economic conditions and to enhance cities' competitiveness as well as their attractiveness (Joss & Molella, 2013; Viitanen & Kingston, 2014).

However, different initiatives include different terms to designate what cities want to achieve. To get an indication of this, one can examine scholarly articles in the Scopus database. Doing so reveals that the following numbers of articles that have the terms designated below in their title:

Smart cities: 4490
Sustainable cities: 741
Eco-cities: 279
Digital cities: 256
Green cities: 195
Low carbon cities: 189
Resilient cities: 140
Knowledge cities: 104
Intelligent cities: 68
Ecological cities: 48
Liveable cities: 41
Inclusive cities: 33
Circular cities: 30

In each of the mentioned articles, different concepts are presented that summarize efforts made by cities (government, companies, institutions, and citizens) to achieve sustainable development. Based on a detailed review of each concept, it is found that many have some points in common (sometimes several, sometimes very few), but clearly they are different concepts from each other. For example, comparing the concepts of resilient cities and smart cities could suggest that they are talking about very related and even interchangeable concepts; however, from reviewing the smart city projects it is evident that it has basically governance and IT content. Relatedly, the term resilience, as mentioned by the United Nations Office for Disaster Risk Reduction (UNISDR), focuses on the prevention and management of disasters, and strengthening governments, communities, and citizens, in order to achieve physical and economic results.

5.5 Can a City Then Belong to a Smart City Initiative and at the Same Time Be an Eco-City and Strive to Be a Sustainable City?

To answer this question, we must remember that the term sustainable development, coined in the 1980s, became a very important paradigm receiving political-normative recognition. The term is maintained to date as the basis for the Sustainable Development Goals (UN, 2015). What is clear is that sustainable development has environmental, economic, and social dimensions, but how exactly the "triple bottom line" is to be understood, and how it applies to urban contexts needs to be analyzed. Due to the lack of consensus on applications, all the initiatives of the cities have room to develop. But, it is necessary to ask if we are dissipating the efforts of a city in several different directions, perhaps in a non-integrated manner.

Cities have historically been the places where innovative activities originate, which are explained by the new needs of the population. For example, at present, people avoid walking long distances to obtain information and instead have incorporated the use of the app to find information related to the nearest banks, the tap station with its respective price or in which pharmacies you can find a certain medication. The United Nations estimates that by 2030, 60% of the population will live in cities (United Nations, 2014), and that there will be more economic, social, and environmental demands that must be planned today.

The challenge is specifically in those cities that have grown without planning since this disorganized growth will have important negative consequences on the environment and finally on the citizens themselves (Annez & Buckley, 2009); therefore, there are major sustainability challenges at the level of housing construction and road construction (David, 2017).

A double approach is therefore required to meet these demands: to achieve smart cities and at the same time achieve sustainable cities, that is, an innovative city that uses information and communication is needed. Technologies (ICT) means to improve the quality of life, the efficiency of operations, urban services, and competitiveness while ensuring that it meets the needs of the present and future population with respect to economic, social, environmental, and cultural (UNECE, 2015). It is also important to be able to measure the performance of these innovative interventions (Webb, Hawkey, & Tingey, 2016). For decades, different rankings have been generated that allow comparing countries and even cities in relation to the implementation of sustainable activities. This information is crucial for authorities and private investors, since it allows us to be certain of which interventions are working best. A first measurement tool was United Nation's (UN) City Prosperity Index (UN-HABITAT, 2015) which was created in 2012 and which was also supported by a conceptual proposal called "The Wheel of Urban Prosperity" (UN-Habitat, 2013, p.12). Some basic concepts and indicators were established in this initiative:

5.6 Productivity

A prosperous city contributes to economic growth and development, generating income, employment, and equal opportunities that provide adequate living standards for the entire population.

1. Economic Strength—City Product Per Capita
2. Employment—Unemployment Rate

5.7 Infrastructure

A prosperous city deploys the infrastructure, physical assets and amenities—adequate water, sanitation, power supply, road network, information and communications technology, etc.—required to sustain both the population and the economy, and provide better quality of life.

1. Housing Infrastructure—Improved Shelter—Access to Improved Water
2. Social Infrastructure—Physicians Density
3. ICT—Internet Access
4. Urban Mobility—Traffic Fatalities

5.8 Quality of Life

Prosperous cities provide amenities such as social services, education, health, recreation, safety, and security required for improved living standards, enabling the population to maximize individual potential and to lead fulfilling lives.

1. Health—Life Expectancy at Birth—Under-Five Mortality Rate
2. Education—Literacy Rate—Mean Years of Schooling
3. Safety and Security—Homicide Rate

5.9 Equity and Social Inclusion

A city is only prosperous to the extent that poverty and inequalities are minimal. No city can claim to be prosperous when large segments of the population live in abject poverty and deprivation. This includes reducing the incidence of slums and new forms of poverty and marginalization.

1. Economic Equity—Gini Coefficient—Poverty Rate
2. Social Inclusion—Slum Households—Youth Unemployment
3. Gender Inclusion—Equitable Secondary School Enrolment

5.10 Environmental Sustainability

The growth of cities and their economic development do not destroy or degrade the environment; instead, the city's natural assets are preserved for the sake of sustainable urbanization. Well-planned cities promote environmental sustainability.

1. Air Quality—PM2.5 Concentration
2. Waste Management—Waste Water Treatment
3. Energy—Share of Renewable Energy—CO_2 Emissions

5.11 Governance and Legislation

Cities are best able to combine sustainability and shared prosperity through effective urban governance and transformational leadership, deploying appropriate and effective policies, laws, and regulations, and creating adequate institutional frameworks with strong local institutions and sound institutional arrangements.

1. Participation—Voter Turnout
2. Institutional Capacity—Days to Start a Business

Outcomes of index measurement showed five types of solid prosperity (UN-HABITAT, 2015):

Cities with very solid prosperity index (more than 80 points)

Oslo, Copenhagen, Stockholm, Helsinki, Paris, Vienna, and Melbourne.

Cities with solid prosperity index (between 70 and 79 points)

Montreal, Toronto, Sydney, Berlin, Milan, Amsterdam-Utrecht, Brussels, Tokyo, Manchester, and Prague.

Cities with moderately solid prosperity index (between 60 and 69 points)

Buenos Aires, Mexico City, Lima, Almaty, Ciudad Obregon, Guadalajara, Medellin, Panama City, and Guayaquil.

Cities with moderately weak prosperity index (between 50 and 59 points)

Sao Paulo Jakarta, Bangkok, Ulaanbaatar, Guatemala City, Manila, Quito, Abha, Yerevan, and Fortaleza.

Cities with weak and very weak prosperity index (below 49 points)

Nairobi, Cape Town, Kathmandu, Accra, Mekelle, Kampala, Dar es Salaam, Lagos, Karachi, and Addis Ababa.

5.12 How to Manage a City with a Sustainable Approach?

In many countries, the citizen is part of the decisions that occur in their city, their participation is active and relevant, having facilities through dynamic means to send their criticisms and suggestions. Then when the citizen gets his opinion and his opinion is reflected in tangible improvements in his city, a virtuous circle is gener-

Table 5.1 Indicators standards on sustainable cities according UN SDG

Standard	Indicators
Sustainable development goals Goal 11: Make cities and human settlements inclusive, safe, resilient, and sustainable (UN, 2018)	UN SDG targets 11.1, 11.2, 11.3, 11.4, 11.5, 11.6, 11.7, 11a, 11b, 11c, 1.4, 6.3
ITU-T Y.4902/L.1602 Key performance indicators related to the sustainability impacts of information and communication technology in smart sustainable cities (ITU, 2008a)	Environmental sustainability, productivity, quality of life, equity and social inclusion, physical infrastructure
ITU-T Y.4903/L.1603 Key performance indicators for smart sustainable cities to assess the achievement of sustainable development goals (ITU, 2008b)	Economy, environment, society, and culture
ISO 37120: 2018 Sustainable development of communities—indicators for city services and quality of life (ISO, 2018)	Economy, education, energy, environment and climate change, finance, governance, health, housing, population and social conditions, recreation, safety, solid waste, sport and culture, telecommunication, transportation, urban/local agriculture and food security, urban planning, wastewater, water

ated as it will generate that citizen and his environment have confidence in their authorities and increase their participation, such as what happens in Dublin, Ireland, where there is an area called "Smart Dublin office" which is a unit shared between the four Dublin local authorities to coordinate and promote its smart city mission (Cardullo & Kitchin, 2018). In addition, it can be expected that even citizens have a co-creation function of sustainable cities (Gutiérrez, Amaxilatis, Mylonas, & Muñoz, 2018). For example, through initiatives such as crowdsourcing, participation can be obtained to propose improvements in the city. When crowdsourcing leads to an economic retribution for participation, it is called crowdfunding (Hong & Ryu, 2019; Özdemir, Faris, & Srivastava, 2015; Stiver, Barroca, Minocha, Richards, & Roberts, 2015).

Some indicators standards have been developed to monitor sustainable cities efforts. Table 5.1 describes some indicators.

5.13 Does the Implementation of Sustainable Cities Have Pitfalls?

In literature, some articles show potential pitfalls in the implementation of sustainable cities. For instance, Saiu (2017) states that there can be three main pitfalls: the idea of the city as a business, oversimplification of urban complexity, and the quest for the ideal community.

5.13.1 Pitfall by the Idea of the City as a Business

Project Lavasa. Critics of the project say the plan violates a host of statutes and laws (corruption) (Thekaekara, 2010).

Project Masdar. The Masdarian understanding of sustainability interprets urban development purely in economic terms (Cugurullo, 2016).

Project Tianjin. The involvement of private capital can help relieve local government's financial burden, but also introduces advanced technology and management experience to the construction (Aylett, 2014).

Project Caofeidian. The "eco-city" was made possible through huge bank loans. Once it was half-built, these loans were halted and many projects suspended due to the rising cost of raw materials and a lack of government support (Sabrie, 2014).

Project Dongtan. The scale of Dongtan was perhaps too small to attract economic development and activities and the cost of construction to achieve the target was very high (Wu, 2012).

5.13.2 Pitfall by the Oversimplification of Urban Complexity

Project Bo01. A functional planning ideology in the efforts made to plan for a general "user," while socio-economical stratification and differentiation of the population were not (Kärrholm, 2011).

Project Hammarby. The City of Stockholm did not involve residents in the planning process and then tried to influence their attitudes (Mazhouni, 2015).

Project Sociopolis. In view of the inefficiency of the strategy of letting time pass by, in this case the most reasonable alternative would be a change of use, recovering its original one or adapting it to implement a zone of urban agricultural gardens, as it partially was in the initial project (Díaz, 2016).

Project Valdespartera. Insufficient transport options from eco-city to the center of Zaragoza. This encourages residents to use private cars instead of social transport (Fédorova, 2016).

Project Dongtan. The growth stage of the eco-city is vulnerable to complex problems, including radical power changes (Zhuang, 2015).

5.13.3 Pitfall by and Quest for the Ideal Community

Project Bo01. Bo01 was initially planned to be a heterogeneous and socially sustainable area, but at one point the city chose to consider the question of integration on the scale of the municipality, claiming that the city of Malmo needed more wealthy taxpayers (Kärrholm, 2011).

Project Sociopolis. No expense was spared in hiring the cream of the architectural star system. In 2013, with the development finished, barely 22% was built (2800 houses on its 35 hectares; 5 out of 18 planned towers) but most of them remained vacant (Díaz, 2016).

Project Dongtan. Displaced farmers were not likely to be able to afford housing at the eco-city site, even with 20% of dwelling units designated as affordable housing (ADB, 2010).

Project Tianjin. The majority of the housing is targeted at households with above average incomes (Wippel, Bromber, & Krawietz, 2014).

Project Lavasa. The risk existed of producing exclusionary urban spaces. This would largely exclude indigenous and rural populations who were displaced from their land to build India's planned hill city (Datta, 2012).

Closing Remarks

All efforts to build and maintain a sustainable city are valuable from a social, economic, environmental, and cultural point of view. Although there are different terminologies and different global initiatives to achieve sustainable cities, it is still a challenge to achieve maximum involvement by the different authorities at the national level and especially at the local level. We can think of a sustainable city when in this city there is a harmonious interface between vehicular transport and the circulation of bicycles; when citizens report all traffic lights that do not work or do not have an intelligent system for their operation; and when accessing medical appointments and deliveries of medications to patients is given partially and outside of the times the medications are needed. A day to day sustainable city needs to be maintained, and it is mission of companies along with governments and universities, as well as citizens, to lead this process of change.

References

Annez, P. C., & Buckley, R. (2009). Urbanization and growth: Setting the context. In M. Spence, P. C. Annez, & R. M. Buckley (Eds.), *Urbanization and growth. Commission on growth and development* (pp. 1–45). Washington, DC: The World Bank.

Asian Development Bank (ADB). (2010). Urban innovations and best practices, sustainable urban development in the people's Republic of China: Eco-City development—A new and sustainable way forward? Retrieved February 5, 2020, from http://www.adb.org/publications/eco-city-development-new-and-sustainable-way-forward

Aylett, A. (2014). *Progress and challenges in the urban governance of climate change: Results of a global survey*. MIT: Cambridge, MA.

Cardullo, P., & Kitchin, R. (2018). Being a 'citizen' in the smart city: Up and down the scaffold of smart citizen participation in Dublin, Ireland. *GeoJournal, 84*(1), 1–13.

Cugurullo, F. (2016). Urban eco-modernisation and the policy context of new eco-city projects: Where Masdar City fails and why. *Urban Studies, 53*(11), 2417–2433.

Datta, A. (2012). India's Ecocity? Environment, urbanisation, and mobility in the making of Lavasa. *Environment and Planning C: Politics and Space, 30*, 982–996.

David, D. (2017). Environment and urbanization. *The International Encyclopedia of Geography, 24*(1), 31–46. https://doi.org/10.1002/9781118786352.wbieg0623

Díaz, G. F. (2016). The vacant city: Is there anything we can do? The case of València. *International Journal of Sustainable Development and Planning, 11*, 930–938.
Fédorova, K. (2016). Sustainability, urbanization and civilizations: Focus on Spain. *Reflexión Política, 18*, 42–56.
Gutiérrez, V., Amaxilatis, D., Mylonas, G., & Muñoz, L. (2018). Empowering citizens towards the co-creation of sustainable cities. *IEEE Internet of Things Journal, 5*(2), 668–676.
Habitat III. (2017). New Urban Agent. Retrieved February 5, 2020, from http://habitat3.org/wp-content/uploads/NUA-English.pdf
Hong, S., & Ryu, J. (2019). *Crowdfunding public projects: Collaborative governance for achieving citizen co-funding of public goods. Government Information Quarterly, 36*(1), 145–153.
ISO. (2018). *ISO 37120:2018. Sustainable cities and communities – Indicators for city services and quality of life* (2nd ed.).
ITU. (2008a). Key performance indicators related to the sustainability impacts of information and communication technology in smart sustainable cities. Retrieved February 5, 2020, from https://www.itu.int/itu-t/recommendations/rec.aspx?rec=12662
ITU. (2008b). Key performance indicators for smart sustainable cities to assess the achievement of sustainable development goals. Retrieved February 5, 2020, from https://www.itu.int/rec/T-REC-L.1603/en
Joss, S., & Molella, A. P. (2013). The eco-city as urban technology: Perspectives on Caofeidian International Eco-City (China). *Journal of Urban Technology, 20*(1), 115–137.
Kärrholm, M. (2011). The scaling of sustainable urban form: A case of scale-related issues and sustainable planning in Malmö, Sweden. *European Planning Studies, 19*, 97–112.
Mazhouni, A. (2015). The 'Policy Mix' for sustainable urban transition: The city district of Hammarby Sjöstad in Stockholm. *Environmental Policy and Governance, 25*, 288–302.
Özdemir, V., Faris, J., & Srivastava, S. (2015). Crowdfunding 2.0: The next-generation philanthropy: A new approach for philanthropists and citizens to co-fund disruptive innovation in global health. *EMBO Reports, 16*(3), 267–271.
Sabrie, G. (2014). Caofeidian, the Chinese Eco-City that became a ghost town-in pictures. Retrieved February 5, 2020, from https://www.theguardian.com/cities/gallery/2014/jul/23/caofeidian-chinese-eco-city-ghost-town-in-pictures
Saiu, V. (2017). The three pitfalls of sustainable city: A conceptual framework for evaluating the theory-practice gap. *Sustainability, 9*(12), 2311.
Stiver, A., Barroca, L., Minocha, S., Richards, M., & Roberts, D. (2015). Civic crowdfunding research: Challenges, opportunities, and future agenda. *New Media & Society, 17*(2), 249–271.
Thekaekara, T. (2010). City without Soul. Retrieved February 5, 2020, from http://www.tehelka.com/2010/02/city-without-soul
UN. (2015). Sustainable development goals. Retrieved February 5, 2020, from https://sustainabledevelopment.un.org/?menu=1300
UN. (2018). Sustainable development goals. Retrieved February 5, 2020, from https://sustainabledevelopment.un.org/sdg11
UNECE. (2015). Key performance indicators for smart sustainable cities to assess the achievement of sustainable development goals. Retrieved February 5, 2020, from https://www.itu.int/rec/T-REC-Y.4903-201610-I/en
UN-Habitat. (2013). *State of the world's cities 2012/2013: Prosperity of cities*. London: Earthscan.
UN-Habitat. (2015). *The city prosperity initiative: 2015 global city report*. Nairobi.
United Nations. (2014). *World urbanization prospects: The 2014 revision, highlights (ST/ESA/SER.A/352)*. New York. https://doi.org/10.4054/DemRes.2005.12.9.
Viitanen, J., & Kingston, R. (2014). Smart cities and green growth: Outsourcing, democratic and environmental resilience to the global technology sector. *Environment and Planning A: Economy and Space, 46*(4), 803–819.
Webb, J., Hawkey, D., & Tingey, M. (2016). Governing cities for sustainable energy: The UK case. *Cities, 54*, 28–35. https://doi.org/10.1016/j.cities.2015.10.014

WHO. (2016). Shanghai declaration on promoting health in the 2030 Agenda for Sustainable Development. Retrieved February 5, 2020, from https://www.who.int/healthpromotion/conferences/9gchp/shanghai-declaration.pdf?ua=1

Wippel, S., Bromber, K., & Krawietz, B. (Eds.). (2014). *Under construction: Logics of urbanism in the Gulf Region*. Farnham: Ashgate Publishing.

World Bank. (2017). Systems of cities. Retrieved February 5, 2020, from http://documents.worldbank.org/curated/en/877591468163481401/pdf/518600WP0Urban10Box342051B01PUBLIC1.pdf

World Bank. (2018). Inclusive cities. Retrieved February 5, 2020, from http://www.worldbank.org/en/topic/inclusive-cities#2

Wu, F. (2012). China's Eco-cities. *Geoforum, 43*, 169–171.

WUF9. (2016). Kuala Lumpur Declaration on cities 2030. Retrieved February 5, 2020, from http://wuf9.org/kuala-lumpur-declaration

Zhuang, Y. (2015). Confucian ecological vision and the Chinese eco-city. *Cities, 45*, 142–147.

Part II
Social Sustainability in Cities

Chapter 6
Informality and City, Denied Concomitants

Marté Sánchez-Villagómez

Abstract Labor and organizational tasks that today are typified as informal were not always so; in fact, work and organization are intrinsic to the development of human societies. Talking about informality does not have to have a pejorative character since history shows that there is an evolution of productive activities and that at the same time, depending on the context and resources, they are the only possible activities to be carried out for the survival of individuals.

Keywords Informality · City · Informal · Sustainable cities · Development

6.1 The City, a Heterogeneous Scenario

According to Plattner (1991), based on the theory of the central place of Christaller, with the development of the mercantile activities new population concentrations emerged, and other pre-existing ones gained importance depending on their location, but above all, urban centers (new and old) were articulated to a commercial network, whose administrative center was installed in cities strategically located for maritime or land commercial activity, if applicable.

The new urban centers, in constant and ascending commercial dynamics, managed to establish a clear difference between the activities (productive, commercial, and service) that developed in their own entrails from those that had their center in rural areas, but articulated to the network urban commercial. The process of industrialization strengthened the point of break between the rural and the urban, finally and with the passing of time the differences between both spaces were geographically configured; several decades ago, authors like Hobsbwan (1973), Lefebvre (1980), Wolf (1993), and Hobsbwan and Ranger (2012) pointed out the process that allowed to consolidate the differences between the rural and the urban, the supremacy of the new city as a multiple scenario and space par excellence for the citizen's exercise.

M. Sánchez-Villagómez (✉)
Universidad de San Martín de Porres, Lima, Peru
e-mail: msanchezv1@usmp.pe

A. Alvarez-Risco et al. (eds.), *Building Sustainable Cities*,
https://doi.org/10.1007/978-3-030-45533-0_6

The city, according to Lefebvre (1969) and Sennett (2013), in good account is the expression of each economic order; in this sense, roughly, three types of city can be historically indicated:

- Ancient, essentially political and linked to the possession of slaves.
- Medieval, mainly focused on commercial activity, banking and the growth of artisanal activity; nevertheless, its political character persists.
- Modern, arises with the consolidation of industrialization; its main characteristics were machine and mass production as the basis of economic activity, and the emergence of a new leadership (bourgeoisie).

On the other hand, it is essential to define the word city; therefore, it must be done on the basis of his permanence among humanity. Given the context, the city should be understood as the space where there is a social, economic, and political dynamic, whose function is focused on the preservation of wealth, knowledge, techniques, and artistic works. It is not just a new place to live. In this regard, recently UNESCO (2014) and the United Nations (2017) understand the relevance of cities, their livability, and the contribution that cultural activities bring to the gross domestic product (GDP). The first of the aforementioned supraorganisms recognizes the role of informal activity in favor of the aforementioned macroeconomic indicator, while the second international organization only recognizes the contribution of formal cultural activities.

The cities are not patrimony of the denominated contemporary urban areas, nor it is due to assume that the population conglomerates have their origin in the spaces before mentioned. The irrefutable is that the cities achieved a dynamic without equal within the urban context, the same that allowed them to differentiate themselves from other habitable scenarios. In that sense, not only did they bring together people, but also they were a place of concentration of power in all its manifestations. Therefore, urban development as a model of human habitat went hand in hand with the political process of affirming and consolidating sovereign States. It should be noted that this process became much more evident during the twentieth century.

It is also true that all the aforementioned caused administrative, political, and economic centralism that, most of the time, fell on the city designated as political and administrative capital of the country, thus sharpening the distance not only territorial but also geographical between urban and rural spaces. According to Remy (2009), the conception of the urban was attributed to higher levels of development than to the typified spaces of rural areas; in fact, the latter were perceived as a kind of obstacle to the progress of the former. The representation of the city as a symbol of progress and access to services was confronted with another equally powerful image, but in the opposite sense, the rural was presented as a synonym of backwardness and absolute lack of state services; this antagonism was installed in the majority of the population of the countries.

It was during the second half of the twentieth century, also known as the post-World War II period, that cities became the scenarios of great transformations; many situations converged for it; for example, the strengthening of the national states, and the productive change that turned clearly towards an industrial model. In this context, the companies proliferated and concentrated their activity in/for the cities, the population in general accentuated their migratory flow from rural to urban areas, and migrants settled in the cities, forcing an expansion of their geographical borders. The conflicts that this implied between migrant and local population have been registered by Elías and Scotson (2016). However, all groups of migrants indistinctly claimed "right to the city," needless to say, the migrants minimal access to coverage of services and basic infrastructure.

For Aruj (2008), the migratory process undertaken by the population since the end of the nineteenth century can be typified in four moments: transoceanic (from the late nineteenth century to mid twentieth century), internal (result of the economic crisis of 1930), cross-border (consequence of the internal sociopolitical conflicts of each country in the decades of 1960–1970), and international migration in a context of globalization (from the last two decades of the twentieth century to the present). The last migratory moment has mobilized, according to the International Organization for Migration—IOM (2018), an estimated 244 million people outside its territorial borders, where you can see a true global diaspora. The main destinations of international migration are European Union, with 76 million migrants; Asia, with 75 million; and North America who registers 54 million. It should be noted that these three destinations are home to 84% of the world diaspora within their territories.

The same IOM's report points out as a relevant fact that more than 70% of international migrants are of working age. On the other hand, it is known that population censuses record the demographic history of each country and/or region of the world. In that sense, they have been an important instrument to warn that the rural population began a continuous and growing migratory route to the cities. This confirms that the process of population mobilization on a global scale has followed the same course as internal migration, that is, mobilizing population from spaces with less development towards areas of greater advancement, from spaces categorized as rural towards others typified as urban.

The Department of Economic and Social Affairs (DESA—ONU 2014) of the United Nations noted that 56% of the world's population lives in urban spaces, its projection for the year 2050 estimates that the figure would rise to 66%. For the European Union, Eurostat (2015) reports that 28% of its population lives in rural areas, just over 40% reside in urban areas, while the remaining 32% live in intermediate cities; that in fact they are so urban by the level of services rendered as the second, but without achieving the political or economic relevance of the aforementioned urban areas. In the case of Latin America, CEPAL (1999) notes that during the 1970s, the subcontinent recorded that 57% of the total population lived in urban areas; 25 years later, the urban population grew to 73% of the total population of the subcontinent already mentioned.

The migratory processes, internal and international, have been and are studied from different perspectives. Critics of the process because of the negative effects it produces in a sector of the population have argued that the myth of development associated with the growth and advancement of cities has not been equitable for all those who live in it. Wacquant (2007) states that economic growth within the capitalist system did not benefit everyone, that there are population sectors within the city that were trapped in precarious, poorly paid jobs, while other sectors of the population achieved stable employment and personal success. Delgado (2010) analyzes the case of the city of Barcelona, argues that the Catalan city—from the Olympic Games of 1992 onwards—was redesigned as a business city, one capable of excluding and expelling the citizen or visitor considered insolvent; the author points out that behind the "spectacle city" is a project that thinks in terms of the market and not of citizen coexistence. Although it seems contradictory, for the aforementioned reason, is that cities such as Barcelona attract huge migration—international and international—in search of work; in reality, it is a workforce that moves around the interior of a country, a region, and/or the world in search of job opportunities that in their home city they would not have easily.

From another perspective, diverse investigations have indicated the benefits that migration grants not only to the migrant and his or her family. The positive effects of it are perceived and impact even in the country of origin. The remittances sent by the migrants to their families inject capital to start businesses; they are and generate a constant source of income in the countries of origin of the population in international migration. According to the World Bank, in 1990 migrants sent approximately 29,000 million US dollars to low and middle income countries. This figure doubled before the year 2000. For 2016, remittances reached 429,000 million dollars. Currently, in global terms, remittances triple the amount of official assistance allocated for development. Nowadays remittances are an economic engine for a large number of families in underdeveloped or developing countries.

Until now, the strong relationship between migration to cities and the search for job opportunities is evident, although it is not the only one. Undoubtedly, it is one of the main reasons for immigrants, what is not said is how much of the money sent to the countries of origin was obtained in labor activities without any recognition of rights or labor benefits, labor informality clearly. Self-employment is another source of direct income for migrants, very often associated with the ambulatory sale of products and/or services. For example, the formal tourist guide service coexists with the informal service offered through social networks, this is not a secret; the same happens with the souvenirs or bags for women that are offered in the subway accesses of the cities, just to mention some of them: Paris, Rome, Barcelona, or Manhattan. In summary, much of the money that is sent as remittances to the countries of origin comes from informal activities and activities that take place in the so-called global or cosmopolitan cities of the world.

6.2 Ruminations Around Informality

Labor and organizational activities that today are typified as informal were not always so; in fact, work and organization are intrinsic to the development of human societies. Authors such as Berdan (1991) and Plattner (1991) have pointed out that trade and the market predate the organizational forms with State presence; that is to say, that the aforementioned activities existed in pre-capitalist societies. The consolidation of the nation-state model brought with it a greater capacity to manage and regulate the activities within the administered territory. Likewise, it managed to concentrate power—unlike the systems that preceded it—which progressively facilitated the governments of the day to regulate productive and labor activities.

In general terms, and in accordance with international standards, it was the government administration that created the legal frameworks; the same ones that worked as a watershed when pointing out the permitted activities of those that were proscribed. It was thus that a set of economic activities, traditionally accepted, ended up being informalized and outside the normative frameworks established by the national authority. In the sense indicated in the previous paragraph, it is understood that certain ways of exercising productive or commercial work ended up outside of formality; therefore, the regulations established by the States delineated new frontiers between what was allowed and what was left aside. It should be remembered that human groups, in terms of standards of coexistence and work, have been regulated by pre-existing organization systems at the appearance of the sovereign State. This new form of social and political order, because of its scope, is more complex than its predecessors.

Taking into account what has been pointed out up to this point, and in opposition to what De Soto (1986) suggests, it is maintained that the informality category cannot be applicable to different eras of the development of humanity; its use—as a category of analysis—serves to account for a social fact based on the validity of the State-nation model of organization going forward. It was Hart's study (1970), conducted in Ghana, which gave an initial meaning to the term. The aforementioned author found that a part of the population of the interior of the country migrated to Accra (capital of Ghana) and failed to be inserted into employment neither by the state administration nor by the business organizations based in that city. The response of the migrants was rapid, they undertook a part-time business activity; in fact, they were small businesses that complemented their income, in other cases and to a lesser extent they were the main source of their stipends.

In 1973, Hart published "Informal Income Opportunities and Urban Employment in Ghana," it was the article that for the first time used the informal word to indicate a small-scale economic activity that was exercised outside the state regulation, installed in urban and urban-marginal contexts, and generally carried out by the migrant population. These characteristics associated informality with migration and socioeconomic marginality in the cities. Therefore, the informal activity was expressed as a response to the lack of job opportunities of a group of poor and marginal migrant population in the cities. Seen in this way, Hart was the proponent of the theoretical proposal of duality, the same that posed a coexistence of two parallel sectors within the economy: one, formal, and another, informal.

Hart (1973) saw a new situation in internal migration, population displacement from traditional to modern areas, from rural to urban areas, in good account of the countryside to the city. This new population installed in Accra, faced with the lack of employment, was conditioned to create their own, the self-employment of the migrant focused on construction, housing, and transportation; labor activities traditionally neglected by the State. These were activities that the dynamics of growth—accelerated and chaotic—of the cities relegated until they became irrelevant for the State, therefore, neglected by the governments of the day; situation to which Matos (2004)—for the decade of 1980—called "crisis of the State" and whose consequence was not other than a popular overflow that found its own cause. As you can see, it is a positive look at the informal entrepreneurship of migrants in the city. It should be noted that this chapter does not aim to account for the set of theories about informality; nevertheless, if that is the interest, a rigorous balance can be consulted in Chen (2012). Seen this way, informality in its beginnings was associated with poverty and marginality of certain population sectors; it was expressed as a survival strategy for many migrants in the cities. The fact that Hart's research was conducted in a country that was part of the so-called poor Africa left its mark on future studies within the theme; therefore, it was not difficult to extend the idea that informality is associated with poverty; their conclusions were extrapolated to a large part of the developing countries—in that political context they were called third-worldists—for sharing the same entrepreneurial potential in the face of lack of employment and common economic difficulty.

Hart (1970, 1973) identified that entrepreneurs and informal workers contributed to the Ghana economy in different ways. For example, the purchase of heavy vehicles, as a form of self-sustaining work by small non-formal companies, allowed greater communication between the capital city and other territories within the country, which played in favor of greater national integration despite the poor state of roads. Likewise, it facilitated the distribution of processed foods in formal companies, the distribution of household appliances made by small entrepreneurs allowed the incorporation of technology into households. He also points out that the self-construction of houses oriented for lodging or rent allowed to stimulate the expansion of companies in other areas of the interior of the country. All these contributions indicated here become relevant only if we take into consideration that the State implemented a modernizing project, one that failed to consolidate or that failed as a promise of a better quality of life, social advancement, and progress for all; nevertheless, the internal mobilization of the population in the face of adversity in the city opted for an entrepreneurial response that allowed generating its own source of employment.

From America, the Regional Employment Program for Latin America and the Caribbean (PREALC)—ILO considered that the informal economy was linked to a set of activities that do not require specialized worker qualifications, undertakings that are carried out with scarce economic capital, unprofitable and low productivity; in good account, given the characteristics of the activities, its realization depends on few qualifications and general work skills. Today it is known that the type of enterprises lacking economic capital have their main support in social capital. Waisgrais and Sarabia (2008) warn that the poverty scheme linked to informality does not include a point that is relevant; this characterization of the informal includes self-employed labor, family work, domestic service, but excludes professional and technical personnel. That said, the concept of work informality is extended to another group of workers with specialized qualifications, far removed from any link to poverty and low productivity of their activities, which forces us to think about the concept beyond the forms of survival of generalist auto workers/employees informally due to their low job qualifications.

Although consensus has not been achieved, research on informality in general has enriched the debate; one of the contributions has been the information and the casuistry that have analyzed to give account of the alluded phenomenon. Many have been the efforts, all valuable, to formulate a definition that reflects the economic dynamics included within the category of analysis. For more than four decades, various scholars have elaborated approaches as diverse as the contents of the term itself. Precisely, that has been—it has not stopped being—the difficulty to arrive at a consensus of what should be understood by informality. Hart's neologism proliferated worldwide, since then research on the subject is constant and evidence the concern for a more complete definition of a complex and heterogeneous activity; authors such as Tokman (1978, 2001), Bromley (1978), De Soto (1986, 2004), Castells and Portes (1989), Portes (1995), Gërxhani (2004), Tanaka (2010), Schneider (2012), and Chen (2012). All of them, among many others, have helped to clarify the contents that the term informality contains.

One of the characteristics of informal activity is to have the city as its main scenario, where it germinates and intensively develops this activity, work, and business. Clichevsky's research (2001) elaborates a geography of the city, emphasizing that informal dwellings, of self-construction, usually serve as homes and work spaces for families that derive their income from informal activities. However, the growth and expansion of the city, coupled with the demand for housing from social sectors with greater economic resources, manage to exert pressure on precarious areas located within the redeveloped city. For the author mentioned above, the situation alluded to above occurs as a result of an unequal distribution of income within the cities. As we can see the urban problem is broad, here we are interested in the informality. The truth is that the current cities are large urban spaces, in them the logic of the market is impregnated in the daily life of its inhabitants; the mercantile centers (market as a place) are places where a set of commercial interactions, formal and informal, of goods and services are carried out.

6.3 Informality

Conceptually, the term informality is one of the most difficult to frame, given its versatility, use in different circumstances, and its value. The classic concept of informality circumscribed the problem to the low productivity of the units responsible for economic activity; today is not enough to account for the content of the word alluded to in the business world, much less if one thinks of labor informality.

The term informality, roughly, can be used to refer a work activity, a productive organization and/or service, to certain institutional practices; it can also allude to a structural condition of the prevailing economic model. The truth is that any business situation or work activity that is outside the margins of state regulation, in general, can be typified as informal. In conclusion, the aforementioned term encompasses a very broad content, therefore difficult to delimit; his employment attempts to account for a heterogeneous economic reality, one that ranges from the irregular working condition of the worker and/or the business organization to the structural impact of the current economic system.

Informality, as was erroneously thought, does not discriminate between qualified (professional or technical) and unskilled labor, nor does it do so between small and large companies; much less must be assumed that in the developed countries there is no room for it. On the contrary, there are manifestations and forms according to each of the circumstances and/or cases where informality is evidenced. Some authors have pointed out that:

> When the regulation of the State is very effective and widespread, as in many industrialized countries, the situation is different. In these cases, informal activities are more hidden and, as we have seen, they are generally inserted in narrower social networks. Therefore, however well organized the official registration system is, it is more likely that it will not be able to detect a significant proportion of informal activity. (Portes and Haller, 2004: 24)

Macdonald (1994) showed that in the field of developed economies informality indirectly fulfilled, continues to fulfill, a role of economic complementarity for certain population sectors. The aforementioned author takes as an example the case of the city of Cincinnati—State of Ohio. In this city, as a survival strategy, the most vulnerable and marginal population appealed to the social security benefits for unemployment of the social security that they associated in addition to informal and temporary jobs that allowed them to inject additional money to their precarious family economy. It is quite clear that informality serves as a support to the economy of marginal populations, disaffiliated from the economic project and its benefits. In developed countries, informal activity does not disappear, it only adds new forms to existing ones, in certain circumstances its identification and analysis becomes complex. These are new situations that make it more difficult to account for and differentiate formality from informality, be it labor and/or business; their limits are narrower, in other cases they intersect and there are many circumstances in which they collaborate among themselves. The binomial formality–informality is real, but has been treated as a duality of irreconcilable opposites that resulted in the following idea; formalization processes proposed by the State at some point, sooner or later, will eradicate informality, with which each and every one of its manifestations will be corrected.

Neither regularization nor deregulation policies have been effective recipes for this task, therefore, it seems that the aforementioned binomial is not constituted by radically opposed parts, but is based on the complementarity of these, where the practice of one generates adjustments or modifications in the other. If one takes into account, and as an antecedent, that informality has been demonized, criminalized, and penalized by different governments and business sectors from different parts of the world; it is logical that it is difficult to think, even more to assume, that the relationship between formality and informality is complementary. What for decades has been thought of as disjunction cannot, right away, be accepted as a conjunction, much less from the exercise of state power. It is a dilemma!

The actions that they take and execute, central, regional, and local government, will depend on the situation in which the informal activity is located; therefore, its role can be presented as tolerant or repressive in the face of informality. In the first situation, the risk manifests itself before the possible weakening or loss of the legal order. This would occur if the actions taken do not correspond to the magnitude of the infraction committed. In the second scenario, the effect of economic "cushioning" that part of informal activities allow certain population sectors could be eliminated. "An excess of tolerance can put into question the credibility of the legal system and the willingness of formal companies and taxpayers to continue fulfilling their obligations. On the contrary, an attitude that is too repressive would help to eliminate the" cushion "that constitutes informal activities or, what is worse, it can impel them to hide even more, depriving the authorities of all kinds of control and information about them. The systematic concealment of information to government agents has proven to be, by far, the most effective instrument that civil society has to resist authoritarian regimes"(Portes and Haller 2004: 43).

Regarding the use of the sector and informal employment categories, there is an international consensus that is reflected in the document prepared by the ILO (2013), in this sense we have to:

- Informal sector; for Ramos (2015), it is a main source of employment generation; the ILO (2013) argues that it is also for the production of goods and services, which has a positive effect on the consumption of all social strata. Under the proposal, the informal sector is made up of production units (companies) dedicated to producing goods and/or providing services, it is generally made up of small-scale units with an embryonic organization; despite this, they manage to create jobs and generate income to the participants in said activity. However, employment relationships do not involve formal guarantees of employment contracts; on the contrary, these relationships are based on personal, family, and/or social ties. Castells and Portes (1989) pointed out that the difference between formal and informal activity is associated with the way in which products are processed or exchanged and not with the characteristics of the finished product. From this perspective, informality as an activity lies in the way of organizing for the production of goods or services. Something different is the typification of companies from being or not registered as a business organization.
- Labor informality; Longhi (1998), understands that informal work is defined as non-compliance, absolute or partial, of a set of rules, emphasizing the normative conditions in which work activity occurs. For the ILO (2013), the informality of employment has two components: one, employment in the informal sector; two, informal employment. The first case refers to the business sector within the informal sector that provides employment without social or labor benefits, while the second refers to the type of employment that takes place in both the formal and informal sectors.

It should be noted that the definitions proposed by the ILO (2013) allow for the definition of two categories, sector and informal work, which the supranational body referred to above understands as central and decisive when it comes to talking and measuring informality.

6.4 Informality and Sustainable Cities

Gelman and Du (2018) recently stated that the majority of the population worldwide do not know the number of people who work informally, let alone the contribution that this activity makes to national economies. Mainly, urban areas are spaces where a large population is concentrated around the economic, political, social, and cultural dynamics that cities develop within them.

A set of postulates, coming from economists and urban planners among others, is concentrated around the following thesis. Labor and business informality will disappear as development increases. The truth is that the inverse correlation formulated as more development, equal to less informality (+ D = −I), has not given the expected results. In the same direction, Perry et al. (2008) argued in a publication for the World Bank that informality is a transitory stage in the development process; or what is similar, informality represents underdevelopment. Nowadays, the informal sector is large, heterogeneous, persistent, and the work associated with it does not exclude any country. In each and every one of the national spaces, informal activities take place on a daily basis.

Informal activities are a source of work for many people, according to ILO (2018) it is estimated that at the global level informal employment represents 61.2% of total employment. This means that 2000 million workers have informal employment. In some continents, informality is a primary source of employment, for example, in the case of Africa, informality represents 85.8% of the total employment in the continent. Gelman and Du (2018) report figures on how informal activity affects countries, regions, and sub-regions. For example, in Mexico, workers in informal employment generate 30% of GDP (gross domestic product), in India the figure rises to 46%, in West Africa it reaches 50% of GDP, while in Latin America the figure represents 25% of GDP. Given this evidence, it is impossible to continue denying economic value or disgusting informal activity.

As pointed out by Matos (2004), faced with the crisis of the State, there was a popular overflow, that is, a large migrant population—through self-employment, informal employment, and/or employment in the informal sector—found or created their own employment. The work, although precarious, prevented them from falling into the criminal. From another perspective, to this process, De Soto (1986) called it the other path, a metaphor that highlights the entrepreneurial attitude of many migrants in the city. These are people who were attracted by the modernizing project of cities, but also those who were expelled by the persistent economic crisis in rural areas, so they migrated to the cities in search of better economic and employment opportunities and of social life for them and their families.

For the Peruvian case—where De Soto conducted the research cited in the previous paragraph—said internal migration was a consequence of the crisis of the landowner economic system and the rise of the productive matrix based on industry. On the other hand, the political violence unleashed at the beginning of 1980—especially—in the rural areas of the country, made the migration to the urban centers take on a particularity: its compulsive character. The result was a greater population concentration in the cities, which generated an oversupply of unskilled labor that formal employment could not absorb. It is evident that the population growth exerts more pressure on the offer of formal employment; the disadvantage arises when said labor force does not manage to be integrated labor neither by the State nor by the companies of the formal system. What happens to people who cannot access a job that recognizes labor rights and assigns social benefits?

A large part of them ends up being employed in the informal sector, another mechanism has been the modality of independent worker; that is to say, workers who work from their homes, in urban transport or in ambulatory commerce among the most used modalities. It is not about making an apology for informality, only to recognize that its economic dynamics are important, both in figures reported and in labor employed. The informal activity allows income to families, generates employment, and has a significant participation when measuring the GDP of the countries. On the other hand, Valenzuela-Aguilera and Monroy-Ortiz (2014) have pointed out that the work generated by the informal activity helps the youths employed by it reject illicit roads or give in to organized crime. Studies have shown that, in certain circumstances, when informal activity is besieged and deteriorated as a result of violent and excessive government regulation, the workforce disaffiliated from work ends up strengthening illegal economic circuits. According to Véliz and Díaz (2014), informality is contributing to the economic growth of cities; also, as already pointed out, it prevents certain more vulnerable labor, social, and age sectors from succumbing to the criminal. It is not an ethical device, nor a moral reserve; informal activity, such as employment or self-employment, becomes economically less favored for the least advanced before falling into absolute precariousness and being exposed to criminal and criminal practices.

The prevailing discourses around informality have maintained that the activities carried out from the informal sector are of low productivity given the inadequate use of resources when producing, that informal business organizations are uncompetitive in the market. Other discourses, the most subjective ones, point to it as an illegal activity and the people who work in this situation are perceived as delinquents. In summary, under this perspective, informal work and business activity is presented as a burden for national economies. These speeches are collected and reflected in the governmental plans and practices of the cities, generating an adverse regulatory environment that makes it very difficult for many people to earn their daily living honestly. On the contrary, they are stigmatized and in many cases, they are criminalized for being informal. The most urgent challenge for urban workers involved in labor informality is that they are not recognized as legitimate actors of the cities; therefore, they do not have citizen participation, much less representation capacity in the city; the organizations of workers or self-employed of the informal activity have been invisible when it comes to a positive look for political and economic purposes, the opposite happens when the perspective is negative because they become the example of all the bad, they are vile beings that they attempt against the future of the nation.

On the other hand, the United Nations (2017) has indicated that cities must face challenges and point out opportunities for sustainable, sustained, and inclusive economic growth, using local economies and accepting the contribution of the informal economy in this economic process. In the same way, it suggests a progressive transition of workers and economic units to formality; it is an approach that seeks a balance between incentives and compliance measures aimed at guaranteeing the preservation and improvement of livelihoods, without neglecting the legislation, policies, practices, and specific priorities of each national environment. The new urban agenda, sponsored by the United Nations, in the sense referred to here, is a commitment to formalization; one that safeguards the integrity of people in conditions of informality and minimally guarantees the continuation of the productive units that are support for the life of many people in a situation of informal work.

The World Resources Institute (WRI Mexico), has pointed out examples of inclusive policies of collaboration between municipal administrations and organizations of informal workers in the world. In this way, Gelman and Du (2018) present some case studies; the first shows as an organization of informal workers evidence before the relevant government body the need for public transport to make viable and sustainable the connection between the periphery of the city (place where they live by relocation) with an important market where they usually make their Business transactions in the city of Bangkok—Thailand. The second case, Bhubaneshwar, India, presents an agreement between the municipal authority and street vendors organized within an inclusive city model. Agreement that allowed the inhabitants to continue benefiting from the products sold by the street vendors, safe areas for sale on the public highway for the ambulantes designated by the municipal authority, and free of confiscation of goods and without harassment. In exchange, the municipal authority was able to point out, through the same agreement, the areas that were restricted for ambulatory trade. Both cases show a model that has allowed the parties involved (municipal authorities—organized informal workers) to obtain benefits. As can be seen, many studies elaborate theoretical approaches that explain informality and its persistence, few that show field information of concrete realities; the vast majority of investigations elaborate theories to explain the persistence, characteristics, and growth of informal activity in the world. In recent years, the organization Women in Informal Employment: Globalizing and Organizing (WIEGO) has carried out a global investigation about economic informality, having as target population three groups of informal workers (home workers, street vendors, and recyclers). The model and the research methodology has been applied in 10 cities of the world; under a conception of inclusive cities, the Monitoring Studies of the Informal Economy (EMEI in Spanish) were developed. In this sense, the works consulted on economic informality within the aforementioned framework are Acosta and Ortiz (2013) for Colombia—Bogotá, Ogando et al. (2013) for Brazil—Belo Horizonte, Castellanos (2014) for Peru—Lima, Mahadevia et al. (2014) for India—Ahmedabad, and Naqeeb et al. (2014) for Pakistan—Lahore.

The EMEI investigations, indicated in the previous paragraph, are longitudinal studies of the urban informal economy, all—beyond their specificities—point to the same purpose. On the basis of a set of false myths, a negative and isolated image of the informal economy has been constructed; it has been said that it is disconnected from the formal economy, which is a lag of surviving premodern economies. It has also been argued that informal workers do not intend to formalize and therefore hide their activities from government regulations. The EMEI studies provide and show a set of evidences that, in each case, begin to build a different perspective on work and informal activity. They also point out the importance that the informal economy has for many as it is a source of work in the cities.

Finally, the experiences outlined here lead to the following climax. It requires the collaboration and participation of government agencies: central, regional, and municipal, supranational organizations, and associations of informal workers to find solutions that provide benefits to all parties involved. This implies the visibility, recognition, and incorporation of informal workers organized as legitimate social actors within the cities; it also presupposes consensus, training, processes, and deadlines for the formalization of activities carried out informally, but above all it means including them in decision-making, listening to them, and inserting their opinions in the urban policies developed by the municipal and national governments, respectively.

Closing Remarks

It is time to review the policies and practices of citizen security that have focused on outpatient trade, among many other forms of informal work, as a criminal activity to be fought without truce; it is about stopping criminalizing unstructured labor activities. Harassment, eviction, and even relocations—which do not guarantee the viability and connection of the place with the rest of the city—have been unsuccessful answers for decades. It is time to make a different approach to people who live in/informality. It is about supporting the labor initiative, providing access to goods and spaces to develop their activities, improve public services. Only then will formalization as a process of change succeed and without social victims to regret. Smart cities are those that fulfill global commitments and those assumed with their citizens, those that promote inclusion, equity, and social justice within their space. These are the cities called to guarantee the fulfillment of the Sustainable Development Goals (SDG), especially Objective 11: Sustainable Cities and Communities.

References

Acosta, T., and Ortiz, R. (2013). Informal economy monitoring study: Recyclers and recyclers from Bogotá, Colombia. [*Estudio de monitoreo de la economía informal: Recicladoras y recicladores de Bogotá, Colombia*]. Manchester, RU: WIEGO.

Aruj, R. (2008). Causes, consequences, effects and impact of migration in Latin America. [Causas, consecuencias, efectos e impacto de las migraciones en Latinoamérica]. *Papeles de población*, 14(55), 95–116. Retrieved February 5, 2020, from http://www.scielo.org.mx/scielo.php?script=sci_arttext&pid=S1405-74252008000100005&lng=es&tlng=es

Berdan, F. (1991). Commerce and markets in precapitalist states. [*Comercio y mercados en los Estados precapitalistas*]. En: *Antropológia económica* (pp. 235–284). Plattner (comp). México DF, México: Alianza editorial. México.

Bromley, R. (1978). Organization, regulation and exploitation in the so-called 'urban informal sector': The street traders of Cali, Colombia. *World Development, 6*(9/10), 1161–1171.

Castellanos, T. (2014). *Estudio de monitoreo de la economía informal: Vendedoras y vendedores ambulantes de Lima Metropolitana, Perú*. Manchester, RU: WIEGO.

Castells, M., and Portes, A. (1989). World underneath: The origins, dynamics, and effects of the informal economy. In Portes, A., Castells, M. y Benton, L. (Eds.), *The informal economy: Studies in advanced and less advanced developed countries*. Baltimore: Johns Hopkins University Press.

CEPAL. (1999). Latin America: Urban and rural population projections, 1970–2025. [*América Latina: proyecciones de población urbana y rural, 1970–2025*]. En: *Boletín Demográfico*. Año 32, N° 63—Enero. Santiago, Chile. Publicaciones de las Naciones Unidas.

Chen, M. (2012). *The informal economy: Definitions, theories and policies*. Cambridge, MA: Women in Informal Employment: Globalizing and Organizing (WIEGO).

Clichevsky, N. (2001). Habitat informal en América Latina: entre la permisividad. El desalojo y la regularización. *Territorios* Vol. 6 (enero–junio), 15–30.

De Soto, H. (1986). The other path. The informal revolution. [*El Otro sendero. La revolución informal*]. Lima, Perú: Instituto Libertad y Democracia.

De Soto, H. (2004). The mystery of capital. [*El misterio del capital*]. Bogotá, Colombia: Planeta.

Delgado, M. (2010). The liar city. Fraud and misery of the 'Barcelona model'. [*La ciudad mentirosa. Fraude y miseria del `modelo Barcelona'*]. Madrid, España: Los libros de la catarata.

DESA—ONU. (2014). More than half of the population lives in urban areas and will continue to grow. [*Más de la mitad de la población vive en áreas urbanas y seguirá creciendo*]. Nueva York, EEUU. Retrieved February 5, 2020, from http://www.un.org/es/development/desa/news/population/world-urbanization-prospects-2014.html

Elías, N., and Scotson, J. (2016). Established and marginalized. A sociological research on community problems. [*Establecidos y marginados. Una investigación sociológica sobre problemas comunitarios*]. México DF, México: Fondo de Cultura Económica.

Eurostat. (2015). *People in the EU: Who are we and how do we live?* Luxembourg: Publications Office of the European Union. https://doi.org/10.2785/406462.

Gelman, V., and Du, J. (2018). Informal work helps cities work: Stories from Thailand, India and Colombia. [*El trabajo informal ayuda a que las ciudades funcionen: Historias de Tailandia, India y Colombia]* [Blog]. Blog WRI México. Retrieved February 5, 2020, from https://wrimexico.org/bloga/el-trabajo-informal-ayuda-que-las-ciudades-funcionen-historias-de-tailandia-india-y-colombia-4

Gërxhani, K. (2004). The informal sector in developed and less developed countries: A literature survey. *Public Choice, 120*(3/4), 267–300.

Hart, K. (1970). Small-scale entrepreneurs in Ghana and development planning. *Journal of Development Studies, 6*(4), 104–120. https://doi.org/10.1080/00220387008421338

Hart, K. (1973). Informal income opportunities and urban employment in Ghana. *The Journal of Modern African Studies* Vol 11, pp. 61-89. https://doi.org/10.1017/S0022278X00008089.

Hobsbwan, E. (1973). Around the origins of the industrial revolution. [*Entorno a los origenes de la revolución industrial*]. Buenos Aires, Argentina: Siglo veintiuno editores.

Hobsbwan, E., & Ranger, T. (2012). *La invención de la tradición*. Barcelona, España: Crítica.

ILO. (2013). Informality measurement: Statistical manual on the informal sector and informal employment. [La medición de la Informalidad: *Manual estadístico sobre el sector informal y el empleo informal*]. Turín, Italia: Centro Internacional de Formación de la OIT.

ILO. (2018). Women and men in the informal economy: A statistical overview. [*Mujeres y hombres en la economía informal: un panorama estadístico*]. Ginebra, Suiza: OIT.

IOM. (2018). World Migration Report. [*Informe sobre las migraciones en el mundo*]. Ginebra, Suiza: Organización Internacional para las Migraciones (OIM).

Lefebvre, H. (1969). The right to the city. [*El derecho a la ciudad*]. Barcelona, España: Ediciones península.

Lefebvre, H. (1980). The urban revolution. [*La revolución urbana*]. Madrid, España: Alianza editorial.

Longhi, A. (1998). Work and the informal economy: Discussion of theoretical approaches. [El trabajo y la economía informal: Discusión de enfoque teóricos]. *Serie documento de trabajo* (34). Retrieved February 5, 2020, from http://cienciassociales.edu.uy/wp-content/uploads/2013/archivos/DocTrab34.pdf

MacDonald, S. (1994). *Professional academic writing in the humanities and social sciences*. Carbondale, EEUU: Southern Illinois University press.

Mahadevia, D., Mishra, A., & Vyas, S. (2014). *Informal economy monitoring study: Home-based workers in Ahmedabad, India*. Manchester, UK: WIEGO.

Matos, J. (2004). Popular overflow and state crisis. Twenty years later. [*Desborde Popular y crísis del Estado. Veinte años después*]. Lima, Perú: Fondo Editorial del Congreso de la República del Perú.

Naqeeb, B., Saigol, R., & Azhar, U.-L. (2014). *Informal economy monitoring study: Home-based Workers in Lahore, Pakistan*. Manchester, UK: WIEGO.

Ogando, A., Brito, M., Oliveira, A., and Dias, S. (2013). Informal economy monitoring study: Waste pickers in Belo Horizonte, Brazil. [*Estudo de Monitoramento de Economia Informal: Catadoras e Catadores em Belo Horizonte, Brasil*]. Manchester, RU: WIEGO.

ILO. (2013). Informality measurement: Statistical manual on the informal sector and informal employment. [La medición de la Informalidad: *Manual estadístico sobre el sector informal y el empleo informal]*. Turín, Italia: Centro Internacional de Formación de la OIT.

ILO. (2018). Women and men in the informal economy: A statistical overview. [*Mujeres y hombres en la economía informal: un panorama estadístico*]. Ginebra, Suiza: OIT.

Perry, G., Maloney, W., Arias, O., Fajnzylber, P., Mason, A., & Saavedra-Chanduvi, J. (2008). *Informalidad: escape y exclusion*. Bogota, Colombia: Banco Mundial—Mayoy ediciones S.A.

Plattner, S. (1991). Markets and shopping centers. [*Mercados y centros mercantiles*]. En: *Antropológia económica* (pp. 235–284). México DF, México: Alianza editorial.

Portes, A. (1995). Around informality: Essays on theory and measurement of the unregulated economy. [*En torno a la informalidad: ensayos sobre teoría y medición de la economía no regulada*]. México DF, México: Flacso.

Portes, A., and Haller, W. (2004). Informal economy. [*La economía informal*]. Santiago, Chile: Naciones Unidas.

Ramos, A. (2015). Informal sector, informal economy and informality. [Sector informal, economía informal e informalidad]. *Revista Iberoamericana para la investigación y el desarrollo educativo, 6*(11), 1–12.

Remy, M. (2009). Cities, cities and rural people. [Las urbes, las ciudades y la población rural]. Revista Argumentos. N° 2, Mayo 2009. Retrieved February 5, 2020, from http://revistaargumentos.iep.org.pe/articulos/las-urbes-las-ciudades-y-la-poblacion-rural/

Schneider, F. (2012). *The shadow economy and work in the shadow: What do we (not) know?* En: *IZA Discussion Paper N° 642*. Deutschland: Univeristy of Bonn.

Sennett, R. (2013). Crafts, technology and new ways of working. [*Artesanía, tecnología y nuevas formas de trabajo]*. Barcelona, España: Centre de Cultura Contemporània de Barcelona (CCCB)—Katz.

Tanaka, V. (2010). The 'informal sector' and the political economy of development. *Public Choice, 145*, 295–317.

Tokman, V. (1978). An exploration into the nature of the informal-formal sector relationship. *World Development* 6(9/10), 1065-1075. https://doi.org/10.1016/0305-750X(78)90063-3.

Tokman, V. (2001). Relations between the formal and informal sectors. An exploration of its nature. [Las relaciones entre los sectores formal e informal. Una exploración sobre su naturaleza]. *Revista Economía* 24 (48), pp. 17–74. Lima, Perú: PUCP.

UN. (2017). New urban agenda. [*Nueva agenda urbana*]. Quito, Ecuador: Editada por la secretaría de Habita III.

UNESCO. (2014). UNESCO indicators of culture for development: methodological manual. [*Indicadores UNESCO de cultura para el desarrollo: manual metodológico*]. París, Francia: UNESCO—AECID.

Valenzuela-Aguilera, A., & Monroy-Ortiz, R. (2014). Formal/informal/illegal: The three circuits of space economy in Latin America. [Formal/Informal/Ilegal: Los Tres Circuitos de la Economía Espacial en América Latina]. *Journal of Latin American Geography, 13*(1), 117–135. https://doi.org/10.1353/lag.2014.0009

Véliz, J., & Díaz, S. (2014). The phenomenon of informality and its contribution to economic growth: The case of the city of Guayaquil. [El fenómeno de la informalidad y su contribución al crecimiento económico: el caso de la ciudad de Guayaquil]. *Journal of Economics, Finance & Administrative Science, 19*(37), 90–97. https://doi.org/10.1016/j.sjefas.2014.09.001

Wacquant, L. (2007). The damned of the city. Ghetto, peripheries and state. [Los condenados de la ciudad. Gueto, periferias y Estado]. Buenos Aires, Argentina: Siglo veintiuno editores.

Waisgrais, S., & Sarabia, M. (2008). Heterogeneidad social y productiva: caracterización del trabajo informal en el Gran Buenos Aires. *Aportes a una nueva visión de la informalidad laboral en la Argentina*, 179–230.

Wolf, E. (1993). *Europe, and the people without history. [Europa y la gente sin historia].* Buenos Aires, Argentina: Fondo de Cultura Económica.

Chapter 7
Health Literacy, Pharmaceutical Care, and Population Health

Aldo Alvarez-Risco, Shyla Del-Aguila-Arcentales, Santiago Diaz-Risco, M. Chandra Sekar, and Coralia Mesa-Gomez

Abstract Providing health to the population has different dimensions. One may think that giving access to medical consultation is sufficient and it is even thought that ensuring that patients receive the prescribed medication after the medical appointment is sufficient. The lack of optimal results with the use of medications has been evidenced for at least 40 years; in other words, the therapeutic goals are not being met in a large percentage of patients, so it is necessary to measure the level of understanding of patients in health matters and at the same time, it is urgent to implement pharmaceutical services in the health system of many so that they can contribute to achieving therapeutic goals, especially those with chronic diseases. The evidence in several countries provides academic, clinical, and financial support for its immediate implementation.

Keywords Health literacy · Health · Pharmaceutical care · Patients · Clinical pharmacy · Pharmacy practice

---

A. Alvarez-Risco (✉)
Facultad de Ciencias Administrativas y Recursos Humanos, Universidad de San Martín de Porres, Lima, Peru
e-mail: aldo.alvarez@redsaf.org

S. Del-Aguila-Arcentales
Escuela Nacional de Marina Mercante "Almirante Miguel Grau", Callao, Peru

Universidad Nacional de la Amazonia Peruana, Iquitos, Peru
e-mail: sdelaguila@enamm.edu.pe

S. Diaz-Risco
Centro de Fertilidad Cajamarca, Cajamarca, Peru

M. C. Sekar
University of Findlay, Findlay, OH, USA
e-mail: sekar@findlay.edu

C. Mesa-Gomez
Ministerio de Salud Pública de Cuba, La Habana, Cuba
e-mail: coraliamesa@infomed.sld.cu

A. Alvarez-Risco et al. (eds.), *Building Sustainable Cities*,
https://doi.org/10.1007/978-3-030-45533-0_7

7.1 Health Literacy (Sorensen, 2013; Thakkar, 2018)

Health literacy has been defined in multiple different ways:

> The cognitive and social skills which determine the motivation and ability of individuals to gain access to understand and use information in ways which promote and maintain good health. (Nutbeam, 1998)
>
> The constellation of skills, including the ability to perform basic reading and numeral tasks required to function in the healthcare environment. (Ad Hoc Committee on Health Literacy for the Council on Scientific Affairs AMA, 1999)
>
> The personal, cognitive, and social skills which determine the ability of individuals to gain access to, understand, and use information to promote and maintain good health. (Nutbeam, 2000)
>
> The individuals' capacity to obtain, process, and understand basic health information and services needed to make appropriate health decisions. (Institute of Medicine, 2004)
>
> The ability to make sound health decision(s) in the context of everyday life–at home, in the community, at the workplace, the healthcare system, the market place, and the political arena. It is a critical empowerment strategy to increase people's control over their health, their ability to seek out information, and their ability to take responsibility. (Kickbusch, Wait, Maag, & Banks, 2006)
>
> The wide range of skills, and competencies that people develop to seek out, comprehend, evaluate, and use health information and concepts to make informed choices, reduce health risks, and increase quality of life. (Zarcadoolas, Pleasant, & Greer, 2009)
>
> An individual's possession of requisite skills for making health-related decisions, which means that health literacy must always be examined in the context of the specific tasks that need to be accomplished. The importance of a contextual appreciation of health literacy must be underscored. (Paasche-Orlow & Wolf, 2007)
>
> The ability to read, filter, and understand health information in order to form sound judgments. (European commission, 2007)
>
> The capacity to obtain, interpret, and understand basic health information and services and the competence to use such information to enhance health. (Pavlekovic, 2008)
>
> The ability to access, understand, evaluate, and communicate information as a way to promote, maintain, and improve health in a variety of settings across the life course (Rootman & Gordon-El-Bihbety, 2008)
>
> The knowledge, skills, and abilities that pertain to interactions with the healthcare system. (Ishikawa & Yano, 2008)
>
> A process that evolves over one's lifetime and encompasses the attributes of capacity, comprehension, and communication. The attributes of health literacy are integrated within and preceded by the skills, strategies, and abilities embedded within the competencies needed to attain health literacy. (Mancuso, 2008)
>
> The ability of citizens to make sound decisions concerning health in daily life—at home, at work, in health care, at the market place, and in the political arena. (Kickbusch & Maag, 2008)
>
> The degree to which individuals have the capacity to read and comprehend health-related print material, identify and interpret information presented in graphical format (charts,

graphs, and tables), and perform arithmetic operations in order to make appropriate health and care decisions. (Yost et al., 2009)

The ability to understand and interpret the meaning of health information in written, spoken, or digital form and how this motivates people to embrace or disregard actions relating to health. (Adams et al., 2009)

The ability to derive meaning from different forms of communication by using a variety of skills to accomplish health-related objectives. (Adkins, 2009)

The degree to which individuals and groups can obtain process, understand, evaluate, and act upon information needed to make public health decisions that benefit the community. (Freedman et al., 2009)

The degree to which an individual has the capacity to obtain, communicate, process, and understand basic health information and services to make appropriate health decisions. (DPC, 2010)

7.1.1 *What Is the Level of Health Literacy?*

Information about health literacy is limited in global literature. However, Sørensen et al. (2015) measured the level of health literacy in eight European countries (Austria, Bulgaria, Germany, Greece, Ireland, the Netherlands, Poland, and Spain). This study used a scale: insufficient, problematic, sufficient, and excellent. At least 1 in 10 (12%) respondents showed insufficient health literacy and almost 1 in 2 (47%) had limited (insufficient or problematic) health literacy. The patients' demands for services in health care depend on their health literacy (HLS-EU Consortium, 2008; Ngoh, 2009). For example, the low health literacy in many countries of Latin America is an important barrier for the development and implementation of pharmaceutical care.

Health literacy impacts healthcare outcomes in multiple ways: improved knowledge of risks (Baker, 2006; Mancuso, 2008; Nutbeam, 2000; Speros, 2005), health status (Lee, Arozullah, & Choc, 2004; Mancuso, 2008; Speros, 2005), health outcomes and costs (Baker, 2006; Institute of Medicine, 2004; Kickbusch & Maag, 2008; Mancuso, 2008; Manganello, 2008), lower healthcare costs (Speros, 2005), self-care (Paasche-Orlow & Wolf, 2007), adherence (Mancuso, 2008), shorter hospitalization (Speros, 2005), and patient–provider interaction (von Wagner, Steptoe, Wolf, & Wardle, 2009); and these in turn have a direct impact on the prevalence of drug-related problems.

7.2 Drug-Related Problems

The Pharmaceutical Care Network Europe (PCNE) defined a DRP as "An event or circumstance involving drug therapy that actually or potentially interferes with desired health outcomes" (PCNE, 2018). Other terms used to designate DRPs have been "drug-therapy problems" (Cipolle, Strand, & Morley, 1998), "pharmaceutical care issues" (Krska et al., 2002), "pharmacotherapy failures," and "negative clinical outcomes" (Fernandez-Llimos, Tuneu, Baena, Garcia-Delgado, & Faus, 2004). About DRP classification, PCNE made the most comprehensive classification of DRP which has been revised several times, the current one being version 8.02 (PCNE, 2018). The primary domains are as follows:

Problems	
P1	Treatment effectiveness
P2	Treatment safety
P3	Others

Causes	
C1	Drug selection
C2	Drug form
C3	Drug dose
C4	Treatment duration
C5	Dispensing
C6	Drug use process
C7	Patient related
C8	Other

Planned interventions	
I0	No intervention
I1	At prescriber level
I2	At patient level
I3	At drug level
I4	Other

Intervention acceptance	
A1	Intervention accepted
A2	Intervention not accepted
A3	Other

Status of the DRP	
O1	Problem status unknown
O2	Problem solved
O3	Problem partially solved
O4	Problem not solved

7.3 Pharmaceutical Care

Pharmaceutical care is the pharmacist's contribution to the care of individuals in order to optimize medicines use and improve health outcomes (PCNE) (Alves da Costa, Van Mil, & Alvarez-Risco, 2019).

Pharmaceutical care is the main tool for improving of health literacy and, in this way, ensuring the effectiveness and safety of the pharmacotherapy. For authors, "Pharmaceutical Care is the pharmacist's contribution to the care of individuals in order to optimize medicines use and improve health outcomes" (Alves da Costa, Van Mil, & Alvarez-Risco, 2019).

7.4 Why the Pharmaceutical Care Has an Impact Over Health System Sustainability?

In order to contribute to the sustainability of a country's health system, it is necessary to optimize the services provided to citizens and focus on achieving appropriate endpoints. So that, patients can achieve greater access to medical appointments, get correct diagnosis in a timely manner, and even receive all the medications that are prescribed by the physician provider. However, the desired therapeutic outcome will only result if patient takes the medication correctly.

Are there then problems with the way patients use medications?

Several reports clearly demonstrate that there are multiple problems with the use of medications: lack of effectiveness and safety due to lack of adherence in patients with several diseases as diabetes (Alqarni, Alrahbeni, & Al Qarni, 2019; Jannoo & Khan, 2019; Peres, Pereira, Martinez, Viana, & de Freitas, 2018), hypertension (Lor, Koleck, Bakken, Yoon, & Navarra, 2019; Mackenzie & MacDonald, 2019; Sabio, 2018), asthma (Desager, Vermeulen, & Bodart, 2018; Sleath et al., 2018; Vasbinder et al., 2016), and HIV (Brittain et al., 2018; Kim, Lee, Park, Bang, & Lee, 2018; Nichols et al., 2019).

Pharmaceutical care addresses this global problem by focusing on ensuring improving clinical outcomes for patients. Is there evidence in the medical literature about pharmaceutical care activities that are able to improve outcomes in patients?

VOSviewer analysis shows the research leaders in pharmaceutical care (Fig. 7.1).

Also, there is evidence that pharmaceutical care generates impact on economic, clinical, and humanistic outcomes in several diseases.

Fig. 7.1 Conglomerate of authors with greater co-citation in Scopus 1998–2018. Source: Scopus, 02/20/2019. Only authors who had at least 5 citations were considered

7.5 Economic Outcomes

Adibe, Aguwa, and Ukwe (2013), Alvarez-Risco et al. (2014), Obreli-Neto et al. (2015), Alvarez-Risco, Del-Aguila-Arcentales, and Stevenson (2015a), Upadhyay, Ibrahim, Mishra, Alurkar, and Ansari (2016), Alvarez-Risco and Del-Aguila-Arcentales (2016a), Karapinar-Çarkıt et al. (2017), Vazin et al. (2018), Mateti, Nagappa, Attur, Nagaraju, and Rangaswamy (2018), de Souza Cazarim et al. (2018), Alvarez-Risco, Del-Aguila-Arcentales, & Diaz-Risco (2018a).

7.6 Clinical Outcomes

Alvarez-Risco, Lilja, Juarez-Eyzaguirre, and Tafur Valderrama (2006), Peña Marin, Junchaya Yllescas, Cerron Sauce, & Alvarez Risco (2007), Alvarez-Risco et al. (2007), Alvarez-Risco, Zegarra Arellano, Matos Valerio, Mejía Acosta, & Solis

Tarazona (2013), Alvarez-Risco et al. (2015b, 2015c), de Souza Cazarim et al. (2016), Alvarez-Risco et al. (2016b, 2016c), Mejia-Acosta et al. (2016), Dilokpattanamongkol, Tangsujaritvijit, Suansanae, and Suthisisang (2017), Silva-Villanueva, Alvarez-Risco, Del-Aguila-Arcentales, and Sanchez-Parra (2017), Alvarez-Risco, Lu, Del-Aguila-Arcentales, and Yu (2017), Vazin et al. (2018), Sakthong and Sangthonganotai (2018), Leguelinel-Blache et al. (2018), Lu et al. (2018), Alvarez-Risco et al. (2018b), Del-Aguila-Arcentales, Alvarez-Risco, and Diaz-Risco (2018), Alvarez-Risco et al. (2018c).

7.7 Humanistic Outcomes

Alvarez-Risco et al. (2002), Alvarez-Risco et al. (2006), Mohammed, Moles, and Chen (2016), Anum, Anto, and Forson (2017), Malik, Khan, Hussain, and Hashmi (2017), Chandrasekhar, Pradeep, Geoji, George, and Athira (2018), Sakthong and Sangthonganotai (2018).

7.8 Cases of Pharmaceutical Care in Health System

United States
Medicare is the federal health insurance program for:

a. People who are 65 or older.
b. Certain younger people with disabilities.
c. People with end-stage renal disease (permanent kidney failure requiring dialysis or a transplant, sometimes called ESRD).

In health system of the USA, pharmaceutical care is described as Medication Therapy Management (MTM) and is now covered under Medicare, specifically in Part D of Medicare. MTM process included Personal Medication Record, Medication-Related Action Plan, Intervention and/or Referral, and Documentation and Follow-up. According to CMS (2018), some examples of drug therapy problem recommendations made as a result of MTM services include, but are not limited to:

- Needs additional therapy
- Unnecessary drug therapy
- Dosage too high
- Dosage too low
- More effective drug available
- Adverse drug reaction
- Medication Non-compliance/Non-adherence

CMS (2018) explained that some examples of drug therapy problem resolutions made as a result of MTM recommendations include, but are not limited to:

- Initiate drug
- Change drug (such as product in different therapeutic class, dose, dosage form, quantity, or interval)
- Discontinue or substitute drug (such as discontinue drug, generic substitution, therapeutic substitution, or formulary substitution)
- Medication compliance/adherence

Evidence shows the benefits of MTM as have been described by Murali et al. (2016) and Alshehri, Barner, Brown, and Rush (2015).

United Kingdom

In health system of United Kingdom, pharmaceutical care is described as Medicine Use Review (MUR) which is an Advanced Service within the NHS Community Pharmacy Contractual Framework. According to the Pharmaceutical Services Negotiating Committee (PSNC, 2018), MUR process is a way to:

a. Improve patients' understanding of their medicines
b. Highlight problematic side effects and propose solutions where appropriate
c. Improve adherence
d. Reduce medicines wastage, usually by encouraging the patient only to order the medicines they require

Also, they mentioned that MUR is not:

a. A full clinical review
b. An agreement about changes to medicines
c. A discussion about the medical condition beyond that which is needed to achieve the above objectives
d. A discussion on the effectiveness of treatment based on test results

Evidence shows the benefits of MUR as have been described by Latif, Pollock, and Boardman (2011) and Hatah et al. (2014).

Closing Remarks

Everything described in this chapter demonstrates the value of pharmaceutical care in improving patient outcome in a cost-effective manner. Despite all the evidence, several countries have not yet implemented pharmaceutical care in their health systems. This leads to suboptimal outcomes for a high percentage of chronic patients as they continue to use drugs inappropriately, resulting in poor healthcare outcome and increased health expenditure. Health authorities should formulate policies and plans based on scientific evidence to ensure continued sustainability of the health system while improving patient outcome.

References

Ad Hoc Committee on Health Literacy for the Council on Scientific Affairs AMA. (1999). Health literacy: Report of the council on scientific affairs. *Journal of the American Medical Association, 281*(6), 552–557.

Adams, R. J., Stocks, N. P., Wilson, D. H., Hill, C. L., Gravier, S., Kickbusch, I., et al. (2009). Health literacy. A new concept for general practice? *Australian Family Physician, 38*(3), 144–147.

Adibe, M. O., Aguwa, C. N., & Ukwe, C. V. (2013). Cost-utility analysis of pharmaceutical care intervention versus usual care in management of Nigerian patients with type 2 diabetes. *Value in Health Regional Issues, 2*(2), 189–198.

Adkins, R., & Corus, C. (2009). Health literacy for improved health outcomes: Effective capital in the marketplace. *Journal of Consumer Affairs, 43*(2), 199–222.

Alqarni, A. M., Alrahbeni, T., & Al Qarni, A. (2019). Adherence to diabetes medication among diabetic patients in the Bisha governorate of Saudi Arabia–a cross-sectional survey. *Patient Preference and Adherence, 13*, 63–71.

Alshehri, A. M., Barner, J. C., Brown, C. M., & Rush, S. K. (2015). Medication and health problems and recommendations from MTM services provided by third-year pharmacy students in community pharmacies. *Currents in Pharmacy Teaching and Learning, 7*(1), 70–77.

Alvarez-Risco, A., & Juarez-Eyzaguirre, J. (2002). Atención Farmacéutica en pacientes con insuficiencia cardiaca. *Revista O.F.I.L., 12*(2), 32–38.

Alvarez-Risco, A., Lilja, J., Juarez-Eyzaguirre, J., & Tafur Valderrama, E. (2006). Seguimiento Farmacoterapéutico y el impacto de la Internet en los pacientes. *Revista O.F.I.L., 16*(4), 57–61.

Alvarez-Risco, A., & van Mil, J. W. (2007). Pharmaceutical care in community pharmacies: Practice and research in Peru. *Annals of Pharmacotherapy, 41*(12), 2032–2037. https://doi.org/10.1345/aph.1K117

Alvarez-Risco, A., Zegarra Arellano, E., Matos Valerio, E., Mejía Acosta, N., & Solis Tarazona, N. (2013). Campaña de atención farmacéutica como estrategia de implementación de los servicios farmacéuticos: Experiencia Perú. *Pharmaceutical Care España, 15*(1), 50–52.

Alvarez-Risco, A., Conteh-Barrat, M., Dawson, J., Lourenço, L., Sousa Pinto, G., Besançon, J., & Wang, L. N. (2014). Enfermedad por el virus del Ébola: información y guías para farmacéuticos y personal de farmacia. Documento de Salud FIP.

Alvarez-Risco, A., Del-Aguila-Arcentales, S., & Stevenson, J. G. (2015a). Pharmacists and mass communication for implementing pharmaceutical care. American Journal of Pharmacy Benefits, 7(3), e125–e126.

Alvarez-Risco, A., & Del-Aguila-Arcentales, S. (2015b). Atención farmacéutica, ecofarmacovigilancia y su aporte en la creación de ciudades sostenibles. *Revista. O.F.I.L., 25*(3), 183–186.

Alvarez-Risco, A., & Del-Aguila-Arcentales, S. (2015c). Errores de prescripción como barrera para la Atención Farmacéutica en establecimientos de salud públicos: Experiencia Perú. *Pharmaceutical Care España, 17*(6), 725–731.

Alvarez-Risco, A., & Del-Aguila-Arcentales, S. (2016a). Sostenibilidad en salud en las farmacias comunitarias. *Revista O.F.I.L., 26*(2), 147–148.

Alvarez-Risco, A., Quiroz-Delgado, D., & Del-Aguila-Arcentales, S. (2016b). Pharmaceutical care in hypertension patients in a Peruvian hospital. *Indian Journal of Public Health Research & Development, 7*(3), 186–191.

Alvarez-Risco, A., Turpo-Cama, A., Ortiz-Palomino, L., Gongora-Amaut, N., & Del-Aguila-Arcentales, S. (2016c). Barreras para la implementación de la Atención Farmacéutica en establecimientos farmacéuticos de Cusco. *Perú. Pharmaceutical Care España, 18*(5), 194–205.

Alvarez-Risco, A., Lu, Y. F., Del-Aguila-Arcentales, S., & Yu, P. W. (2017). Barreras para la provisión de Atención Farmacéutica en farmacias de Tainan, Taiwán. *Pharmaceutical Care España, 19*(2), 58–68.

Alvarez-Risco, A., Del-Aguila-Arcentales, S., & Diaz-Risco, S. (2018a). Pharmacovigilance as a tool for sustainable development of healthcare in Peru. *Pharmacovigilance Review, 10*(2), 4–6.

Alvarez-Risco, A., Del-Aguila-Arcentales, S., & Diaz-Risco, S. (2018b, April). Dispensing process as a cornerstone of improving pharmacovigilance. *Industrial Pharmacy*, (57), 17–21.

Alvarez-Risco, A., Solís-Tarazona, Z., Del-Aguila-Arcentales, S., & Diaz-Risco, S. (2018c). Saving obtained with the implementation of unit dose drug dispensing system in hospitals and specialized institutes in Lima, Perú [Ahorro obtenido con la implantación del Sistema de Dispensación de Medicamentos en Dosis Unitaria (SDMDU) en hospitales e institutos especializados de Lima, Perú]. *Revista O.F.I.L., 28*(4), 319–322.

Alvarez-Risco, A., Quiroz-Delgado, D., & Del-Aguila-Arcentales, S. (2016). Pharmaceutical care in hypertension patients in a Peruvian Hospital. *Indian Journal of Public Health Research & Development, 7*(3), 186–191.

Alvarez-Risco, A., Turpo-Cama, A., Ortiz-Palomino, L., Gongora-Amaut, N., & Del-Aguila-Arcentales, S. (2016). Barreras para la implementación de la Atención Farmacéutica en establecimientos farmacéuticos de Cusco, Perú. *Pharmaceutical Care España, 18*(5), 194–205.

Alves da Costa, F., Van Mil, F., & Alvarez-Risco, A. (2019). *The pharmacist guide to implementing pharmaceutical care*. Springer Natural Science Publishing.

Anum, P. O., Anto, B. P., & Forson, A. G. (2017). Structured pharmaceutical care improves the health-related quality of life of patients with asthma. *Journal of Pharmaceutical Policy and Practice, 10*(1), 8.

Baker, D. W. (2006). The meaning and the measure of health literacy. *Journal of Internal Medicine, 21*, 878–883.

Brittain, K., Remien, R. H., Mellins, C. A., Phillips, T. K., Zerbe, A., Abrams, E. J., et al. (2018). Determinants of suboptimal adherence and elevated HIV viral load in pregnant women already on antiretroviral therapy when entering antenatal care in Cape Town, South Africa. *AIDS Care, 30*(12), 1517–1523.

Chandrasekhar, D., Pradeep, A., Geoji, A. S., George, A. E., & Athira, V. (2018). Impact of intensified pharmaceutical care on health related quality of life in patients with stroke in a tertiary care hospital. *Clinical Epidemiology and Global Health, 6*(4), 198–202.

Cipolle, R. J., Strand, L. M., & Morley, P. C. (1998). *Pharmaceutical care practice*. New York: McGraw-Hill.

CMS. (2018). CY 2019 Medication Therapy Management Program Guidance and Submission Instructions. Retrieved February 5, 2020, from https://www.cms.gov/Medicare/Prescription-Drug-Coverage/PrescriptionDrugCovContra/Downloads/Memo-Contract-Year-2019-Medication-Therapy-Management-MTM-Program-Submission-v-040618.pdf

de Souza Cazarim, M., de Freitas, O., Penaforte, T. R., Achcar, A., & Pereira, L. R. L. (2016). Impact assessment of pharmaceutical care in the management of hypertension and coronary risk factors after discharge. *PloS One, 11*(6), e0155204.

de Souza Cazarim, M., & Pereira, L. R. L. (2018). Cost-effectiveness analysis of pharmaceutical care for hypertensive patients from the perspective of the public health system in Brazil. *PloS One, 13*(3), e0193567.

Del-Aguila-Arcentales, A., Alvarez-Risco, A., & Diaz-Risco, S. (2018). Barriers for implementation of cosmetovigilance in Peru: Structural equation modeling using partial least square. *Ars Pharmaceutica, 59*(1), 21–26.

Desager, K., Vermeulen, F., & Bodart, E. (2018). Adherence to asthma treatment in childhood and adolescence – A narrative literature review. *Acta Clinica Belgica, 73*(5), 348–355.

Dilokpattanamongkol, P., Tangsujaritvijit, V., Suansanae, T., & Suthisisang, C. (2017). Impact of pharmaceutical care on pain and agitation in a medical intensive care unit in Thailand. *International Journal of Clinical Pharmacy, 39*(3), 573–581.

DPC. (2010). The Patient Protection and Affordable Care Act, 2010. Retrieved February 5, 2020, from https://www.dpc.senate.gov/healthreformbill/healthbill04.pdf

European Commission. (2007). Together for Health: A strategic approach for the EU 2008–2013: White Paper. OOPEC.

Fernandez-Llimos, F., Tuneu, L., Baena, M. I., Garcia-Delgado, A., & Faus, M. J. (2004). Morbidity and 444 mortality associated with pharmacotherapy. Evolution and current concept of drug-related 445 problems. *Current Pharmaceutical Design, 10*, 3947–3967.

Freedman, D. A., Bess, K. D., Tucker, H. A., Boyd, D. L., Tuchman, A. M., & Wallston, K. A. (2009). Public health literacy defined. *American Journal of Preventive Medicine, 36*(5), 446–451.

Hatah, E. (2014). A systematic review and meta-analysis of pharmacist-led fee-for-services medication review. *British Journal of Clinical Pharmacology, 77*(1), 102–115.

HLS-EU Consortium. (2008). *The European health literacy project Grant Agreement 2007113*. Luxembourg: European Agency for Health and Consumers.

Institute of Medicine. (2004). *Health literacy: A prescription to end confusion*. Committee on Health Literacy, Board on Neuroscience and Behavioral Health.

Ishikawa, H., & Yano, E. (2008). Patient health literacy and participation in the health-care process. *Health Expectations, 11*(2), 113–122.

Jannoo, Z., & Khan, N. M. (2019). Medication adherence and diabetes self-care activities among patients with type 2 diabetes mellitus. *Value in Health Regional Issues, 18*, 30–35.

Karapinar-Çarkıt, F., van der Knaap, R., Bouhannouch, F., Borgsteede, S. D., Janssen, M. J., Siegert, C. E. H., et al. (2017). Cost-effectiveness of a transitional pharmaceutical care program for patients discharged from the hospital. *PloS One, 12*(4), e0174513.

Kickbusch, I., & Maag, D. (2008). Health literacy. In H. Kris & Q. Stella (Eds.), *International encyclopedia of public health* (Vol. 3, pp. 204–211). Academic Press.

Kickbusch, I., Wait, S., Maag, D., & Banks, I. (2006). *Navigating health: The role of health literacy. Alliance for health and the future*. International Longevity Centre.

Kim, J., Lee, E., Park, B. J., Bang, J. H., & Lee, J. Y. (2018). Adherence to antiretroviral therapy and factors affecting low medication adherence among incident HIV-infected individuals during 2009–2016: A nationwide study. *Scientific Reports, 8*(1), 3133.

Krska, J., Jamieson, D., Arris, F., McGuire, A., Abbott, S., Hansford, D., et al. (2002). A classification 442 system for issues identified in pharmaceutical care practice. *International Journal of Pharmacy Practice, 10*, 91–100.

Latif, A., Pollock, K., & Boardman, H. F. (2011). The contribution of the Medicines Use Review (MUR) consultation to counseling practice in community pharmacies. *Patient Education and Counseling, 83*(3), 336–344.

Lee, S. D., Arozullah, A. M., & Choc, Y. I. (2004). Health literacy, social support, and health: A research agenda. *Social Science & Medicine, 58*, 1309–1321.

Leguelinel-Blache, G., Castelli, C., Roux-Marson, C., Bouvet, S., Andrieu, S., Cestac, P., et al. (2018). Impact of collaborative pharmaceutical care on in-patients' medication safety: Study protocol for a stepped wedge cluster randomized trial (MEDREV study). *Trials, 19*(1), 19.

Lor, M., Koleck, T. A., Bakken, S., Yoon, S., & Navarra, A. M. D. (2019). Association between health literacy and medication adherence among Hispanics with hypertension. *Journal of Racial and Ethnic Health Disparities, 6*(3), 517–524.

Lu, M., Zhou, Y., Wang, B., Hu, Z., Du, Y., Liu, S., et al. (2018). Impact of multidisciplinary collaborative pharmaceutical care on knowledge, adherence, and efficacy of hormone therapy in climacteric women. *Patient Preference and Adherence, 12*, 1273.

Mackenzie, I. S., & MacDonald, T. M. (2019). Identifying poor adherence to antihypertensive medications in patients with resistant hypertension. *British Journal of Clinical Pharmacology, 85*(1), 5.

Malik, M., Khan, A., Hussain, A., & Hashmi, A. (2017). Assessment of health-related quality of life among Asthmatic patients: The need for structured pharmaceutical care delivery system in Pakistan. *Journal of Pharmacy and Bioallied Sciences, 9*(4), 272.

Mancuso, J. M. (2008). Health literacy: A concept/dimensional analysis. *Nursing & Health Sciences, 10*(3), 248–255.

Manganello, J. A. (2008). Health literacy and adolescents: A framework and agenda for future research. *Health Education Research, 23*(5), 840–847.

Mateti, U. V., Nagappa, A. N., Attur, R. P., Nagaraju, S. P., & Rangaswamy, D. (2018). Cost-effectiveness of pharmaceutical care on patients undergoing maintenance hemodialysis – A multicenter randomized controlled study. *Postgraduate Medicine, 130*(7), 621–626.

Mejia-Acosta, N., Alvarez-Risco, A., Solis-Tarazona, Z., Matos-Valerio, E., Zegarra-Arellano, E., & Del-Aguila-Arcentales, S. (2016). Adverse drug reactions reported as a result of the implementation of pharmaceutical care in the Institutional Pharmacy DIGEMID-Ministry of Health. *Pharmaceutical Care España, 18*(2), 67–74.

Mohammed, M. A., Moles, R. J., & Chen, T. F. (2016). Impact of pharmaceutical care interventions on health-related quality-of-life outcomes: A systematic review and meta-analysis. *Annals of Pharmacotherapy, 50*(10), 862–881.

Murali, A. B., Boban, B., Shanmughan, A. K., Marimuthu, K., Sreelatha, A. R., & Xavier, A. (2016). Medication therapy management (MTM): An innovative approach to improve medication adherence in diabetics. *Drug Metabolism and Personalized Therapy, 31*(3), 151–155.

Ngoh, L. N. (2009). Health literacy: A barrier to pharmacist–patient communication and medication adherence. *Journal of the American Pharmacists Association, 49*(5), e132–e149.

Nichols, J. S., et al. (2019). High prevalence of non-adherence to antiretroviral therapy among undisclosed HIV-infected children in Ghana. *AIDS Care, 31*(1), 25–34.

Nutbeam, D. (1998). Health promotion glossary. *Health Promotion International, 13*, 349–364.

Nutbeam, D. (2000). Health literacy as a public health goal: A challenge for contemporary health education and communication strategies into the 21st century. *Health Promotion International, 15*(3), 259–267.

Obreli-Neto, P. R., Marusic, S., Guidoni, C. M., Baldoni, A. D. O., Renovato, R. D., Pilger, D., et al. (2015). Economic evaluation of a pharmaceutical care program for elderly diabetic and hypertensive patients in primary health care: A 36-month randomized controlled clinical trial. *Journal of Managed Care & Specialty Pharmacy, 21*(1), 66–75.

Paasche-Orlow, M. K., & Wolf, M. S. (2007). The causal pathways linking health literacy to health outcomes. *American Journal of Health Behavior, 31*(1), S19–S26.

Pavlekovic, G. (2008). *In health literacy*. Programmes for Training on Research in Public Health for South Eastern Europe.

Peña Marin, J., Junchaya Yllescas, A., Cerron Sauce, M., & Alvarez Risco, A. (2007). Conocimiento de los factores relacionados con la gastritis en pacientes ambulatorios de la sierra de Perú. *Revista O.F.I.L., 17*(4), 35–39.

Peres, H. A., Pereira, L. R. L., Martinez, E. Z., Viana, C. M., & de Freitas, M. C. F. (2018). Heart failure is associated with non-adherence to pharmacotherapy in elderly with type 2 diabetes mellitus in public health system Brazilians. *Diabetes & Metabolic Syndrome: Clinical Research & Reviews, 13*(2), 939–946.

Pharmaceutical Care Network Europe (PCNE). (2018). PCNE Drug related problem classification. Retrieved February 5, 2020, from http://www.pcne.org/working-groups/2/drug-related-problem-classification

PSNC. (2018). What is the medicines use review & prescription intervention service? Retrieved February 5, 2020, from https://psnc.org.uk/services-commissioning/advanced-services/murs/murs-the-basics

Rootman, I., & Gordon-El-Bihbety, D. (2008). *A vision for a health literate Canada*. Ottawa, ON: Canadian Public Health Association.

Sabio, R. (2018). Arterial hypertension and adherence to treatment: The gap between clinical trials and reality. *Revista Cubana de Salud Pública, 44*(3), 1–4.

Sakthong, P., & Sangthonganotai, T. (2018). A randomized controlled trial of the impact of pharmacist-led patient-centered pharmaceutical care on patients' medicine therapy-related quality of life. *Research in Social and Administrative Pharmacy, 14*(4), 332–339.

Silva-Villanueva, M., Alvarez-Risco, A., Del-Aguila-Arcentales, S., & Sanchez-Parra, G. (2017). Impact of pharmaceutical care in adherence of HIV patients in Hospital San Pablo de Coquimbo, Chile. *Pharmaceutical Care España, 19*(1), 3–15.

Sleath, B., Gratie, D., Carpenter, D., Davis, S. A., Lee, C., Loughlin, C. E., et al. (2018). Reported problems and adherence in using asthma medications among adolescents and their caregivers. *Annals of Pharmacotherapy, 52*(9), 855–861. https://doi.org/10.1177/1060028018766603

Sorensen, K. (2013). Health literacy: The neglected European public health disparity. Retrieved from https://inthealth.mumc.maastrichtuniversity.nl/sites/intranet.mumc.maastrichtuniversity.nl/files/inthealth_mumc_maastrichtuniversity_nl/e-reader_health_literacy_phd_thesis_kristine_sorensen_0.pdf

Sørensen, K., Pelikan, J. M., Röthlin, F., Ganahl, K., Slonska, Z., Doyle, G., et al. (2015). Health literacy in Europe: Comparative results of the European health literacy survey (HLS-EU). *European Journal of Public Health, 25*(6), 1053–1058.

Speros, C. (2005). Health literacy: Concept analysis. *Journal of Advanced Nursing, 50*(6), 633–640.

Thakkar, U. (2018). Health literacy awareness in school students. Retrieved from http://www.aiir-journal.com/uploads/Articles/1534427798Final%20Seminar%2025th%20April%202018.pdf

Upadhyay, D. K., Ibrahim, M. I. M., Mishra, P., Alurkar, V. M., & Ansari, M. (2016). Does pharmacist-supervised intervention through pharmaceutical care program influence direct healthcare cost burden of newly diagnosed diabetics in a tertiary care teaching hospital in Nepal: A non-clinical randomised controlled trial approach. *DARU Journal of Pharmaceutical Sciences, 24*(1), 6.

Vasbinder, E. C., Belitser, S. V., Souverein, P. C., van Dijk, L., Vulto, A. G., & van den Bemt, P. M. (2016). Non-adherence to inhaled corticosteroids and the risk of asthma exacerbations in children. *Patient Preference and Adherence, 10*, 531.

Vazin, A., Karimzadeh, I., Karamikhah, R., Oveisi, Z., Mohseni, S., Keykhaee, M., et al. (2018). Clinical and economical impacts of guideline implementation by the pharmaceutical care unit for high cost medications in a referral teaching hospital. *BMC Health Services Research, 18*(1), 815.

von Wagner, C., Steptoe, A., Wolf, M. S., & Wardle, J. (2009). Health literacy and health actions: A review and a framework from health psychology. *Health Education & Behaviour, 36*(5), 860–877.

Yost, K. J., Webster, K., Baker, D. W., Choi, S. W., Bode, R. K., & Hahn, E. A. (2009). Bilingual health literacy assessment using the Talking Touchscreen/la Pantalla Parlanchina: Development and pilot testing. *Patient Education and Counseling, 75*(3), 295–301.

Zarcadoolas, C., Pleasant, A., & Greer, D. S. (2009). *Advancing health literacy: A framework for understanding and action* (Vol. 45). John Wiley & Sons.

Chapter 8
Equity

Andrew J. Chapman

Abstract As more and more people make cities their home, the issue of social sustainability in these cities is being brought to the fore. Although a key component of sustainability, social aspects are rarely addressed in a quantitative manner, leading to an evaluation of sustainability which is not comprehensive. This chapter outlines the concept of social equity, both singularly, and as an important part of sustainability evaluation. Linking these evaluations and the concept of stakeholder engagement in the energy policy making process, social equity and sustainability are prioritized in terms of engendering robust, socially sustainable policy making. By linking an academic, quantitative approach of social equity and policy making in an inclusive process, policy making becomes more effective and fair, particularly for those sectors of society which are not typically represented in the traditional policy making process.

Keywords Social equity · Distribution · Sustainability · Social factors · Participation

8.1 Introduction

Currently, approximately half of the world's population lives in cities, and, due to economic development and the gathering of people and businesses to sustain economic activities, over two-thirds of the population will choose to live in a city by 2050 (OECD, 2012). While the positive aspects of cities are easily understood through their contribution to employment opportunities, bolstering economic prosperity and the provision of critical services, cities are also responsible for excess of two-thirds of global energy-related greenhouse gas emissions (Seto et al., 2014). This level of emissions is expected to grow to three-quarters by 2030, and to be

A. J. Chapman (✉)
International Institute for Carbon Neutral Energy Research, Kyushu University, Fukuoka, Japan
e-mail: chapman@i2cner.kyushu-u.ac.jp

A. Alvarez-Risco et al. (eds.), *Building Sustainable Cities*,
https://doi.org/10.1007/978-3-030-45533-0_8

dominated by developing nations and emerging Asian economies (OECD, 2014). As more and more people make cities their home, especially in developing nations where social inequity is an ongoing issue, it is important to consider the social sustainability of cities, and this chapter focuses on social equity as one of the key measures of social sustainability.

In simple terms, social equity describes the "fairness" of a society; however, this is a subjective term, not understood broadly, and the subject of some conceptual confusion (Chapman, McLellan, & Tezuka, 2016b). Often, comparative expressions are used to give an image of "equity." Some examples include the distribution of income, i.e., the gap between rich and poor, or equal opportunity to participate, irrespective of race, income level, gender, or other subjective factors. Often social equity is difficult to quantify, as different stakeholders have different views of a fair or just system. In order to overcome some of these limitations, this chapter focuses on social equity aspects pertinent to the energy system, critical to social sustainability in cities.

8.2 The Importance of Social Equity Within Sustainability Evaluations

Generally speaking, sustainability is made up of economic, environmental, and social aspects, each of which is critical to achieving sustainable development and a sustainable society. In terms of the energy system, economic and environmental factors are well understood, and their expression is relatively uniform, in terms of dollars, tons of carbon dioxide, etc., making them fungible and interchangeable. However, social aspects are not so easily quantified and are often considered in qualitative terms, i.e., how something makes us feel, opinions with regard to fairness, etc. Building on recent advances in sustainability and energy system evaluation approaches such as energy justice, the social aspects of sustainability can also be quantified in terms of distributive, recognition, and procedural outcomes (Jenkins, McCauley, Heffron, Stephan, & Rehner, 2016). Distributive outcomes can be quantified in terms of who benefits from, and who pays for various energy policy, technology deployment, and overall system approaches, allowing us to observe changes in the gap between high- and low-income households. Recognition outcomes go further still, identifying groups within benefitting and burden bearing sectors of society in terms of not only their income level, but also where they live, their race, gender, beliefs, and other factors which may also affect social equity. Finally, procedural outcomes measure the ability of stakeholders to participate in the policy development process, demonstrating the openness (or otherwise) of policy making processes in a given jurisdiction. This notion of "energy justice," although relatively new in academic terms, seeks relevance in developing practicable policy outcomes specific to energy issues. Energy justice is underpinned by the precedential research streams of environmental justice and climate justice and builds on their success in terms of broad recognition and grass root origins, toward environmental improvement and normative policy impacts (Heffron, McCauley, & Sovacool, 2015).

8.3 Socially Equitable Cities

As more and more people move into cities, the improvement of social equity in urban areas becomes increasingly more important. Cities are home to rich and poor, home owners, renters, and are generally more cosmopolitan than their rural counterparts. With such a broad stakeholder group, the definition and achievement of consensus for a single definition of equity is likely impossible; however, through direct stakeholder engagement, a city-level "ideal" can be approximated. Based on this ideal, and a series of demographic, economic, and environmental factors, the level of social equity and the distribution of burdens and benefits can be quantified. Building on these findings, city level policies can be derived, and actions defined to improve social equity outcomes in line with stakeholder's desires. Obviously, individual lifestyle choices require the consumption of resources to sustain, and when these resources are limited in nature, imbalances can emerge between sectors of society, often dependent on income or other socio-economic factors. Furthermore, consumption of limited resources and energy to sustain our lifestyles can engender flow-on impacts such as the generation of social ills including pollution and the depletion of critical materials. Ideally, benefits and burdens would be shared equitably, however, in the case of the environment, and the depletion of limited resources, those who benefit most do not necessarily bear the burden that their lifestyles entail (Johnson, 2012).

In order to provide a definition for social equity and a benchmark for its evaluation, the concept of "vertical equity" is applied (a pertinent example being a progressive tax system), and as such, social equity is said to be improved when the benefits of energy policies are shared fairly, and the gap between lower and higher income households is reduced. This concept is applied to ensure that each energy policy engenders a user pays approach and that households are not restricted from societal participation due to their household means or other demographic or exogenous aspects.

The process for the evaluation and redress of social equity issues in cities is detailed in Fig. 8.1 and requires policy makers to engage openly with stakeholders and to objectively measure economic and environmental factors common to sustainability assessments.

As described above, the first stage of this process (adapted from Chapman et al., 2016b) is to define social equity, in this case in terms of energy justice, vertical equity, and the notion of closing the gap between rich and poor (in terms of income, employment opportunity, participation, etc.)

The second stage of the process requires an identification of social equity issues being experienced, as well as a city level ideal. Equity issues can be identified by investigating current policies and their pitfalls, while the city level ideal is best iden-

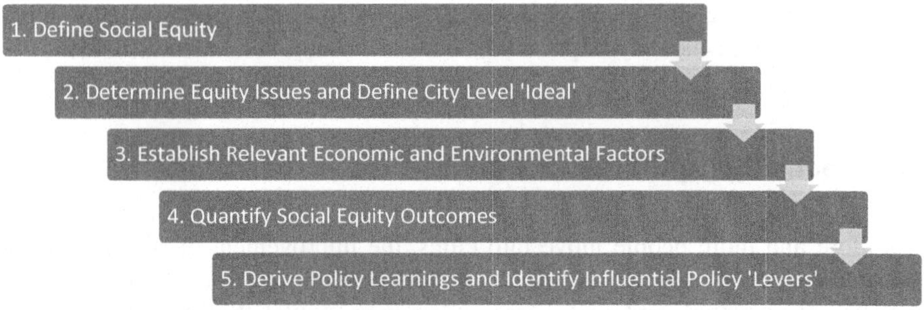

Fig. 8.1 City level social equity evaluation and improvement framework

tified through a survey of stakeholders to establish key social equity factors requiring redress and an appropriate weighting (i.e., how important these factors are to the stakeholder group; see Chapman, McLellan, & Tezuka, 2018 for a regional level example). A survey is a very useful tool as it can (1) identify whether policy maker's preconceptions of "important issues" match with the local stakeholder's ideals, (2) clarify the importance of each social equity factor, essential for matching outcomes with expectations, and (3) uncover any specific local or cultural factors of importance, unique to the region being investigated.

Third, in order to quantify each of the factors identified, the underpinning economic and environmental factors need to be determined. In terms of social equity, income distribution needs to be understood in each case, while the underpinning environmental factors vary. Some common examples include PM2.5 or greenhouse gas emissions and their distribution, the number of jobs gained or lost due to changing policies, and the ability to participate in policy initiatives based on other economic factors such as home ownership.

Finally, following the quantification of social equity outcomes, analysis can be undertaken to derive policy learning, in terms of identifying the policy levers which improve or impair social equity levels. These levers will be different in each area and depending on social equity factor importance will have a different level of impact on future outcomes. For example, in a society such as Japan which has limited fossil fuel resources and an aging society, issues such as employment and economic development may be prioritized, whereas in Australia or the USA, both resource rich nations, low electricity prices, and participation may be considered more important. Additional city specific priorities may also emerge within nations. A recent investigation of the transition from fossil fuels to renewable energy in Australia, for example, found that the deployment of different renewable technologies and the primary fossil fuel source engaged in each region weighed heavily on social equity outcomes. Furthermore, the socio-economic status and diversity in each region also had impacts on health, employment, and electricity price outcomes (Chapman et al., 2018). One of the key lessons learned through the ongoing investigation of social equity at the city level is that there is no "one size fits all" approach, and that national and regional policies will need to be tailored to address the emerging issue of societal disparity.

8.4 Improving Equity in Cities Through a Policy Approach

It is natural to assume that the people who live in cities will not want to reduce their level of comfort or change their lifestyles voluntarily, and in most cases, people are looking to improve their level of well-being. As people act in their own interests, a centralized policy approach is likely to be the most effective method of addressing social equity issues for current and future generations, cognizant of socio-economic variety within a given jurisdiction. As a greater proportion of people choose to live in cities in the future, the identified issues of energy use and greenhouse gas emissions will become further concentrated in urban areas. It is likely that these issues will continue to be dealt with through a top-down policy approach. To date, energy policy approaches which aim to reduce emissions have included carbon trading schemes, feed-in tariffs for renewable energy, and energy efficiency regulations, among others. Each of these approaches generally employs a financial reward or penalty, associated with certain actions, i.e., a cost for producing carbon dioxide, or a reward for generating clean energy, etc. By introducing financial rewards (or penalties) for certain behaviors, policy can engender a societal level change which can positively impact on environmental outcomes. These approaches link economic and environmental aspects of sustainability, with some attention given to adjusting social behaviors; however, they are often developed and implemented without considering the social equity impacts that they subsequently engender.

Through a detailed investigation of eight Organization for Economic Cooperation and Development (OECD) nations' approach to energy policy making processes, it was found that energy policy tools employed are often misaligned with energy policy goals, particularly regarding the social aspects of sustainability (Chapman, McLellan, & Tezuka, 2016a). Policy targets are often stated in terms of economic or environmental outcomes, with social outcomes not assessed until the final, evaluation stage of the policy cycle which may occur years after policy implementation in some cases. The UK, for example, commits to an Annual Energy Statement in line with an independent committee review of their renewables target, while Australia is legislated to have a biennial review by an independent authority. The US energy review is on a quadrennial schedule, while Japan suggests a "periodic" review of their Long-term Energy Supply and Demand Outlook, to occur at least once every three years.

In order to improve the social equity and sustainability outcomes of energy policy implementation in cities, based on the OECD policy making and priority setting evidence, a modified policy cycle is proposed, incorporating a pre-evaluation phase between the policy formulation and decision-making steps, as shown in Fig. 8.2.

As shown, the inclusion of the policy pre-evaluation phase is proposed in order to measure policy approaches against the three pillars of sustainability (economic,

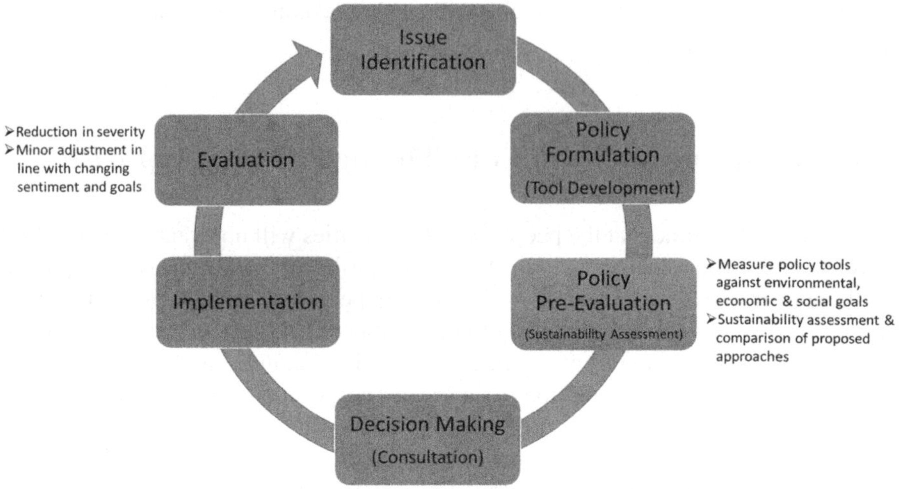

Fig. 8.2 Sustainability conscious policy cycle (from Chapman et al., 2016a)

environmental, and social), and to allow an objective comparative analysis of a variety of policies. In doing so, it is anticipated that the previously radical evaluation and re-working phase of policy making (often due to unanticipated negative social equity outcomes) can be reduced in severity, and that policy can undergo relatively minor, incremental change in line with stakeholder's sustainability and social equity ideals.

Precedential scholarship has identified sustainable economic growth, energy security, and low or reduced energy prices as energy policy priorities for OECD nations. In line with these goals, social equity is also considered broadly, using various terms, ranging from the consideration of social issues during energy policy development through to the specific identification of disadvantaged groups and targeted support needs analysis. The need to reduce the current public burden level, as well as the burden level for future generations (intergenerational equity) has also been identified as a priority issue (Chapman et al., 2016a). Furthermore, it has been noted that not only is the gap between rich and poor at its highest level in 30 years, but also that this level of inequality and wealth concentration limits investment opportunities and impairs economic growth (OECD, 2015).

Closing Remarks

Ideally, social equity can be improved through a combination of stakeholder engagement, quantitative academic analysis of social equity factors, and the application of these findings to policy development and implementation processes. Through the development of sustainability and social equity aware energy policy, government goals for the future energy system can be met, while also achieving sustainability and social equity goals, leading to improved sustainable economic growth, and enhancing social equity outcomes for cities and regions.

References

Chapman, A., McLellan, B., & Tezuka, T. (2016a). Strengthening the energy policy making process and sustainability outcomes in the OECD through policy design. *Administrative Sciences, 6*(3), 9. https://doi.org/10.3390/admsci6030009

Chapman, A. J., McLellan, B., & Tezuka, T. (2016b). Proposing an evaluation framework for energy policy making incorporating equity: Applications in Australia. *Energy Research and Social Science, 21*, 54–69. https://doi.org/10.1016/j.erss.2016.06.021

Chapman, A. J., McLellan, B. C., & Tezuka, T. (2018). Prioritizing mitigation efforts considering co-benefits, equity and energy justice: Fossil fuel to renewable energy transition pathways. *Applied Energy, 219*, 187–198. https://doi.org/10.1016/j.apenergy.2018.03.054

Heffron, R. J., McCauley, D., & Sovacool, B. K. (2015). Resolving society's energy trilemma through the energy justice metric. *Energy Policy, 87*, 168–176. https://doi.org/10.1016/j.enpol.2015.08.033

Jenkins, K., McCauley, D., Heffron, R., Stephan, H., & Rehner, R. (2016). Energy justice: A conceptual review. *Energy Research and Social Science, 11*, 174–182. https://doi.org/10.1016/j.erss.2015.10.004

Johnson, J. (2012). Environmental justice. In R. Chadwick (Ed.), *Encyclopedia of applied ethics* (2nd ed., pp. 124–132). Amsterdam: Elsevier/Academic Press.

OECD. (2012). *Compact city policies: A comparative assessment*. Paris: OECD Publishing.

OECD. (2014). *Cities and climate change: National governments enabling local action*. Paris: OECD Publishing.

OECD. (2015). *In it together: Why less inequality benefits all*. Paris: OECD Publishing.

Seto, K.C., Dhakal, S., Bigio, A., Blanco, H., Delgado, G. C., Dewar, D., et al. (2014). Human settlements, infrastructure and spatial planning. In *Climate change 2014: mitigation of climate change*. Contribution of Working Group III to the Fifth Assessment Report of the Intergovernmental Panel on Climate Change. Cambridge and New York, NY: Cambridge University Press.

Chapter 9
Food Insecurity

Arístides Vara-Horna and Aldo Alvarez-Risco

Abstract Food insecurity is a global problem and has different causes and effects. Usually it is thought that it is exclusively related to poverty as the main cause but other aspects such as violence against women are also reported in the literature; likewise, food insecurity directly impacts the worsening of diseases such as HIV or cancer. A comprehensive approach is necessary to ensure the food security of the population, through multidisciplinary strategies and with approaches based on research results.

Keywords Food · Food insecurity · Food security · Agriculture · FIES

9.1 Introduction

During the World Food Summit held in Rome in 1996, the member countries of the United Nations Organization UN reaffirmed their commitment to eradicate hunger. Four years later, they raised 8 Millennium Development Goals, where the first was to eradicate extreme poverty. Thanks to the commitment of the member countries of the General Assembly of the United States, until 2015, the population suffering from hunger had been halved, yet 800 million people still face malnutrition in the world. Therefore, the UN developed the Sustainable Development Goals (SDG) by 2030, which also included eradicating hunger.

In recent years, the measurement of hunger or food security has been part of many debates among specialists. However, to know the prevalence worldwide, in 2013 the Food and Agriculture Organization of the United Nations (FAO) launched the project "Voices of Hunger" to provide updated information on food insecurity that are relevant to policy and practice. A methodology was developed to measure the severity of food insecurity experienced by individuals or households, so that it could be compared between countries through a direct interview in relationship to

A. Vara-Horna · A. Alvarez-Risco (✉)
Facultad de Ciencias Administrativas y Recursos Humanos, Universidad de San Martín de Porres, Lima, Peru
e-mail: aldo.alvarez@redsaf.org

A. Alvarez-Risco et al. (eds.), *Building Sustainable Cities*,
https://doi.org/10.1007/978-3-030-45533-0_9

personal, and as experiences, as a new global standard for measuring the experience of food insecurity validated at the international level and as a global and national monitoring tool. The FIES has been applied in 140 countries since 2014, through the Gallup World Survey applied to a representative sample of adults nationwide. The instrument has managed to determine a similar trend of global food insecurity. The FIES offers reliable information about the severity of people's food insecurity, quickly and at a lower cost, through 8 questions about the access adequacy of adults (15 years of age or older) to food, asking them direct questions about their experiences. Through FIES, FAO plays the role of monitor of the severity of food insecurity in 150 countries, funded by the United Kingdom and Belgium. In addition, FAO offers technical assistance to implement the survey in the countries considering contextual variables associated with food insecurity and the potential causes, consequences, and experiences. In effect, national institutions can use FIES to obtain evidence based on data about the distribution and severity of food insecurity to develop and implement policies that contemplate the right to adequate food.

9.2 Food Insecurity

Poverty is commonly defined as the lack of what is necessary to ensure material well-being, particularly food, but also housing, land, and other assets, which leads to hunger and physical deprivation (World Bank, 2000). It is considered as a determinant of food insecurity (FAO, 2018). Food insecurity is an indicator of social inequity associated with the lack of regular and permanent access to food, in sufficient quantity and quality. It represents a concern for its long-term access (Silva et al., 2017) and its main determinants are poverty (lack of money or other resources) and social inequalities (Sperandio, 2015). Indeed, food insecurity represents a global problem that transgresses the human right to adequate food. According to the last report (FAO, 2018) on the security of food and nutrition in the world, since 2015 there has been an increase in hunger in the world, since people suffering from chronic food deprivation has increased to almost 821 million in 2017, of around 804 million in 2016, mainly due to climatic and economic factors. The experiences of food insecurity are similar around the world, its severity begins when there is a concern for the ability to obtain food, in the case of not having the capacity increases the level with the decrease in the quality and variety of the food, continuing with a decrease in the amount of food consumed and culminate in experiences of hunger (FAO, 2016). During 2017, the percentage of food insecurity worldwide reached 10.7%, in Africa 29.8%, in Asia 6.9%, in Latin America 9.8%, and in North America and Europe 1.4% (FAO, 2018).

9.3 The Measurement of Food Insecurity

Although the right to adequate food was declared in 1949, efforts to measure the severity of food insecurity and hunger began to develop in 1990 in the USA, as a result of a great need for measurement by policymakers at all levels of government (Carlson, Andrews, & Bickel, 1999). In 1996, during the World Food Summit in Rome, discussions on that universal right were resumed (Ballard, Kepple, Cafiero, & Schmidhuber, 2014). In effect, the concepts underlying the measurement of food security based on experience have a long history based on ethnographic studies to understand the experience of hunger. However, research in the USA revealed that the experience of food insecurity is characterized by uncertainty and anxiety regarding access to food and changes in the quality of the diet, such as a less balanced and more monotonous diet; gravity that increases when the amount of food consumed decreases as portion sizes are reduced, the severity when meals are omitted and people stop eating (FAO, 2017).

9.3.1 The FIES scale

In September 2015, the 193 Member States of the United Nations (UN) adopted the 2030 Agenda for Sustainable Development, to update the Millennium Development Goals, and, in March 2016, the UN Statistical Commission developed a framework of 230 global indicators, to monitor the objectives and measure progress towards the achievement of the 17 Sustainable Development Goals (SDGs). Of the total indicators, FAO assumed the responsibility of 21 of them, related to the second SDG (2.1 Indicators that measure people's access to food), whereby countries commit themselves to end hunger and guarantee the access of all people, particularly the poor and people in vulnerable situations, including babies, to sufficient safe and nutritious food.

To provide a better information, FAO launched the project "Voices for hunger" and a new methodology called "Scale of Experiences of Food Insecurity" or Food Insecurity Experiences Scale (FIES), adapted from the Latin American and Caribbean Scale of Security Alimentaria (Escala Latinoamericana y Caribeña de Seguridad Alimentaria—ELCSA), whose origins are derived from the United States Household Food Security Survey Module, the Brazilian Food Insecurity Scale, the scale of access to food insecurity in households (HFIAS), and a similar scale adapted for Colombia. Through the FIES, FAO offers information on the prevalence of experiences of food insecurity worldwide, provides technical assistance to countries, and fulfills its role as monitor receiving and communicating the data to the Department of Economic and Social Affairs of the UN (UNDESA).

FIES has become a reliable and valuable tool for determining the prevalence of moderate or severe food security in the population, at the individual and household level, applied in more than 140 countries (see Table 9.1), providing reliability,

Table 9.1 Implementation of FIES worldwide

Status	Countries
Use of its own national scale of food security based on experiences	USA, Bolivia, Brasil, Canada, Colombia, Ecuador, Mexico, Guatemala, Corea del Sur, Filipinas and Sri Lanka
FIES or similar scale already included in national surveys	Bangladesh, Botswana, Burkina Faso, Chile, Dominican Republic, El Salvador, Ethiopia, Ghana, Israel, Lesotho, Malawi, Malaysia, Marshall Islands, Namibia, Palestine, Rwanda, Sierra Leone, St. Lucia, Sudan, Swaziland, Uganda, Vietnam, Zimbabwe, and Cote d'Ivoire
FIES already included in national surveys and government plans to collect data through FIES on a regular basis	Indonesia, Jordan, Kenya, Pakistan, and Seychelles
Plans to include the FIES in national surveys	Afganistán, Benin, Capo Verde, Chad, Guinea, Guinea-Bissau, Kiribati, Mauritania, Mali, Micronesia, Nicaragua, Niger, Nigeria, Samoa, Senegal, Solomon Islands, Togo, Tokelau, Tonga, Vanuatu

Source: (FAO, 2018)

quickness, at low cost, and adapted in more than 200 local languages and dialects. This scale has the capacity to produce results and effects of change according to the level of their progress in relation to the second SDG and subsequent development of governmental policies on food security (FAO, 2016).

9.4 Description of Items

FIES is based on a well-founded concept of the experience of food insecurity structured into three levels: uncertainty/concern, changes in the quality of food, and changes in the amount of food, from which an underlying scale of severity is derived (food insecurity at the family or individual level). The FIES is composed of 8 questions (the first three are subjective and the remaining are objective) (see Table 9.2), with yes/no dichotomous answers, on access to adequate food in a reference period from 30 days to 12 months. It is a scale of statistical measurement designed to calculate a range of the severity of food insecurity and is analyzed together as a scale, not as separate items. Each question refers to a different experience and is associated with a different level of severity of insecurity. Unlike other scales of measurement related to the FIES, the latter not only considers the experiences of food insecurity, but also the quality of the compromised diet and the reduced amount of food and psychosocial elements associated with anxiety or uncertainty with respect to the ability to procure enough food. Table 9.3 describes the conceptual definition of the items of the FIES.

Table 9.2 Survey module on the scale of food insecurity (during the last 12 months)

N°	Standard label	Questions	
		Reference unit—individual	Reference unit—home
1	Worried	Are you worried about not having enough food to eat due to lack of money or other resources?	Are you or someone else in your household worried about not having enough food to eat because of lack of money or other resources?
2	Healthy	Still thinking about the last 12 months, was there ever a time when you were not able to eat healthy and nutritious foods due to lack of money or other resources?	Still thinking about the last 12 months, was there ever a time when you or someone else in your household could not eat healthy and nutritious foods due to lack of money or other resources?
3	Few foods	Was there ever a time when you ate a small variety of foods due to lack of money or other resources?	Was there ever a time when you or someone else in your household ate a small variety of foods due to lack of money or other resources?
4	Skipped	Was there ever a time when you had to stop having breakfast, lunch, or dinner because there was not enough money or other resources to get food?	Was there ever a time when you or someone else in your household had to stop having breakfast, lunch, or dinner because there was not enough money or other resources to get food?
5	Ate less	Still thinking about the last 12 months, was there ever a time when you ate less than you thought you should eat for lack of money or other resources?	Even in the last 12 months, was there ever a time when you or someone else in your household ate less than you thought you should eat for lack of money or other resources?
6	Ran out	Was there ever a time when your home has run out of food due to lack of money or other resources?	Was there ever a time when your home has run out of food due to lack of money or other resources?
7	Hungry	Was there ever a time when you felt hungry but did not eat because there was not enough money or other resources to get food?	Was there ever a time when you or someone else in your household felt hungry but did not eat because there was not enough money or other resources to get food?
8	Whole day	Was there ever a time when you stopped eating all day for lack of money or other resources?	Was there ever a time when you or someone else in your household stopped eating for a day because of a lack of money or other resources?

Source: (FAO, 2018)

In addition, when analyzing the scale jointly, it is considered a quantitative tool to measure the prevalence of moderate and severe levels of food insecurity (see Table 9.4) in a given population, using statistical methods to estimate the error (confidence intervals) around the measurements, and its reliability and validity is formally evaluated and compared between countries.

Among the main features of the estimation of FIES parameters are highlighted (FAO, 2017):

Table 9.3 Conceptual definition of the items of the FIES

Questions	Conceptual definition
You are worried about not having enough food to eat because of lack of money or other resources	The question refers to a state of worry, anxiety, fear, or fear due to the fact that there is not enough food or it is over due to insufficient money or other resources to obtain food, due to circumstances that affect the ability to obtain food. (Loss of employment, insufficient food production, insufficient food availability for hunting and gathering, deterioration of social relationships, loss of usual food or assistance, or environmental or political crises.) It is not necessary that the respondent has lacked sufficient food or has been effectively deprived of them to answer this question positively
You have not been able to eat healthy and nutritious foods for lack of money or other resources	It inquires about the ability to obtain the foods that the respondent or respondent considers healthy or good for him/her, those that allow him/her to enjoy good health, or those that integrate a nutritious and balanced diet (for lack of sufficient money or other resources) to obtain food. The question refers to the quality of the diet and not the amount of food that is eaten
You have eaten little variety of food for lack of money or other resources	Inquire about the quality of the diet and not the amount of food that is eaten. It implies that the lack of money or resources, and not traditional habits or other circumstances (i.e., health or religion), are the reason for limiting the variety of foods
You have had to stop having breakfast, lunch, or dinner because there was not enough money or other resources to obtain food	Find out about the experience of having to skip a main meal (for example, breakfast, lunch, or dinner depending on the norm regarding the number and times of meals in the culture in question) carries out normally (for lack of sufficient money or other resources to obtain food). This question refers to an insufficient amount of food
You have eaten less than you thought you should eat for lack of money or other resources	It refers to a food consumption lower than what should be in the opinion of the respondent, even if a meal has not been skipped (due to the lack of money or other resources to obtain food at home). The answer depends on the respondent's own opinion about how much he thinks he should eat. The question refers to the amount of food consumed and not the quality of the diet
Your home has run out of food due to lack of money or other resources	It refers to experiences of affective lack of food in the home due to lack of money, other resources, or any other means to obtain food
You have felt hungry but did not eat because there was not enough money or other resources to obtain food	This question has as its object the physical experience of suffering from hunger, and, specifically, of being hungry and not being able to eat enough (due to lack of money or resources to obtain food)
You have stopped eating a whole day for lack of money or other resources	This question inquires about a specific behavior: not eating anything all day (due to lack of money and other resources to obtain food)

Note: None of the questions refer to special diets for losing weight or fasting for health or religious reasons.
Source: (FAO, 2015)

Table 9.4 Gravity of food insecurity along a continuous gravity scale

Levels	Definition
Food safety	When there is no food restriction of any nature, not even fear or worry about the lack of future food
Mild food insecurity	When there is concern about the ability to obtain food
Moderate food insecurity	When the quality and variety of foods are compromised
Serious food insecurity	When you reduce the amounts, skip meals and/or go hungry

Source: (FAO, 2017)

- In FIES, the items and respondents (individuals or households) are positioned on the same underlying scale of severity of food insecurity.
- The probability that a respondent answers "yes" to a FIES item depends on the distance along the scale between the severity of the respondent and the severity of the item. The more severe the food insecurity of a respondent, in relation to that of the item, the greater the probability that he will respond affirmatively.
- The relative position of the items and respondents in the severity scale is expressed through their respective parameters. Both the elements and the parameters of the respondent are estimated according to the response patterns given to the eight FIES questions.
- The raw score of a respondent (a whole number with a value between zero and eight) is the sum of the affirmative answers given to the eight FIES questions. For the data that passes the validation tests, the raw score itself is already an ordinal measure of gravity, with lower gross scores corresponding to less severe food insecurity. The raw score is the basis for estimating the respondent's parameter, which provides a measure of the severity of food insecurity and allows the production of comparable food insecurity measures across countries and contexts.
- The order of the FIES items in terms of the severity they reflect is not constant, but is revealed through an analysis of the specific data collected. The relative severity of a given experience of food insecurity depends on the frequency with which people respond affirmatively to that element, and more severe experiences are reported less frequently. This is similar to a relatively difficult test question that causes a smaller proportion of correct answers than the easier ones.
- In different countries and subpopulations, the same item may be associated with a different level of severity due to nuances in translation or actual differences in the way food insecurity is experienced and managed in diverse cultures and media systems lifetime. The implication of the specificity of the scale is that the parameters cannot be automatically compared through the FIES applications. However, this does not prevent them from being formally compared.
- Comparability can be achieved by calibrating the scales in a common metric, in a process called equalization. This is done by adjusting all the measures obtained at the national level to the global standard, which is a set of parameter values of items based on the results of more than 140 countries covered by Gallup World Poll in 2014, 2015, and 2016.

Table 9.5 Elements of measurement of food insecurity

Definition	Refers to limited access to food, at the level of individuals or households, due to lack of money or other resources
Measuring elements	**Criteria**
Parameters of severity of the respondent	They are based on the answers to the 8 questions of the FIES. The number of affirmative responses from 0 to 8 is transformed at the gross (ordinal) level (it is not an interval measure, so a linear regression model is not used) The respondent's severity parameter conforms to the VoH global standard metric, so it is intended to represent the same level of severity in all countries
Probability of moderate or severe food insecurity	The values range from 0 to 1 The measurement error is taken into account and can be considered as the proportion of people that represent a true sample of the population with food insecurity, which exceeds the threshold established in the level of severity of the item, and is compared with the global standard of VoH. The value is based on adjusting the respondent's severity parameters to the VoH global standard, so it is intended to be comparable in all countries. The probability of moderate or severe food insecurity for cases with a zero gross score will be presented as zero. The values of the raw score 8 will be based on the standard VoH methods (used to calculate the national prevalence rates).
Probability of severe food insecurity	Values range from 0 to 1 This is similar to the likelihood of moderate or severe food insecurity, except that the threshold is more severe, at the level of severity according to the VoH global reference scale. This element of data will be missing for cases with missing answers to the FIES questions

Source: (FAO, 2015)

9.5 Operational Definition of Characteristics

To calculate internationally comparable estimates of the prevalence of food insecurity, respondents must first be assigned to (in) defined food safety classes by standard thresholds established along the severity scale. Two global standard thresholds are established according to the severity of two specific elements of FIES: ATELESS and WHLDAY, defining the classes of moderate and severe food insecurity, respectively. The matching procedure ensures that these thresholds are mapped to national scales, and respondents are assigned probabilistically to common food insecurity classes, given their gross scores. The prevalence of food insecurity in the population is given by the weighted sum of the specific gross probabilities of the score. Table 9.5 shows the elements of measurement of food insecurity.

9.6 Indicators Based on FIES

Two indicators based on FIES can be used for national and global monitoring purposes (FAO, 2017). The first indicator is an estimate of the sum of the segments of the population with moderate food insecurity and severe food insecurity, and the second only with severe severity.

- FImod + sev: Proportion of the population with moderate or severe food insecurity
- FIsev: Proportion of the population that experiences severe food insecurity.

People who experience moderate levels of food insecurity often consume low quality diets and may have been forced, sometimes during the year, to also reduce the amount of food they would normally eat, while those who experienced severe levels would have spent whole days without eat, due to lack of money or other resources to obtain food. It is expected that the prevalence of severe food insecurity is highly correlated, in all countries, with the prevalence of undernourishment.

9.7 Theory ITR—Rasch model used in FIES

The FIES data is analyzed using item response theory (IRT), a branch of statistics that allows the measurement of unobservable traits through the analysis of responses to surveys and tests, establishing the severity of each. The ITR applies to the FIES due to the intrinsically unobservable characteristic of food security, which can only be measured by examining its observable manifestations (FAO, 2017). The specific IRT model applied to the FIES data is the Rasch model, widely used in health, education, and psychology (Nord, 2014). The Rasch model provides a theoretical basis and a set of statistical tools to assess the suitability of a set of survey questions to build a measurement scale, and compare the performance of a scale in different populations and survey contexts.

The Rasch model postulates that the probability of observing an affirmative response per respondent i to question j is a logistic function of distance, on an underlying severity scale, between the position of the respondent, ai, and that of the item, bj.

$$Prob\left(x_{i,j} = S'\mathrm{i}\right) = \frac{\exp\left(a_i - b_j\right)}{1 + \exp\left(a_i - b_j\right)}$$

By applying the Rasch model to the FIES data, the probability of food insecurity ($p_{i,L}$) can be estimated at any level of food insecurity severity L, for each respondent i, with $0 < p_{i,L} < 1$.

The severity of an item, then, is the level of severity of households that are barely on the threshold of affirming or denying that question. The probability that a household affirms a fair item in the level of severity of the household is 1, which corresponds to a probability of 0.5. The probability that a household affirms an item with a severity parameter one unit lower than that of the household is 1, or approximately 2.7, which corresponds to a probability of 0.73 [i.e., 1 / (1 + 1 / 2.7)]. The probability that the household affirms an item with two units lower than its own measure of severity is 0.88, and for an item with three units lower, it is 0.95.

The Rasch model also provides item adjustment statistics, which evaluate how well each item, each household, and the global data fit the assumptions of the measurement model. Two commonly used statistics are "item infit" and "item outfit," similar to the chi-square statistic that compares the mismatch of each element with the degree of expected maladaptation under the assumptions of the model.

- The "infit" is an adjustment statistic "weighted by information" for each item, so it is sensitive to the responses of households with severity scores in the range close to the severity level of the particular item and calculates are calculated comparing the real answers with the probabilistically expected answers, the acceptable value is between 0.7 and 1.3.
- The "outfit" is sensitive to unexpected responses from households with severities much higher or lower than those of the item, that is, highly unlikely responses (outliers).

Both statistics compare the observed deviations of the responses of the expected deviations according to the assumptions of Rasch, so the expected values of the statistics are 1 and the values above 1.0 indicate items that are less or more consistently related to the underlying condition (food insecurity) measured by the set of items. The analysis of the FIES data includes the following steps:

- Estimation of parameters: calculation of the severity of the food insecurity associated with each item of the survey and each respondent. According to the ITR criteria, the parameters of the FIES respondent range from approximately – 1.0 to + 2.5.
- Statistical validation: the evaluation of whether, according to the quality of the data collected, the measure is valid. Through the Rasch model, the psychometric evaluation is determined through the validation of the theoretical assumptions and once a set of questions has been evaluated in a large sample of a population or subpopulation and the assumptions of the model of measurement, psychometric evaluation in subsequent surveys may not be necessary (Nord, 2014).
- Calculation of food insecurity measures. (1). Individual probabilities: for each individual or household sampled (each case in the data), the probability that the individual/household experiences food insecurity above a given severity level, according to their responses to the FIES items, is calculated. (2). Population prevalence estimates: the probabilities are used to estimate the prevalence of food insecurity at moderate and severe levels in the population.

9.8 Statistical Validation

The statistical validation evaluates the quality of the data collected through the FIES, to test its consistency with the assumptions of the Rasch model. This analysis involves the interpretation of several statistics that reveal if any of the items does not work well in a given context, due to cases with response patterns with a high level of error, pairs of items that may be redundant, or due to the low proportion of the total variance in the population that is explained by the measurement model. It is possible that the severity of the items varies in countries for various reasons such as the nuances of translation, culture, livelihoods, or the management of food shortages, so the methodology of FIES anticipates this possibility and adjusts these differences, when they exist, so that they do not affect the validity of the prevalence estimates and their comparability between countries, for this it makes use of the item response theory and the Rasch model, which in turn employs other statistical methods.

The FIES analyzes its validity and reliability, according to the criteria of the theory of the response to the item, ITR (Cafiero, Melgar-Quiñonez, Ballard, & Kepple, 2014), where it considers:

- The validity of the FIES considers two conditions: (1) that the severity of food insecurity in fact involves the domains that have been taken into account when creating the elements that make up the scales, and (2) that the occurrence of experiences in those domains it can be reliably detected and linked in a significant way, although in a probabilistic sense, with food insecurity
- Reliability is analyzed through the study of associations between the measures obtained with the scale and the classifications that they produce with those obtained using other variables that theoretically are part of the same construct food in safety and/or that would vary in an expected way in different levels of food insecurity. According to the Gallup World Survey standard in more than 150 countries, where the FIES was included since 2014, the preliminary analysis of the results obtained from 20 countries revealed measures the reliability is between 0.69 and 0.78, with a median of 0.73 (Rasch is a measure of the overall fit of the data to the measurement model, with theoretical values ranging from 0 to 1, and 1 indicates a perfect fit).

9.9 Calculation of Parameter Estimates and Evaluation Statistics

According to Cafiero, Viviani, and Nord (2018) to estimate the single parameter Rasch model for dichotomous and polytomic item responses with a maximum of four responses (partial credit), the maximum conditional likelihood or conditional maximum likelihood (CML) weighted method is used, which allows:

- Estimate the item parameters and gross scores of a Rasch model for binary item responses, where the input data must be a 0/1 matrix (1 = yes), which also informs about the residual correlation, the statistics of adjustment and the corresponding standard errors, Rasch reliability, and individual adjustment statistics.

- Calibrate the measurement derived from a scale applied in a context (for example, country) to the metric of a reference scale, or standard (for example, the scale applied in another country, or in the same country, but through a different survey), or any other standard. The main result is the prevalence rate in the country of interest calculated in specific thresholds along the latent trait. Other different contexts that are also analyzed are geographical, linguistic, cultural, etc.
- Perform a Wald test of independence of sampling in the parameters of severity of the item.

9.10 Calculation of Prevalence

If the objective is to estimate the prevalence at the national level, it is possible to analyze and estimate the data based on the gross scores (zero to eight). While if the objective is to compare between countries, the estimated indices should be compared with the Gallup World Survey data, using the same statistical methodology, equating to the global scale of FIES, and the same severity thresholds used by the Voices of Hunger monitor.

The prevalence of food insecurity at a certain level of severity (FIL) in the population is calculated as the weighted sum of the probability of being severely insecure for all respondents (i) in a sample:

$$FI_L = \Sigma P_{i,L} W_i$$

where Wi are the post-stratification weights that indicate the proportion of individuals or households in the national population represented by each record in the sample.

9.11 Bibliometric Outcomes

Table 9.6 shows the evaluation of the publications on "food insecurity" that have more citations. Google Scholar is used between 2010 and 2019.

Table 9.7 shows the academic journals that published more studies on "food insecurity". Scopus is used between 2010 and 2019.

Table 9.8 shows the list of institutions that have published scientific articles on "food insecurity." You can see only universities of Canada and United States of America appear.

The cocitation analysis allowed obtaining three conglomerates of authors who published studies on "food insecurity," which were indexed in Scopus (see Figure 9.1).

Each conglomerate is identified by a characteristic color that groups the authors that are part of it and that were obtained from Scopus. Thus, in brown color, Vraig Gundersen of University of Illinois at Urbana-Champaign, whose research is based

Table 9.6 Publications on "food insecurity" most cited in Google Scholar

N°	Title	Author	Year	Cites
1.	Historical warnings of future food insecurity with unprecedented seasonal heat	Battisti & Naylor	2009	1321
2.	Measuring food insecurity	Barrett	2010	869
3.	Food insecurity is associated with chronic disease among low-income NHANES participants	Seligman, Laraia, & Kushel	2009	816
4.	Exploring mediators of food insecurity and obesity: a review of recent literature	Franklin et al.	2012	299
5.	Food insecurity and weight status among US children and families: a review of the literature	Larson & Story	2011	287
6.	The economics of food insecurity in the United States	Gundersen, Kreider, & Pepper	2011	277
7.	Development and validity of a 2-item screen to identify families at risk for food insecurity	Hager et al.	2010	276
8.	Conceptual framework for understanding the bidirectional links between food insecurity and HIV/AIDS	Weiser et al.	2011	257
9.	Food insecurity and health outcomes	Gundersen & Ziliak	2015	237
10.	Food insecurity as a barrier to sustained antiretroviral therapy adherence in Uganda	Weiser et al.	2010	233
11.	Food insecurity and HIV/AIDS: current knowledge, gaps, and research priorities	Anema, Vogenthaler, Frongillo, Kadiyala, & Weiser	2009	228
12.	HIV/AIDS, undernutrition, and food insecurity	Ivers et al.	2009	227
13.	How much does the Supplemental Nutrition Assistance Program reduce food insecurity?	Ratcliffe, McKernan, & Zhang	2011	219
14.	Position of the American Dietetic Association: food insecurity in the United States	Holben	2010	210
15.	Poverty, food insecurity, and the behavior for childhood internalizing and externalizing disorders	Slopen, Fitzmaurice, Williams, & Gilman	2010	203
16.	Food insecurity: special considerations for women	Ivers & Cullen	2011	188

Source: Google Scholar, 01/08/2019
Search command: "food insecurity"

Table 9.7 Academic journals that published more studies on "food insecurity"

N°	Journal	Scopus
1.	Public Health Nutrition	102
2.	Journal Of Nutrition	91
3.	Journal Of Hunger And Environmental Nutrition	80
4.	Plos One	36
5.	Journal Of Nutrition Education And Behavior	33
6.	Social Science And Medicine	27
7.	BMC Public Health	26
8.	AIDS And Behavior	25
9.	Food Security	25
10.	Ecology Of Food And Nutrition	24

Source: Scopus, 28/07/2019.
Search command in Scopus: TITLE ("food insecurity") OR (food AND insecurity)
Own elaboration

Table 9.8 Institutions that published more studies on "food insecurity"

N°	Institution	Scopus
1.	University of California, San Francisco	96
2.	University of Toronto	70
3.	University of South Carolina	69
4.	Cornell University	53
5.	Harvard Medical School	51
6.	McGill University	49
7.	San Francisco General Hospital and Trauma Center	47
8.	Yale University	41
9.	Massachusetts General Hospital	40
10.	Emory University	39

Source: Scopus, 28/07/2019. Search command in Scopus: TITLE("food insecurity")
Own elaboration

on programs in food insecurity. The celestial node represents Edward Frongillo of University of South Carolina, with research oriented to food insecurity and women. The authors around the main author share similar contents and approach.

Closing Remarks

FIES is a measurement tool that completes the existing set of food and nutrition security indicators and has the capacity to provide updated information on the prevalence of people who struggle every day to have access to a safe and nutritious diet (FAO, 2016), so that national institutions can use the FIES to know the prevalence and severity of food insecurity in different sectors or geographical areas of their population, the causes and consequences for formulating more specific policies, and

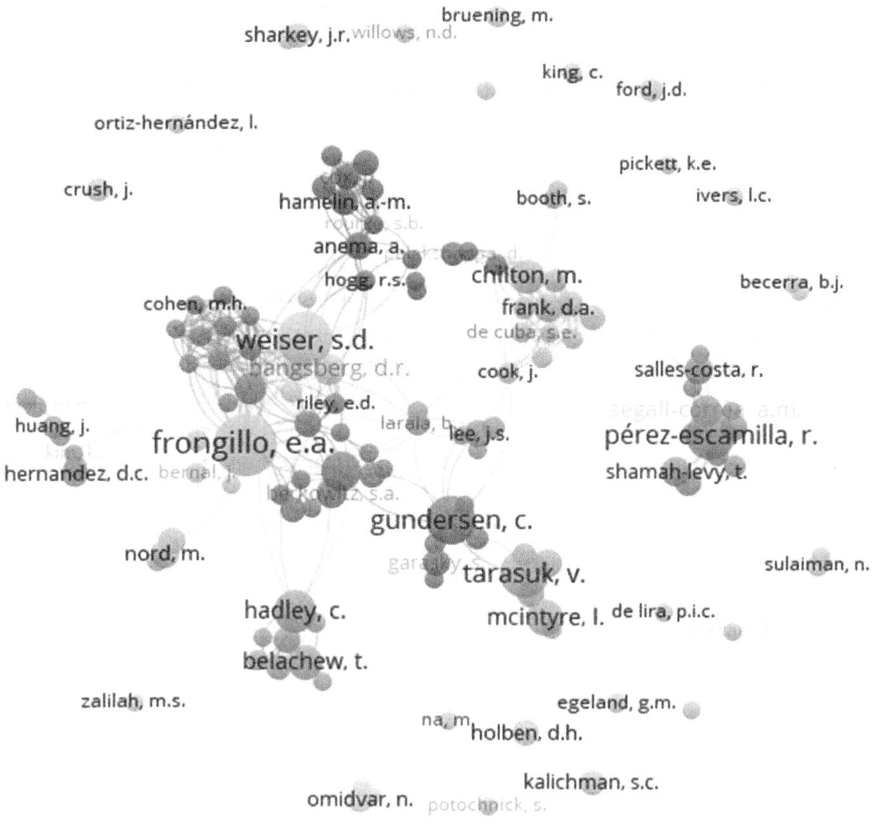

Fig. 9.1 Conglomerate of authors with greater cocitation in Scopus 2010–2019. Source: Scopus, 28/07/2019. Only authors who had at least 5 citations were considered

interventions that allow the effective realization of the right to an adequate diet. FIES makes it easy to estimate the proportion of the population that experiences food insecurity with different levels of severity, which also allows identifying the risk factors of food insecurity in people or households and its consequences in different contexts. Regarding the FIES data analysis, it is necessary to analyze the questions comprehensively, as a scale, not as separate items, to avoid measurement errors; in addition, to compare the results between countries, statistical techniques should be used for the models based on the theory of response to the item (TRI), such as the Rasch model, used for unobservable characteristics in tests of the educational and psychological field. It is recommended to validate the scale in countries with high levels of poverty, in order to establish public strategies that allow them to face this challenge, aligned with the objectives of the second SDG. Likewise, to take a better advantage of the application of FIES and know the prevalence of child food safety, raised by Fram, Bernal, and Frongillo (2015) it is possible to include them in the National Surveys, such as the Multiple Group Indicator Surveys (MICS) or in the Demographic and Health Surveys (DHS), to determine the prevalence of children living in households characterized by food insecurity, as an important step to know data associated with context and culture. As it has been possible to review, food insecurity is a global problem, independent of the socio-economic level of a country, involving the death of many people and a subhuman living condition. For this reason, this chapter has sought to show the available evidence for its measurement but mainly its real conceptual understanding and to see the different areas of impact. Thus, food insecurity is a consequence of violence against women, climate change, and informality, especially which associated with mass migration. However, these same reasons are consequences of food insecurity because it generates violence against women, promotes survival by carrying out productive activities that pollute the environment and promote migration due to lack of food. The authors are committed to the investigation of causes and effects of food insecurity in order to generate management models that allow a correct and complete understanding of the concept, know the causes, and mechanisms involved and their consequences in order to generate successful local, national, regional, and global programs that contribute to the reduction of food insecurity, especially among the most vulnerable populations.

References

Anema, A., Vogenthaler, N., Frongillo, E. A., Kadiyala, S., & Weiser, S. D. (2009). Food insecurity and HIV/AIDS: Current knowledge, gaps, and research priorities. *Current HIV/AIDS Reports, 6*(4), 224–231.

Ballard, T., Kepple, A., Cafiero, C., & Schmidhuber, J. (2014). Better measurement of food insecurity in the context of enhancing nutrition. *Ernahrungs Umschau, 61*(2), 38–41. https://doi.org/10.4455/eu.2014.007

Barrett, C. B. (2010). Measuring food insecurity. *Science, 327*(5967), 825–828.

Battisti, D. S., & Naylor, R. L. (2009). Historical warnings of future food insecurity with unprecedented seasonal heat. *Science, 323*(5911), 240–244.

Cafiero, C., Melgar-Quiñonez, H., Ballard, T., & Kepple, A. (2014). Validity and reliability of food security measures. *Annals of the New York Academy of Sciences, 1331*, 230–248. https://doi.org/10.1111/nyas.12594

Cafiero, C., Viviani, S., & Nord, M. (2018). Weighted Rasch Modeling and Extensions using Conditional Maximum Likelihood., (p21). Retrieved February 5, 2020, from https://cran.r-project.org/web/packages/RM.weights/RM.weights.pdf

Carlson, S., Andrews, M., & Bickel, G. (1999). Measuring food insecurity and hunger in the United States: Development of a national benchmark measure and prevalence estimates. *The Journal of Nutrition, 129*(2), 510–516. https://doi.org/10.1093/jn/129.2.510S

FAO. (2015). The Food Insecurity Experience Scale (FIES). Guidance for translation: Intended meanings of the questions and specific terms-Spanish. (p3). Rome: FAO. Retrieved February 5, 2020, from http://www.fao.org/3/a-be898s.pdf

FAO. (2016). Voices of the hungry: Food Insecurity Experiences Scale. One Metric for the World. FAO. Retrieved February 5, 2020, from http://www.fao.org/3/a-c0053e.pdf

FAO. (2017). The Food Insecurity Experience Scale: Measuring food insecurity through people's experiences. Retrieved February 5, 2020, from http://www.fao.org/3/a-i7835e.pdf

FAO. (2018). Proyecto Voces del Hambre. Escala de seguridad alimentaria basada en la experiencia. Módulos de la encuesta, (p2). Retrieved February 5, 2020, from http://www.fao.org/3/a-bl404s.pdf

Fram, M., Bernal, J., & Frongillo, E. (2015). *The Measurement of Food Insecurity among Children: Review of Literature and Concept Note*. Office of Research Working Paper, 32.

Franklin, B., Jones, A., Love, D., Puckett, S., Macklin, J., & White-Means, S. (2012). Exploring mediators of food insecurity and obesity: A review of recent literature. *Journal of Community Health, 37*(1), 253–264.

Gundersen, C., Kreider, B., & Pepper, J. (2011). The economics of food insecurity in the United States. *Applied Economic Perspectives and Policy, 33*(3), 281–303.

Gundersen, C., & Ziliak, J. P. (2015). Food insecurity and health outcomes. *Health Affairs, 34*(11), 1830–1839.

Hager, E. R., et al. (2010). Development and validity of a 2-item screen to identify families at risk for food insecurity. *Pediatrics, 126*(1), e26–e32.

Holben, D. (2010). Position of the American Dietetic Association: Food insecurity in the United States. *Journal of the American Dietetic Association, 110*(9), 1368–1377.

Ivers, L. C., & Cullen, K. A. (2011). Food insecurity: Special considerations for women. *The American journal of clinical nutrition, 94*(6), 1740S–1744S.

Ivers, L. C., Cullen, K. A., Freedberg, K. A., Block, S., Coates, J., Webb, P., et al. (2009). HIV/AIDS, undernutrition, and food insecurity. *Clinical Infectious Diseases, 49*(7), 1096–1102.

Larson, N. I., & Story, M. T. (2011). Food insecurity and weight status among US children and families: A review of the literature. *American Journal of Preventive Medicine, 40*(2), 166–173.

Nord, M. (2014). *Introduction to item response theory applied to food security measurement: Basic concepts, parameters, and statistics*. Technical Paper (pag. 20). Rome: FAO.

Ratcliffe, C., McKernan, S. M., & Zhang, S. (2011). How much does the supplemental nutrition assistance program reduce food insecurity? *American Journal of Agricultural Economics, 93*(4), 1082–1098.

Seligman, H. K., Laraia, B. A., & Kushel, M. B. (2009). Food insecurity is associated with chronic disease among low-income NHANES participants. *The Journal of Nutrition, 140*(2), 304–310.

Silva, E., Medeiros, D., Martins, P., Sousa Lima, G., Rêgo, M., & Silva, F. (2017). Insegurança alimentar em comunidades rurais no Nordeste brasileiro: faz diferença ser quilombola? *Cadernos de Saúde Pública, 33*(4), 1–14. https://doi.org/10.1590/0102-311X00005716

Slopen, N., Fitzmaurice, G., Williams, D. R., & Gilman, S. E. (2010). Poverty, food insecurity, and the behavior for childhood internalizing and externalizing disorders. *Journal of the American Academy of Child & Adolescent Psychiatry, 49*(5), 444–452.

Sperandio, N. (2015). Prevalência de insegurança alimentar domiciliar e fatores associados em famílias com pré-escolares, beneficiárias do Programa Bolsa Família de Viçosa, Minas Gerais, Brasil. *Epidemiologia e Serviços de Saúde*. https://doi.org/10.5123/S1679-49742015000400016

Weiser, S. D., Tuller, D. M., Frongillo, E. A., Senkungu, J., Mukiibi, N., & Bangsberg, D. R. (2010). Food insecurity as a barrier to sustained antiretroviral therapy adherence in Uganda. *PloS one, 5*(4), e10340.

Weiser, S. D., et al. (2011). Conceptual framework for understanding the bidirectional links between food insecurity and HIV/AIDS. *The American Journal of Clinical Nutrition, 94*(6), 1729S–1739S.

World Bank (2000). World Development Report 2000/2001: Attacking Poverty. Washington, DC: World Bank.

Suggested Reading and Websites

Sinclair, K., Ahmadigheidari, D., Dallmann, D., Miller, M., & Melgar-Quiñonez, H. (2019). Rural women: Most likely to experience food insecurity and poor health in low-and middle-income countries. *Global Food Security, 23*, 104–115.

Sani, S., & Kemaw, B. (2019). Analysis of households food insecurity and its coping mechanisms in Western Ethiopia. *Agricultural and Food Economics, 7*(1), 5.

Lentz, E. C., Michelson, H., Baylis, K., & Zhou, Y. (2019). A data-driven approach improves food insecurity crisis prediction. *World Development, 122*, 399–409.

Osei-Kwasi, H. A., Nicolaou, M., Powell, K., & Holdsworth, M. (2019). "I cannot sit here and eat alone when I know a fellow Ghanaian is suffering": Perceptions of food insecurity among Ghanaian migrants. *Appetite, 140*, 190–196.

Christian, A. K., Marquis, G. S., Colecraft, E. K., Lartey, A., & Soueida, R. (2019). Household food insecurity but not dietary diversity is associated with children's mean micronutrient density adequacy in rural communities across Ghana. *Nutrition, 65*, 97–102.

Patil, S. et al. (2019). Elevated highly sensitive C-reactive protein and d-dimer levels are associated with food insecurity among people living with HIV in Pune, India. Public health nutrition, 1-8.

Atuoye, K. N., Antabe, R., Sano, Y., Luginaah, I., & Bayne, J. (2019). Household Income Diversification and Food Insecurity in the Upper West Region of Ghana. *Social Indicators Research*, 1–22.

Rubio, R., Grineski, S. E., Morales, D. X., & Collins, T. W. (2019). The Role of Parents' Nativity in Shaping Differential Risks of Food Insecurity Among US First Graders. *Maternal and child health journal*, 1–9.

Oberle, M. M., Romero Willson, S., Gross, A. C., Kelly, A. S., and Fox, C. K. (2019). Relationships among Child Eating Behaviors and Household Food Insecurity in Youth with Obesity. Childhood Obesity.

Distel, L. M., Egbert, A. H., Bohnert, A. M., & Santiago, C. D. (2019). Chronic Stress and Food Insecurity: Examining Key Environmental Family Factors Related to Body Mass Index Among Low-Income Mexican-Origin Youth. *Family & community health, 42*(3), 213–220.

Websites

Institute of Research. Universidad de San Martin de Porres. Available in https://www.administracion.usmp.edu.pe/investigacion

Hunger and Health. Available in https://hungerandhealth.feedingamerica.org

The food foundation. Available in https://foodfoundation.org.uk

End Hunger UK. http://endhungeruk.org

Proof. Available in https://proof.utoronto.ca

Chapter 10
Violence Against Women and Sustainable Cities

Arístides Vara-Horna

Abstract Sustainable cities pose challenges that go beyond the environmental dimension. By definition, a sustainable city is "a safe, healthy, attractive, orderly city, with respect to the environment and its historical and cultural heritage, governable, competitive, efficient in its operation and development, so that its inhabitants can live in a comfortable environment, promoting increased productivity, and that can be bequeathed to future generations". In this regard, safety, health, respect for historical and cultural heritage, comfortable habitat, productivity, and intergenerational legacy are elements that transcend the environmental.

Keywords Violence against women · Violence · Gender · Sustainable cities · VAW

10.1 Can Sustainable Cities Be Achieved Without Preventing Violence Against Women?

Peaceful coexistence in cities, for example, requires contexts where all its citizens can fully exercise their rights. Citizen security is usually an important driver within local policies; however, this usually lacks a gender approach. As in most cities, women represent approximately 50% of the population; and although they legally have the same rights and duties, in practice they have serious restrictions based on gender, mainly in the domestic sphere, with serious public repercussions.

Violence against women is a serious chronic and persistent problem in almost all latitudes. WHO estimates that at least 1 in 3 women has been physically and/or sexually assaulted by their partner, bringing enormous impacts on the lives of women, their families, and also for their communities. Is it possible to talk about sustainable societies when half of its citizens—because of the violence suffered—are limited in their resources and capacities for citizen exercise? Can we talk about sustainable societies when women suffer from both public and private insecurity, as

A. Vara-Horna (✉)
Universidad de San Martin de Porres, Lima, Peru
e-mail: avarah@usmp.pe

A. Alvarez-Risco et al. (eds.), *Building Sustainable Cities*,
https://doi.org/10.1007/978-3-030-45533-0_10

well as damage to their health consequences of violence? Can we talk about sustainable cities when within their historical and cultural heritage there are misogynist historical–cultural patterns that despise women and justify violence towards them? Can we talk about sustainable cities when the economic productivity of cities is reduced as a result of violence against women? Can we talk about sustainable cities when the product of violence is perpetuated from generation to generation? A review of the main effects of violence against women, so far documented, will show that thinking of sustainable cities, not including comprehensive prevention of violence against women, is a major mistake.

10.2 Limit Women's Mobility

Violence against women takes many forms. One of the most frequent in the public sphere is sexual violence, which in its most overlapping form is "street sexual harassment" and in its bloodiest form is rape. In this regard, several investigations have shown that sexual violence against women is a very prevalent phenomenon, where 1 in 4 women has been the victim of this type of abuse (WHO).

In 2017, the National Institute of Statistics and Informatics of Peru conducted a National Specialized Survey on Victimization (ENEVIC in Spanish), finding that 29.4% of the population aged 15 and over in Metropolitan Lima was victim of a criminal act; Villa El Salvador being the district with the highest percentage of victimization (38.3%). Analyzing this data, it is found that although insecurity for economic reasons (insecurity in ATMs, banks) is the second reason for insecurity, these do not have significant differences by sex, since both men and women share the same perception. However, analyzing non-institutionalized public spaces, there are significant differences due to gender. Thus, women feel more insecure than men in public spaces (street, markets, and parks), on public transport, and when they walk at night in their neighborhood (see Fig. 10.1a). It should be noted that gender-based violence, in public spaces, rests on the fear of sexual violence. Indeed, as seen in Fig. 10.1b, the fear of sexual violence (harassment, abuse, rape, etc.) is only significantly higher in women. And although, in general, citizen insecurity is a central concern in men and women, it only restricts public activities and women's mobilization. In fact, according to ENEVIC data, a higher percentage of women in Villa El Salvador, unlike men, have stopped going outside, avoid being late home, and avoid taking taxis to get around (Fig. 10.1c).

10.3 Limit Political Participation

The political participation of women is an unavoidable necessity of democratic countries. However, the literature is consistently finding various levels of prevalence of sexism and violence against women in politics. Some studies have found that

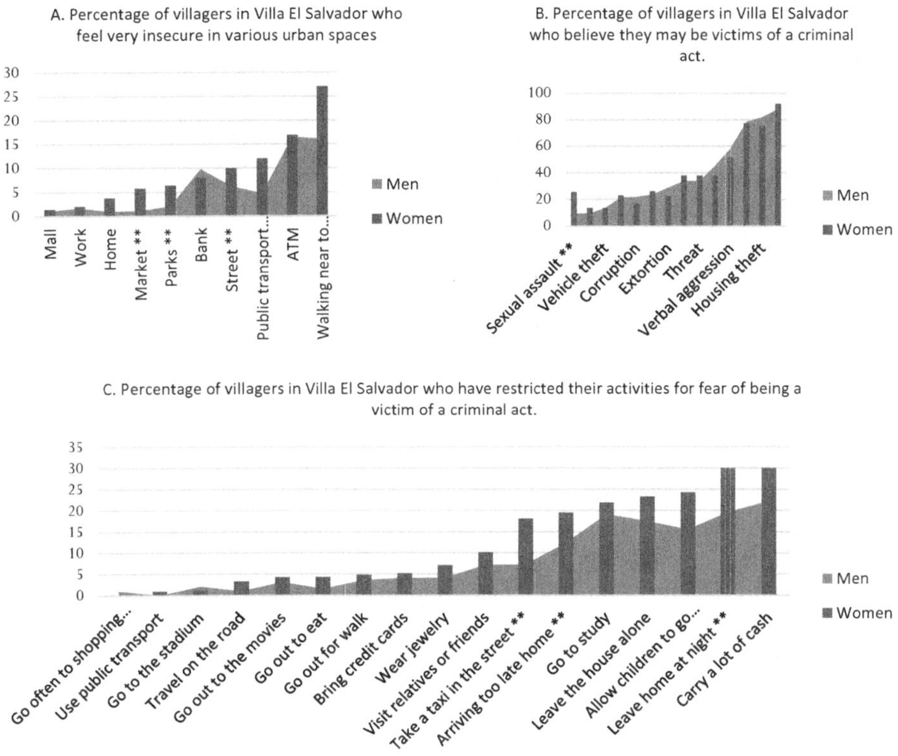

Fig. 10.1 Citizen security indicators with a gender approach in Villa El Salvador. Note: $n = 404$ inhabitants aged 18–65, 48% men, 52% women. (**) Significant differences at 95%. Fig. 10.1b: (**) Only significant difference in sexual assault (X2 = 16,274; gl = 1; $p < 0.001$). Source: INEI—ENEVIC 2017. Elaboration: Arístides Vara-Horna (2019)

both sexism and violence can deter women from participating in politics, because it is a very high cost associated with gender (Haraldsson & Wängnerud, 2019; Huang, 2018; UN Women, 2018). Other studies have found that traditional gender norms can be a serious obstacle for women themselves to participate in politics (Kage, Rosenbluth, & Tanaka, 2018).

In a study conducted in Peru by the United Nations Development Program (UNDP) (Vara-Horna 2019), it has been found that violence against women in politics is an instrumental means that harms physically and emotionally to restrict the political rights of women. In fact, 7.1% of women attacked report having stopped participating in meetings of their community, women's associations, and political parties. The picture is even more depressing in the case of violence against women perpetrated by their partners. 10.6% of women assaulted by their partners reported having stopped participating in meetings of their community, women's associations, and political parties, and the variables that explain it most fall into serious sexual and physical violence: "it has forced her to have sex through blackmail, threats, use of force, weapons, alcohol or drugs" ($r = 0.258$, $p \leq 0.01$), "Has physically forced her to have sex" ($r = 0.287$, $p \leq 0.01$) and "has tried to hang her" ($r = 0.279$, $p \leq 0.01$).

The full exercise of women's political rights should not be an illusion, but should be a guarantee within sustainable cities. However, these results show that women in politics are a very vulnerable group to violence, and that prevention actions, in this case, require special measures, because not only the attacked women cease to participate in politics, but also discourage other women who for fear of suffering the same decline (Campbell & Lovenduski, 2015; National Democratic Institute, 2018; Torres, 2017). Finally, if our political participation is affected, less room is left for positions of political representation occupied by women.

10.4 Limit Your Financial Resources

At the individual level, gender-based violence (GBV) impacts negatively on women's health, increasing their physical and mental morbidity (Cerulli et al., 2012; Coker, Smith, Bethea, King, & McKeown, 2000; Constantino, Sekula, Rabin, & Stone, 2000; Ellsberg et al., 2008; Garcia-Moreno & Watts, 2011; Golding, 1999; Humphreys, & Absler 2011; Nixon et al., 2004; Pico-Alfonso et al., 2006; Plichta, 2004; Sutherland et al., 2002; Vung et al., 2009). Consequently, it causes them double damage: (a) the decrease in their income because their productive capacities and work opportunities are reduced (Adams, 2009; Arias & Corso, 2005; Díaz & Miranda, 2010; Franzway, 2008; IDB, 1997; Karpeles, 2004; Laing & Director, 2001; Laing & Bobic, 2002; Morrison & Orlando, 2004; O'Leary, 2009; Ribero & Sánchez, 2005; Swanberg, Logan, & Macke, 2005; Tennessee Economic Council on Women, 2006; Tolman, 2011) and (b) increase in your out-of-pocket expenses to access health care, justice, and personal protection services (Bonomi et al., 2009; Coker et al., 2000; Fishman et al., 2010; Kruse et al., 2011; Lisboa, Barros, & Cerejo, 2008; Rivera et al., 2007; Vara-Horna, 2015b).

10.5 Perpetuates the Intergenerational Exercise of Violence

Gender-based violence not only affects women, it also impacts their homes, mainly their children or minors in their care. Research has consistently found that violence against women can increase the antisocial behavior of children, decrease their school performance, and affect their health (Agüero, 2013; Ehrensaft et al., 2003; Franklin & Kercher, 2012; Lee et al., 2013; Whitfield et al., 2003; Widom et al., 2014). Vara-Horna (2018, 2019), for example, has found that, in Paraguay and Peru, the children of women attacked by their partners lose school days a year, have higher levels of morbidity, and mothers lose days of domestic care.

School failure and performance problems resulting from violence triple the amount that women must allocate for psychological or pedagogical treatment of their children. The assaulted women are more likely to have to attend school citations for misconduct, and pay more expense as well. The mediate effects of violence on the behavior of children generate opportunity costs for women. In addition, violence also triples the probability of women lending money to address these contingencies, or five times the probability of leaving their home without their care (Vara-Horna, 2019).

On the other hand, violence against women also increases the morbidity of children up to five times as a result of accidents with medical effects (sprains, dislocations, fractures, burns, or severe cuts). Child morbidity increases the likelihood that children will miss school, women will have to ask for permission or be absent from work, or that they will have to borrow money to pay for health expenses (Vara-Horna, 2019).

10.6 Destroy the Productivity of Organizations

In recent years, significant progress has been made in several countries in South America in measuring the costs of violence for large- and medium-sized enterprises (Vara-Horna, 2013, 2015, 2015b, 2018), microenterprises (Vara-Horna, 2015a, 2018), basic education schools (Vara et al., 2017), and universities (Vara-Horna, 2016).

It is now known that violence against women significantly affects the business sector, generating absenteeism, tardiness, increasing staff turnover, decreasing work performance (presenteeism), and producing a negative impact on the entire organization (Henderson, 2000; Hoel, Sparks, & Cooper, 2001; Vara-Horna, 2013, 2015; Vara-Horna, et al., 2015; Yodanis, Godenzi, & Stanko, 2000). Latin American surveys find that between 23.1 and 54.8% of collaborators of large and medium enterprises have been attacked during the last twelve months by their partners or ex-partners (ComVoMujer, 2015; Vara-Horna, 2013, 2015), values similar to which have been reported in international literature (Bureau of Labor Statistics, 2006; Cruz & Klinger, 2011; Philbrick, Sparks, Hass, & Arsenault, 2003; Reeves & O'Learly-Kelly, 2007; Reeves & O'Leary-Kelly, 2009; Vara-Horna, 2013). From the above, it is inevitable that within companies there are cases of women assaulted by their partners (Vara-Horna, 2015). But in the companies there are not only assaulted, there are also collaborators who attack their partners or ex-partners and staff that attests to situations of VcM in their colleagues. For each of them, companies assume significant invisible costs that have an impact on labor productivity and added value (Vara-Horna, 2013, 2015).

Indeed, when women workers are attacked by their partners, companies assume invisible costs, since productivity is affected by delays, absences, decreased performance (presentism), increased staff turnover, and deterioration of staff organizational climate (Brown, 2008; Franzway, 2008; Henderson, 2000; O'Leary, 2009;

Patel & Taylor, 2011; Potter & Banyard, 2011; Rothman, Hathaway, Stidsen, & De Vries, 2007; Soroptimist International of the Americas, 2011; Swanberg et al., 2005; Vara-Horna, 2013, 2015; Yodanis et al., 2000). The judicial restraining orders (in the case of the aggressors), the appointments in the courts, the health care of the assaulted (or accompanying them to avoid being reported, in the case of the aggressors), among other reasons, increase the absence of work from both assaulted women and aggressors (Arias & Corso, 2005; CDC, 2003; Karpeles, 2004; Tennessee Economic Council on Women, 2006; Vara-Horna, 2013). Personnel turnover costs due to the VcM (recruitment, selection, induction, and training costs) are significant depending on the type of employment and industry (Hoel et al., 2001). It has been found that assaulted women are twice as likely to be fired (Adams, 2009; Franzway, 2008; Lisboa et al., 2008; Tolman, 2011). Work performance also decreases significantly, as a result of the distractions and concerns generated by the VcM. The reduction of efficiency and quality of work are frequent indicators of "presentism," which means attending the workplace, but without working at real capacity (Al-Modallal, Hall, & Andreson, 2008; Braaf & Barret-Meyering, 2011; Campbell, 2011; ComVoMujer, 2015; Moe & Myrtle, 2004; O'Leary, 2009; Patel & Taylor, 2011; Swanberg et al., 2005; ; Tolman, 2011Vara-Horna, 2013, 2015; Vara-Horna et al., 2015a). This makes sense because between 60 and 70% of violent women have had difficulties in their work performance and have received sanctions or reprimands at work (Brown, 2008; Potter & Banyard, 2011; Soroptimist International of the Americas, 2011; Swanberg et al., 2005; Vara-Horna, 2013, 2015).

10.7 It Implies Invisible Community Costs

Violence against women generates costs and expenses that many assaulted women cannot bear. Then an invisible subsidy emerges from the social capital of women: Family members, neighbors, and acquaintances provide time, work, and money to cover those needs.

The first measurement of family social support was carried out by Nata Duvvury et al. (2016) in Egypt, when she estimated the costs that family members assume when sheltering women attacked in their homes when they flee from their partners. Duvvury found that many women attacked in Egypt received support and assistance from their relatives and acquaintances, and warned that this is representing a huge invisible social cost. In the investigations carried out with companies, it has been found, in the same way, that knowing assaulted colleagues generates a cost of working time, since the personnel close to them usually comfort and assist them (Al-Modallal et al., 2008; Brandwein & Filiano, 2000; Moe & Myrtle, 2004; O'Leary, 2009; Schmidt & Barnett, 2012; Swanberg et al., 2005; Swanberg, Macke, & Logan, 2006, 2007; Tennessee Economic Council on Women, 2006; Tolman, 2011; Vara-Horna, 2013, 2015, 2018; Zachary et al., 2002).

Vara-Horna (2019) has found for Peru that 1 in 10 women assaulted by their partners or ex-partners has taken refuge in the homes of their relatives or acquaintances, about 29 nights per year, and spending $ 10 the last time. However, he has found that the costs assumed by the community go far beyond the shelter. In the Peruvian district of Villa El Salvador, 2 out of every 7 women attacked report having received, together during the last year, 2.7 million dollars in loans and time equivalent to 5.5 million dollars. These data show the power of sorority and community solidarity. However, we must not forget that this money and time received are debt; that is, at some point they will have to be returned. Many may, but it is unlikely, because violence against women is chronic and maintains an intensity that grows over time. Emergencies will not cease, but social capital is likely to run out and women will lose support from the community. Indeed, a woman may ask the same person for help one or sometimes; and this can help you. But what at first was compassion, will eventually become "guilty," because by ignoring the complex dynamics of violence, the supporter will not understand how difficult it is to break the violent cycle and/or separate from the aggressor, and will tend to blame the woman for her "lack of will, determination or neglect." In other cases, the attacked women will not be able to return the loans granted, so they will tend to isolate themselves voluntarily to avoid the embarrassment of late payment. Consequently, the social network will be weakened.

10.8 Maintains Macho and Misogynist Cultures

There are social norms that favor tolerance towards violence against women. These patterns are attitudes, beliefs, and imaginary prejudices present in the population, which justify and therefore tolerate violence. For example, according to the Peruvian survey of Social Relations called ENARES (INEI, 2015), it finds that 44.3% of people justify the woman being assaulted by her partner if they neglect their children, realizing that these beliefs are still internalized in Peruvian society. These figures should not surprise, sexist social norms are still very valid in the culture of cities.

The most visible expressions of violence against women are feminicide and physical violence; however, this phenomenon has many other manifestations equally harmful to society, such as symbolic violence, understood not as a type of violence, but as a way he continues to think and act that naturalizes and reproduces subordination and abuse, especially towards women. It is a violence normalized in society by customs and customs, and is expressed in different ways, including economic control, control of sociability, mobility, moral contempt, aesthetic contempt, sexual contempt, intellectual disqualification, and disqualification professional. Some living examples that can still be seen in cities are, for example, unrealistic and deteriorated image of the Andean woman, use of the image of women as a sexual object, women as the sole responsible for domestic work, toys that reproduce gender stereotypes, news that justify or relativize violence against women, social pages that reproduce and extol male chauvinism.

Closing Remarks

Violence against women is not only a violation of human rights, it is also a serious obstacle to the development of sustainable cities, causing large economic and social costs. For many centuries, violence against women has been underestimated by society, considering it an inevitable and normal aspect in relations between men and women (Vara-Horna, 2014). The truth is that this is not an inevitable problem, but that it can be eliminated through prevention. A prevention policy can reduce the levels of violence and its effects in the medium and long term, as well as establish efficient paths for development through the full exercise of rights.

References

Adams, A. (2009). *Economic and mental health effects of job instability for low-income survivors of intimate partner violence: Two studies*. Doctoral Thesis, Universidad de Michigan.

Al-Modallal, H., Hall, L., & Andreson, D. (2008). Psychometric properties of a modified version of a worksite harassment tool – Preliminary findings. *AAOHN Journal, 56*(7), 309–316.

Agüero, J. (2013). Causal estimates of the intangible costs of violence against women in Latin America and the Caribbean.

Arias, I., & Corso, P. (2005). Average cost per person victimized by an intimate partner of the opposite gender: A comparison of men and women. *Violence and Victims, 20*(4), 379–391.

Bonomi, A. E., Anderson, M. L., Rivara, F. P., & Thompson, R. S. (2009). Health care utilization and costs associated with physical and nonphysical-only intimate partner violence. *Health Services Research, 44*(3), 1052–1067.

Braaf, R., & Barret-Meyering, I. (2011). *Seeking security: Promoting women's economic wellbeing following domestic violence*. Sydney: Australian Domestic and Family Violence Clearinghouse.

Brandwein, R., & Filiano, D. (2000). Toward real welfare reform: The voice of battered women. *Affilia: Journal of Women and Social Work, 15*(2), 224–243.

Brown, J. (2008). The costs of domestic violence in the employment arena: A call for legal reform and community- based education initiatives. *Virginia Journal of Social Policy and the Law, 16*, 1–45.

Bureau of Labor Statistics. (2006, Octubre). *The survey of workplace violence prevention*.

Campbell, R. (2011). *The financial cost of domestic and family violence*. Sydney: Australian Domestic and Family Violence Clearing House.

Campbell, R., & Lovenduski, J. (2015). What should MPs do? Public and parliamentarians' views compared. *Parliamentary Affairs, 68*, 690–708.

CDC – Center for Disease Control. (2003). *Cost of intimate partner violence against women in the Unites States*. Atlanta: CDC, National Center for Disease Control and Prevention.

Cerulli, C., Poleshuck, E., Raimondi, C., Veale, S., & Chin, N. (2012). "What fresh hell is this?" victims of intimate partner violence describe their experiences of abuse, pain, and depression. *Journal of Family Violence, 27*(8), 773–781.

Coker, A., Smith, P., Bethea, L., King, M., & McKeown, R. (2000). Physical health consequences of physical and psychological intimate partner violence. *Archives of Family Medicine, 9*(5), 451–457.

ComVoMujer. (2015). *Los costos empresariales de la violencia contra las mujeres en Paraguay* [The business costs of violence against women in Paraguay]. Asunción: GIZ and Pacto Global Paraguay.

Constantino, R., Sekula, L., Rabin, B., & Stone, C. (2000). Negative life experiences, depression, and immune function in abused and nonabused women. *Biological Research for Nursing, 1*(3), 190–198.

Cruz, A., & Klinger, S. (2011). *Gender-based violence in the world of work: Overview and selected annotated bibliography*. Geneva: ILO.

Díaz, R., & Miranda, J. (2010). Aproximación del costo económico y determinantes de la violencia doméstica en el Perú. *Psicológica, 29*(29), 30.

Duvvury, N., Ozonas Marcos, M., Gadallah, M., Attia, S., El Adly, N., Maged, W., et al. (2016). *The economic cost of gender based violence survey Egypt 2015*. UNFPA, CAPMAS y NCW.

Ehrensaft, M. K., Cohen, P., Brown, J., Smailes, E., Chen, H., & Johnson, J. G. (2003). Intergenerational transmission of partner violence: A 20-year prospective study. *Journal of Consulting and Clinical Psychology, 71*(4), 741.

Ellsberg, M., Jansen, H. A., Heise, L., Watts, C. H., & Garcia-Moreno, C. W. H. O. (2008). WHO multi-country study on Women's health and domestic violence against women study team. Intimate partner violence and women's physical and mental health in the WHO multi-country study on women's health and domestic violence: An observational study. *Lancet, 371*(9619), 1165–1172.

Franzway, S. (2008). *Framing domestic violence: its impact on women's employment*. Presented on the Annual Conference of the Australian Sociological Association. Australia: University of Melbourne.

Franklin, C. A., & Kercher, G. A. (2012). The intergenerational transmission of intimate partner violence: Differentiating correlates in a random community sample. *Journal of Family Violence, 27*(3), 187–199.

Fishman, P. A., Bonomi, A. E., Anderson, M. L., Reid, R. J., & Rivara, F. P. (2010). Changes in health care costs over time following the cessation of intimate partner violence. *Journal of General Internal Medicine, 25*(9), 920–925.

Garcia-Moreno, C., & Watts, C. (2011). Violence against women: An urgent public health priority. *Bulletin of the World Health Organization, 89*, 2–2.

Golding, J. M. (1999). Intimate partner violence as a risk factor for mental disorders: A meta-analysis. *Journal of Family Violence, 14*(2), 99–132. https://doi.org/10.1023/A:1022079418229

Haraldsson, A., & Wängnerud, L. (2019). The effect of media sexism on women's political ambition: Evidence from a worldwide study. *Feminist Media Studies, 19*(4), 525–541. https://doi.org/10.1080/14680777.2018.1468797

Henderson, M. (2000). *Impacts and costs of domestic violence on the Australian Business/Corporate Sector*. Brisbane: Lord Mayor's Women Advisory Committee. Brisbane City Council.

Hoel, H., Sparks, K., & Cooper, C. (2001). *The cost of violence/stress at work and the benefits of a violence/stress-free working environment*. Report Commissioned by the International Labour Organization (ILO). Manchester: Universidad de Manchester.

Huang, C. (2018). Why low political participation of rural women in China: An interpretation from neo-institutionalism perspective. *Open Journal of Political Science, 8*, 250–262. https://doi.org/10.4236/ojps.2018.83018

Humphreys, C., & Absler, D. (2011). History repeating: Child protection responses to domestic violence. *Child & Family Social Work, 16*(4), 464–473.

INEI (2015). Perú - Encuesta Nacional sobre Relaciones Sociales 2015. Available in https://webinei.inei.gob.pe/anda_inei/index.php/catalog/581 Accessed 30 April, 2020.

Inter-American Development Bank. (1997). *Economic and social progress in Latin America*. Washington: IADB.

Lee, R. D., Fang, X., & Luo, F. (2013). The impact of parental incarceration on the physical and mental health of young adults. *Pediatrics, 131*(4), e1188–e1195.

Kage, R., Rosenbluth, F., & Tanaka, S. (2018). What explains low female political representation? Evidence from survey experiments in Japan. *Politics and Gender*, 1–25. https://doi.org/10.1017/S1743923X18000223

Karpeles, M. (2004). Domestic violence should be workplace concern, too. *Crain's Chicago Business, 27*(40), 11–12.

Kruse, M., Sørensen, J., Brønnum-Hansen, H., & Helweg-Larsen, K. (2011). The health care costs of violence against women. *Journal of Interpersonal Violence, 26*(17), 3494–3508.
Laing, L., & Director, F. C. (2001). Research and evaluation of interventions with women affected by domestic violence. Australian domestic & family violence clearinghouse.
Laing, L., & Bobic, N. (2002). Economic costs of domestic violence. Sydney: Australian domestic and family violence clearinghouse.
Lisboa, M., Barros, P., & Cerejo, S. (2008). *Custodios sociales y económicos de violencia ejercicio contra mulheres en Portugal: dinámicas y procesos socioculturales* [Custos sociais e económicos da violencia exercida contra as mulheres em Portugal: dinámicas e procesos socioculturais]. Lisboa: V Congreesso Português de Sociologia.
Moe, A., & Myrtle, B. (2004). Abject economics: The effects of battering and violence on women's work and employability. *Violence Against Women, 10*(1), 29–55.
Morrison, A., & Orlando, M. B. (2004). The costs and impacts of gender-based violence in developing countries: Methodological considerations and new evidence. *Retrieved March, 5*, 2007.
NDI – National Democratic Institute. (2018). *No party to violence: Analyzing violence against women in political parties*. Preliminary findings from pilots in Cote D'Ivoire, Honduras, Tanzania and Tunisia. NDI.
Nixon, R. D., Resick, P. A., & Nishith, P. (2004). An exploration of comorbid depression among female victims of intimate partner violence with posttraumatic stress disorder. *Journal of Affective Disorders, 82*(2), 315–320.
O'Leary, P. J. (2009). Men who were sexually abused in childhood: Coping strategies and comparisons in psychological functioning. *Child Abuse & Neglect, 33*(7), 471–479.
Patel, D., & Taylor, R. (2011). *Social and economic costs of violence: The value of prevention*. Washington: The National Academies Press.
Philbrick, J. H., Sparks, M. R., Hass, M. E., & Arsenault, S. (2003). Workplace violence: The legal costs can kill you. *American Business Review, 21*(1), 84.
Pico-Alfonso, M., Garcia-Linares, M., Celda-Navarro, N., Blasco-Ros, C., Echeburua, E., & Martinez, M. (2006). The impact of physical, psychological, and sexual intimate male partner violence on women's mental health: Depressive symptoms, posttraumatic stress disorder, state anxiety, and suicide. *Journal of Women's Health, 15*(5), 599–611.
Plichta, S. (2004). Intimate partner violence and physical health consequences: Policy and practice implications. *Journal of Interpersonal Violence, 19*(11), 1296–1323. https://doi.org/10.1177/0886260504269 68
Potter, S., & Banyard, V. (2011). The victimization experiences of women in the workforce: Moving beyond single categories of work or violence. *Violence and Victims, 26*(4), 513–532.
Reeves, C., & O'Learly-Kelly, A. (2007). The effects and costs of intimate partner violence for work organizations. *Journal of Interpersonal Violence, 22*(3), 327–344.
Reeves, C., & O'Leary-Kelly, A. (2009). *Study of the effects of intimate partner violence on the workplace* (pp. 1–57). Fayetteville: Department of Management University of Arkansas.
Ribero, R., & Sánchez, F. (2005). Determinants, effects and costs of domestic violence. *Documento.*, CEDE, 38.
Rivera-Rivera, L., Allen-Leigh, B., Rodríguez-Ortega, G., Chávez-Ayala, R., & Lazcano-Ponce, E. (2007). Prevalence and correlates of adolescent dating violence: Baseline study of a cohort of 7960 male and female Mexican public school students. *Preventive Medicine, 44*(6), 477–484.
Rothman, E., Hathaway, J., Stidsen, A., & De Vries, H. (2007). How employment helps female victims of intimate partner violence: A qualitative study. *Journal of Occupational Health Psychology, 12*(2), 136–143.
Schmidt, M., & Barnett, A. (2012). *Effects of domestic violence on the workplace: A Vermont survey of male offenders enrolled in batterer intervention programs*. Burlington, VT: University of Vermont, Center for Rural Studies.
Soroptimist International of the Americas. (2011). *White paper: Domestic violence as a workplace concern* (pp. 1–13). Philadelphia: Soroptimist.
Sutherland, C. A., Bybee, D. I., & Sullivan, C. M. (2002). Beyond bruises and broken bones: The joint effects of stress and injuries on battered women's health. *American Journal of Community Psychology, 30*(5), 609–636.

Swanberg, J., Logan, T., & Macke, C. (2005). Intimate partner violence, employment, and the workplace: Consequences and future directions. *Trauma, Violence and Abuse, 6*(4), 286–312.

Swanberg, J., Macke, C., & Logan, T. (2006). Intimate partner violence, women, and work: Coping on the job. *Violence and Victims, 21*(5), 561–578.

Swanberg, J., Macke, C., & Logan, T. (2007). Working women making it work: Intimate partner violence, employment, and workplace support. *Journal of Interpersonal Violence, 22*(3), 267–292.

Tennessee Economic Council on Women. (2006). *The impact of domestic violence on the Tennessee economy*. Nashville: State of Tennessee economic council on women.

Tolman, R. (2011). *Impact of intimate partner violence on economic well-being*. Wisconsin: Center of Financial Security.

Torres, I. (2017). *Violence against women in politics: Research on political parties in Honduras*. NDI.

UN Women. (2018). *Violence against women in politics*. Expert group meeting report and recommendations. New York.

Vara-Horna, A. (2013). *Los costos empresariales de la violencia contra las mujeres en el Perú. Una estimación del impacto de la violencia contra las mujeres en relaciones de pareja en la productividad de las empresas peruanas* [The business costs of violence against women in Peru. An estimate of the impact of violence against women in relationships in the productivity of Peruvian companies]. Lima: USMP and ComVoMujer.

Vara-Horna, A. (2015). *Los costos empresariales de la violencia contra las mujeres en Bolivia. Una estimación del impacto invisible para la productividad de la violencia contra las mujeres en relaciones de pareja* [The business costs of violence against women in Bolivia. An estimate of the invisible impact on the productivity of violence against women in relationships]. La Paz: ComVoMujer and USMP.

Vara-Horna, A. (2016). *Impacto de la violencia contra las mujeres en la productividad laboral: Una comparación internacional entre Bolivia, Paraguay y Perú* [Impact of violence against women on labor productivity: An international comparison between Bolivia, Paraguay and Peru]. Lima: GIZ and USMP.

Vara-Horna, A. (2018). *Los costos-país de la violencia contra las mujeres en Paraguay. Una estimación causal-multinivel del impacto económico de la violencia contra las mujeres en relaciones de pareja* [The country costs of violence against women in Paraguay. A causal-multilevel estimate of the economic impact of violence against women in relationships]. Asunción: GIZ.

Vara-Horna, A. (2019). *Los costos económicos de la inacción en la prevención de la violencia contra las mujeres basada en el género en el distrito de Villa El Salvador: 2018* [The economic costs of inaction in the prevention of gender-based violence against women in the district of Villa El Salvador: 2018]. Lima: United Nations Development Program.

Vara-Horna, A. et al. (2015a). *Los costos de la violencia contra las mujeres en las microempresas formales peruanas. Una estimación de su impacto económico* [The costs of violence against women in Peruvian formal microenterprises. An estimate of its economic impact]. Lima: ComVoMujer and USMP.

Vara-Horna, A. et al. (2015b). *Modelo de gestión para prevenir la violencia contra las mujeres en las empresas. Una propuesta integral para involucrar a las empresas en la prevención de la violencia contra las mujeres en relaciones de pareja* [Management model to prevent violence against women in companies. A comprehensive proposal to involve companies in the prevention of violence against women in relationships]. Lima: ComVoMujer and USMP.

Vung, N. D., Ostergren, P. O., & Krantz, G. (2009). Intimate partner violence against women, health effects and health care seeking in rural Vietnam. *European Journal of Public Health, 19*(2), 178–182.

Widom, C. S., Czaja, S., & Dutton, M. A. (2014). Child abuse and neglect and intimate partner violence victimization and perpetration: A prospective investigation. *Child Abuse & Neglect, 38*(4), 650–663.

Whitfield, C. L., Anda, R. F., Dube, S. R., & Felitti, V. J. (2003). Violent childhood experiences and the risk of intimate partner violence in adults: Assessment in a large health maintenance organization. *Journal of Interpersonal Violence, 18*(2), 166–185.

Yodanis, C., Godenzi, A., & Stanko, E. (2000). The benefits of studying costs: A review and agenda for studies on the economic costs of violence against women. *Journal of the Policy Studies Institute, 21*(3), 263–276.

Zachary, M. J., Schechter, C. B., Kaplan, M. L., & Mulvihill, M. N. (2002). Provider evaluation of a multifaceted system of care to improve recognition and management of pregnant women experiencing domestic violence. *Women's Health Issues, 12*(1), 5–15.

Part III
Environmental Sustainability in Cities

Chapter 11
Sustainable Urban Form and Design

Silvia Vásquez-Sánchez, Aldo Alvarez-Risco, and Shyla Del-Aguila-Arcentales

Abstract The relationship between urban design and citizen health will be developed, with emphasis on the basic element of urban structure, the street. The street is the urban space of greater use; it is the first urban space of contact between people and their city. That daily contact necessarily influences the life, just as homes or offices do.

Keywords Sustainable urban form · Sustainable urban form design · Sustainable urban structure

11.1 Introduction

If we visualize at a general level the streets in the city of Lima, we could remember the discomfort we experience when visiting some of them or the feeling of well-being that we experience when walking through others. What makes them belong to one group or the other? How is it or should it be a healthy street? What is the role of planning and urban design in the production of healthy urban life? Do the streets of Lima have the conditions to offer us a healthy urban life? Just like any architectural product, the street can be modeled and designed with relevant technical parameters, and in that direction is the present chapter. Here, the theme of the role of urban design in the health of people will be addressed. For this, the most basic physical

---

Healthy streets: The role of urban design in public health, the case of Lima

S. Vásquez-Sánchez
Pontificia Universidad Católica del Perú, Lima, Peru
e-mail: sylvia.vasquez@pucp.edu.pe

A. Alvarez-Risco (✉)
Facultad de Ciencias Administrativas y Recursos Humanos, Universidad de San Martín de Porres, Lima, Peru
e-mail: aldo.alvarez@redsaf.org

S. Del-Aguila-Arcentales
Escuela Nacional de Marina Mercante "Almirante Miguel Grau", Callao, Peru

Universidad Nacional de la Amazonia Peruana, Iquitos, Peru
e-mail: sdelaguila@enamm.edu.pe

A. Alvarez-Risco et al. (eds.), *Building Sustainable Cities*,
https://doi.org/10.1007/978-3-030-45533-0_11

activity is chosen as a reference: walking. To build a conceptual framework, we must define public health, well-being, quality of life, and the scope of urban planning and design. It will be defined why walking is a benchmark for understanding a healthy street. The international approaches established by the World Health Organization and the trends and ideas developed in Latin America in this regard will be reviewed. About Lima, a review of some cases of interventions carried out in public spaces and the consequences of the present process of urban densification in some districts of Lima will be presented with the possible effects on public health of citizens. We will also review some cases that connect the promotion of the use of the street with health. Urban planning by itself does not necessarily produce sustainable development, and development does not necessarily generate healthy spaces. Bearing this in mind, the balance between urban physical space and human health should be underlined in urban planning and urban design processes. This transversal vision is necessary. The street is the space common to all citizens and must be a space where health can be found and promoted.

11.2 Initial Concepts

"Health is not only the absence of disease, but also the somatic, psychological and social welfare state of the individual and the community" (World Health Organization—WHO, as cited in Navarro, 1998). Throughout its historical review, Navarro (1998) describes a series of approaches to the definition of public health, emphasizing that it is increasingly obsolete the idea that it is from medicine that diseases are prevented and treated: "health public, as defined, is not a branch of medicine, as conceived in Spain but, on the contrary, medicine constitutes a branch of public health" (Navarro, 1998). We therefore have access to a more multidisciplinary public health definition and, therefore, a cross section of several public management sectors to improve the living conditions of the human being.

With respect to the concept of quality of life, Ardila (2003) reviews several definitions in this regard:

- Quality of life is a composite measure of physical, mental, and social well-being, as perceived by each individual and each group, and of happiness, satisfaction, and reward (Levy and Anderson, 1980, p. 7, as cited in Ardila, 2003, p. 163).
- Quality of life is the subjective evaluation of the good or satisfactory character of life as a whole (Szalai, 1980, as cited in Ardila, 2003, p. 163).
- Quality of life is the patient's appreciation of his life and satisfaction with his current level of functioning compared to what he perceives as possible or ideal (Celia and Tulsky, 1990, as cited in Ardila, 2003, p. 163).
- By definition, the quality of life is the subjective feeling of well-being of the individual (Chaturvedi, 1991, as cited in Ardila, 2003, p. 163).
- It is the multidimensional indicator of the material and spiritual well-being of man in a determined social and cultural framework (Quintero, 1992, as cited in Ardila, 2003, p. 163).

In some way, the definitions of quality of life and well-being raise the relationship with the physical environment. Ardilla (2003) points out: "as objective aspects," the quality of life includes:

1. Material well-being
2. Harmonic relations with the environment
3. Harmonic relations with the community
4. Health objectively considered

Urban planning has historically been characterized by prioritizing physical, geographical, and/or strategic political aspects of cities. Cities since their formative beginnings as villages and towns have been linked to their sources of life and resources such as water, or food, as well as access roads, and in other cases to their centers of power such as churches and centers administrative. Environmental management in relation to health is relatively new, basically corresponds to the late twentieth century. The current urban planning includes the management of the environment basically focused on the handling of impacting agents, as well as the physical articulation of urban functions. Thus, urban mobility has become a fundamental aspect of any sustainable urban development plan. The morphological evolution of cities establishes physical pre-existences and the conformation of streets and public spaces, which normally did not have the current vision of sustainability in their origin. In this scenario, urban planners often only see themselves in the possibility of reducing the impacts of inadequate growth. The level of air and sound pollution, for example, can be managed with the definition of adequate standards for human and environmental health and with the application of policies of sustainable mobility, management of green areas, traffic management, etc.

Carmona et al. (2003) indicates that the concept of urban design has evolved from a purely esthetic vision of the definition of the masses of buildings and the space between them to the more physical and socio-cultural approach of making places for the use and enjoyment of the people. Carmona (2003) also points out that urban design is a process that must integrate multidisciplinary approaches because of the complex nature of the problems to be solved. The different actors, local governments, businesses, urban developers, residents, merchants, and even pedestrians all have interests and a role in urban design processes.

In the field of urban policies, there are clear manifestations on the international level. Dr. Jeffrey Koplan (Jackson and Kochtitzky, 2001), former Director of the "U.S. Centers for Disease Control and Prevention," included in a publication referred to the 10 most important public health challenges for the new millennium, which includes at least 4 very significant references to land use and urban design of cities, these are:

- Integrate physical activity in the daily way of people
- Protect and clean the environment
- Recognize the contributions of mental health to the integrity of the health and well-being of people
- Reduce levels of violence in society

In November 2016, in Shanghai China, the Forum Healthy Cities was developed. The meeting that brought together more than 100 mayors from all over the world realizes that health and well-being are at the base of the 2030 Agenda of the United Nations for Development and its Sustainable Development goals. Health is forged at the local level, in the environments of daily life, in the neighborhoods and communities in which people of all ages live, love, work, study, and play. Health for all is not possible without leadership at the local level and the commitment of citizens. The good health of its citizens is one of the strongest and most unequivocal signs that a city has achieved sustainable development. Therefore, health must occupy a central place in the programs of all mayors (WHO, 2016, p. 2).

In the development of the event, commitments were made to prioritize the political option that favors health in all areas of municipal governance. Five governance principles based on the transformation program of the sustainable development goals (SDG) were proposed:

- Integrate health as a fundamental consideration in all policies
- Address all the determinants of health (social, economic, and environmental)
- Promote a strong participation of the community
- Reorient health and social services towards equity
- Evaluate and monitor well-being, the burden of morbidity and the determinants of health. (WHO, 2016, p. 2)

As we observed, the impact of urban living conditions on people's health has been considered from various angles. The planning and urban design must be consistent with the commitments and challenges for healthy cities towards the incentive of physical activity from public policies.

11.3 Aspects to Consider to Encourage Walking from Urban Design

The benefits of walking for health are well known in the medical community. A recent review of them differentiates the physical and mental benefits of walking activity. At the physical level, walking improves physical composition, cardiovascular function, blood pressure, reduces cholesterol level, improves glucose control, among other physical benefits. At the level of benefits on the mental health of walking, it is indicated that there is growing evidence based on the prevention and treatment of mental health such as depression and anxiety based on regular physical activity (Kelly, Murphy, & Mutrie, 2017).

A study developed in 1996 indicates that:

> The positive environmental determinants of physical activity included pleasant landscapes (presence associated with more activity), while the largest barrier perceived was the lack of a safe place to exercise. The research carried out (…) has also indicated that two of the main

> reasons are: The reasons for not exercising are the lack of structures or facilities (such as sidewalks and parks) and fears about safety. In general, (...) they report that the highest levels of perceived safety in the neighborhood are associated with higher levels of physical activity, with the largest differences between racial or ethnic minorities and people over 65. Therefore, people are more likely to use parks, roads, and bicycle paths when they are easy to reach and are safe and well-maintained. (Jackson and Kochtitzky 2001)

From urban design, we have a series of indicators of success of public spaces related to the activities that take place in them. There are multiple studies in this regard, focused on the socio-functional aspects of urban design. According to Ian Bentley (1999), the streets must have the qualities of permeability, variety, versatility, legibility, appropriate image and visual richness, and personalization to guarantee their continuous use, attractiveness, and safety. The streets must have permanent natural vigilance involving people using it and the possibilities of interacting with the street from the private space according to Jane Jacobs (2001), the sidewalks must be safe and the short blocks propitiating multiple options for walking, and the buildings should propitiate looks and encounters with the street to improve the sense of security. Jan Gehl (2011) concentrates more on the activities, characterizes them so that there are necessary, social and optional activities, where all of them together have the potential to generate vitality, but the streets should propitiate it integrating aspects of protection, comfort, and enjoyment. White (1980) points to the behavior of people, the spaces that generate conversations, meetings, permanence of people alone or accompanied, diversity activities, and the agents of the public space that promote them such as kiosks, benches, green areas, where its distribution and the relationship of these different elements generate "triangulations" or encounters between acquaintances and strangers.

All of them have focused in one way or another on the use of the city and the need to design it for life, to build living spaces, meeting spaces. And that is what is generally being fostered in contemporary urban design. The human scale to promote comfort, protection, and enjoyment (Gehl, 2011) of public spaces. Doing the walking exercise is healthy. It was reviewed that the "walkable" streets can be defined by the users in many ways: safe, attractive, comfortable streets, etc. The user can certainly feel persuaded to walk according to the perception he has of the street.

But not all "walkable" streets in the physical sense have a healthy environment. This aspect is fundamental to consider.

Studies (Marshall, Brauer, and Frank, 2009) conducted in neighborhoods of Vancouver, British Columbia, Canada, investigated the relationship defining attributes considered in the promotion of physical activity, or with greater accuracy "walkableness." They are mixed uses, connectivity in streets, and residential density. The results indicate that near the city center exists high levels of nitric oxide concentrations and low O_3 ozone. Areas with lower income levels tend to have higher concentrations of NO, are walkable but have lower concentrations of O_3. The areas of higher income tend to have less contamination of both NO and O_3 agents. The neighborhoods located near the center of the city but not in the center itself have low pollution and high traffic but were those with higher incomes.

11.4 Sustainable Urban Form and Design, Smart Cities, and Big Data

In 1994, the smart city concept started but it was criticized because did not include explicitly the goals of sustainable development in its definition, as well as for lacking the linking with sustainable city concept (Bifulco et al., 2016). Recently, Ahvenniemi (2017) found differences between smart cities and sustainable cities; additionally, Kramers et al. (2014) emphasized that the concept of smart city mentions little about the environmental sustainability performance of the cities. The authors of this chapter think that is really relevant to include in the concept of smart city the factors oriented to environmental sustainability.

Nowadays, it is not available a formal definition of big data. However, different definitions have been used such as Kitchin (2014) who detailed some characteristic features like huge in volume, high in velocity, diverse in variety, exhaustive in scope, fine-grained in resolution, relational in nature, and flexible.

For Bibri (2018), "big data analytics" denotes "any vast amount of data that has the potential to be collected, stored, retrieved, integrated, selected, preprocessed, transformed, analyzed, and interpreted for discovering new or extracting useful knowledge." Firstly, the analytical outcome can be analyzed and visualized in an easy format before their deployment for decision-making purposes (e.g., an improvement of, or a modification in, a process, operation, function, design, strategy, or policy). In relation to sustainable smart and smarter cities, big data analytics refers to a group of sophisticated and dedicated software applications and database management systems run by machines with very high processing power, which can turn a big amount of urban data into practical and useful knowledge for improved, well-informed decision-making and deep insights in relation to various city domains, such as water, transport, mobility, traffic, environment, energy, pollution, land use, waste management, education, healthcare, and others. In this, according to Bibri (2018), big data have some usual components:

- Advanced data mining tasks and algorithms
- Advanced techniques based on data science, fundamental concepts, and computer science methods
- Cloud and fog computing models
- Data mining models
- Computational mechanisms involving such sophisticated and dedicated software applications, and database management systems
- Data processing platforms
- Modeling and simulation approaches, and prediction and optimization methods.

Currently, the analysis of big data is a crucial component of the infrastructure of information and communication technologies (ICT) of smart cities as it contributes to the improvement of sustainability, resilience, efficiency, and quality of life (Kumar & Prakash, 2014; Hashem et al., 2016) through effective decision-making processes. Thus, big data is focused on supporting smart decisions and improving operational functioning, planning, design, and development of urban systems. In the short term, the analysis of big data and its application will create and allow immense possibilities and opportunities for urban management.

All the growth of big data is transforming the way we think, work and, finally, live daily. Thus, *datafication* is a word in vogue at the time of big data and implies that enormous confidence in big data computing and support technologies is generated. But what exactly does the *datafication* entail? Speaking of datafication implies that cities now depend on their data to function correctly in relation to their urban life, especially in relation to sustainable development (Bibri, 2019). In this way, *datafication* implies, for example, taking data, transport data and predict the busiest places according to day and time, considering the avenues that have damaged tracks, avenues where there is a higher rate of vehicle thefts, etc. This applies to the number of households surveyed, number and type of complaints from the municipalities, control of technical reviews of cars, sanitary authorizations of restaurants, and many more activities in one city. Additionally, the analysis of Internet surfing in a city allows the authorities solving hot problems preventing social troubles and reduces time and money.

This leads to a very big challenge since there are various data in a city that are recorded manually (on a sheet of paper), making it impossible to distribute and analyze it. For example, each hospital in a city (ideally in a country) should have an integrated system that allows to know the stock of medicines, both in each hospital pharmacy and in the warehouse so that any shortages that could have an impact can be predicted and anticipated. Another example is in the vials that must be informed to a system that sends alerts to the different drivers, indicating how long the ride will last and the alternate ways they can use, give information about the works period to allow citizens to know and control the management making them feel part of it. Regarding the level of pollution generated by each car in the city, annually, it is expected that the cars are subject to technical review for authorization to operate in the city. Is the information of the technical reviews centralized or is it decentralized and even privatized, making it almost impossible to have a single record of information? By the way, what happens with the other non-standardized vehicles? Thus, various cities in the world have vehicles that do not require a license plate and even their drivers are not required to take a driver authorization. How can we control such activity in the city?

All climate change actions carried out by cities require a transition of urban sustainability in order to efficiently and massively communicate the need for change to successfully achieve green, sustainable, and low carbon emissions (Bulkeley et al., 2011). Specifically, the field of construction contributes greatly to the emissions of greenhouse gases linked to human activities and, at the same time, it has the possibility of generating programs and certifications that promote the reduction of such emissions as it is increasingly being counted with technologies that optimize the construction, generating the planning and construction of ecological buildings as well as the modernization of existing buildings. In this area, it is very important to mention that there are several investigations that are addressing the ecological construction as well as the description of eco-districts, green ventures (called ecopreneurs), and ecological houses (eco-homes) (Frantzeskaki et al., 2017; Pickerill, 2017).

Sustainable cities are also a political objective since cities contribute a high percentage of greenhouse gas emissions compared to non-urban areas, representing 60–80% of energy consumption, more than 75% of consumption of natural resources and the emission of 75% of global carbon emissions. Approximately 30–40% of final energy consumption is used by buildings (Pérez-Lombard et al., 2008). But as much as buildings and cities contribute to greenhouse gas emissions, they are considered a central part of the solution to our climate crisis (van der Heijden, 2014). Most cities face various challenges that force them to plan and carry out plans to mitigate climate change, including the growth of the population, which puts pressure on infrastructures and urban services that are already in difficulties, as well as an increase in total carbon emissions. As a result, various urban climate change initiatives have been developed but it has been relatively slow. An increasing number of cities have begun to act as climate change leaders setting ambitious carbon emission reduction targets and developing a wide range of strategies to achieve these objectives, including ecological and low carbon infrastructure and services. However, the only way to achieve these objectives over time and that can also be controlled efficiently requires data mining, that is, the detailed analysis (almost one by one) of the data obtained in the cities for generating citizens' screens according to their purchases for home, water and electricity consumption, telephony and Internet, as well as private and public green areas.

11.5 Sustainable Urban Forms Focus in Gender Equity

When planning the creation and development of a city must take into account and one of those, who are gaining more followers are the aspects of sex. City planning is changing, based on research that shows that the patterns of behavior of women and men are different, so cities should be built based on these differences. For example, when sports courts are built, they are usually intended for playing football or basketball and, in addition, are usually used mostly by men. Is there equality in this case that affects children, youth, and even adults?

On the other hand, there are more men who drive cars, so women are usually more pedestrian. Thus, an increase and improvement in pedestrian access and transport connections is needed and not only for cars. Is the transport basically oriented to male users?

Public restrooms usually have the same amount of individual spaces but clearly the time to meet the physiological needs of men and women is different. This explains why there is usually a greater queue of people in women's restrooms compared to men's restrooms. Are we building urban areas equally without taking into account gender differences?

Could it be that urban areas with predominance for men and heteronormative families have been designed and continue to be designed?

The issue is not new, since Hayden (1980) in his article "What would a non-sexist city be like" posed this problem. However, in many cities this aspect has not been mentioned, analyzed, or put on the agenda of the authorities that design the cities. Vienna is the pioneer of equity-based city design through the Women-Work-City project in 1999, in which, for example, volleyball and badminton courts were added to the parks and the use of women and men was equated. The term more used by city officials in Vienna for these changes is Fair Shared City. Definitively a greater engagement with daily urban practices provides critical inputs into how claims to the urban area are inherently gendered. A mission of the mayors of the cities is to investigate in depth what the current needs of its inhabitants are and to have on the agenda the need to investigate about equal access to public areas and generate changes in urban design planning. Another crucial aspect that is gaining more and more attention is the great social and economic impact generated by violence against women, to which it is evident that in the case of all the violence carried out in the streets, it is influenced by urban designs. Thus, we have to ask ourselves, how many streets in our cities do not have sufficient lighting and which facilitates the attack on women? How many parks do not have security personnel that prevent theft of children and teenagers' belongings? Is it possible that we are safe when designing means of mass transport where people travel tight, with minimal safety and body care?

Sustainable cities, in relation to their urban design, go beyond the creation of parks and public recreation spaces. It needs drastic changes to make citizens feel safe, full, with easy access to areas, transportation, in other words: to live. What percentage of our streets has names of men and women? These are aspects that we have not taken into account and that need a reassessment. But do you have rankings that allow us to show how cities are progressing and to demonstrate the need for improvements in urban design?

Taking into account Mercer's Quality of Living Reports (Mercer, 2018) it can be seen that there are 10 criteria that are taken into account for this evaluation:

1. Political and social environment (political stability, crime, law enforcement, etc.).
2. Economic environment (currency exchange regulations, banking services).
3. Socio-cultural environment (media availability and censorship, limitations on personal freedom).
4. Medical and health considerations (medical supplies and services, infectious diseases, sewage, waste disposal, air pollution, etc.).
5. Schools and education (standards and availability of international schools).
6. Public services and transportation (electricity, water, public transportation, etc.).
7. Recreation (restaurants, theaters, cinemas, sports and leisure, etc.).
8. Consumer goods (availability of food/daily consumption items, cars, etc.).
9. Housing (rental housing, household appliances, furniture, maintenance services).
10. Natural environment (climate, record of natural disasters).

On urban designs, points 6 and 9 are the necessary objectives to ensure that a city is sustainable. Thus, regarding public services and transportation, it is necessary to ensure access to transport that avoids excessive crowding, a situation of high risk for the personal and bodily safety of women. All of the above must generate that citizens during the electoral election stages for presidents and mayors may demand proposals that are related to this equality in urban designs. Also, have the power to revoke authorities that do not comply with those offers of improvements in urban dimes. But what are the demands that citizens have? We are convinced that it is universities and research centers that must work in depth to obtain information on the social demand for equality in urban designs and transfer the results for the planning of new cities and the redesign of current cities.

Closing Remarks
There is an increasing mismatch between the ways in which everyday rights are framed within the planning and the direct experiences of marginalized urban dwellers. The current trend leads urban designs to prioritize esthetic aspects but there is evidence that the sustainable development of a city and, finally, of a country has a fundamental basis in research, development, and innovation in equitable urban forms, which ensure the progress of families and individuals. The big data is a crucial tool that must be used more and more to analyze the trends of resource use and population displacement to generate concrete urban design policies and actions that generate equity throughout the population of a city.

References

Ahvenniemi, H., Huovila, A., Pinto-Seppä, I., & Airaksinen, M. (2017). What are the differences between sustainable and smart cities? *Cities, 60*, 234–245.

Anderson, L., & Levy, L. (1980). *La tensión psicosocial. Población, ambiente y calidad de vida.* México.

Ardila, R. (2003). Calidad de vida: una definición integradora. *Revista Latinoamericana de psicología, 35*(2), 161–164.

Bentley, I., Alcock, A., & Murrian, P. (1999). Entornos vitales: Hacia un diseño urbano y arquitectónico más humano manual práctico (No. Sirsi) i9780851399676).

Bibri, S. E. (2019). *Big data science and analytics for smart sustainable urbanism: Unprecedented paradigmatic shifts and practical advances*. Germany, Berlin: Springer.

Bibri, S. E. (2018). The IoT for smart sustainable cities of the future: An analytical framework for sensor-based big data applications for environmental sustainability. *Sustainable Cities and Society, 38*, 230–253.

Bifulco, F., Tregua, M., Amitrano, C. C., & D'Auria, A. (2016). ICT and sustainability in smart cities management. *International Journal of Public Sector Management, 29*(2), 132–147.

Bulkeley, H., Castán Broto, V., Hodson, M., & Marvin, S. (Eds.). (2011). *Cities and low carbon transitions*. London: Routledge.

Carmona, M., Heath, T., Oc, T., & Tiesdell, S. (2003). Urban spaces-public places: The dimensions of urban design.

Celia, D. F., & YTulsky, D. S. (1990). Measuring the quality of life today: Methodological sapects. *Oncology, 4*, 29–38.

Chaturvedi, S. K. (1991). What's important for quality of life to Indians—In relation to cancer. *Social Science & Medicine, 33*(1), 91–94.
Frantzeskaki, N., Castán Broto, V., Coenen, L., & Loorbach, D. (Eds.). (2017). *Urban sustainability transitions*. New York: Routledge.
Gehl, J. (2011). Creating places for people. An urban design protocol for australian cities.
Hashem, I. A. T., et al. (2016). The role of big data in smart city. *International Journal of Information Management, 36*, 748–758.
Hayden, D. (1980). What would a non-sexist city be like? Speculations on housing, urban design, and human work. *Signs, 5*(Suppl. 3), S170–S187.
Jacobs, J. (2001). Morte e Vida de Grandes Cidades–tradução Carlos S. Mendes Rosa, 3.
Jackson, R. J., & Kochtitzky, C. (2001). *Creating a healthy environment. The impact of the built environment on public health*. Washington, DC: Sprawl Watch Clearinghouse.
Kelly, P., Murphy, M., & Mutrie, N. (2017). The health benefits of walking. *Walking: Connecting Sustainable Transport with Health*, 61–79.
Kitchin, R. (2014). The real-time city? Big data and smart urbanism. *GeoJournal, 79*, 1–14.
Kramers, A., Höjer, M., Lövehagen, N., & Wangel, J. (2014). Smart sustainable cities–exploring ICT solutions for reduced energy use in cities. *Environmental Modelling & Software, 56*, 52–62.
Kumar, A., & Prakash, A. (2014). The role of big data and analytics in smart cities. *International Journal of Science and Research, 6*(14), 12–23.
Marshall, J. D., Brauer, M., & Frank, L. D. (2009). Healthy neighborhoods: Walkability and air pollution. *Environmental Health Perspectives, 117*(11), 1752–1759.
Mercer. Vienna tops mercer's 20th quality of living ranking. (2018). Retrieved February 5, 2020, from https://www.mercer.com/newsroom/2018-quality-of-living-survey.html
Navarro, V. (1998). *Concepto actual de la salud pública. Martínez, F., Castellanos, PL, Navarro, V., Salud Pública* (pp. 49–54).
Pérez-Lombard, L., Ortiz, J., & Pout, C. (2008). A review on buildings energy consumption information. *Energy and Buildings, 40*(3), 394–398.
Pickerill, J. (2017). Critically interrogating eco-homes. *International Journal of Urban and Regional Research, 41*(2), 1468–2427.
Quintero, G. (1992). Comunicación personal a J. *Grau, 1996*.
Szalai, A. (1980). *The meaning of comparative research on the quality of life. The quality of life* (pp. 7–21).
van der Heijden, J. (2014). *Governance for urban sustainability and resilience: Responding to climate change and the relevance of the built environment*. Cheltenham: Edward Elgar.
WHO (2016) Consenso de Shangai sobre Ciudades Saludables. Retrieved April 30, 2020, from https://www.who.int/healthpromotion/conferences/9gchp/healthycity-pledge/en/

Chapter 12
Sustainable Transportation in Cities

Aldo Alvarez-Risco, Shyla Del-Aguila-Arcentales, and Marc A. Rosen

Abstract Transportation is an important aspect of cities, but it is also a significant user of energy and source of pollution in cities. Sustainable transportation necessitates the existence of effective, clean, efficient, and socially acceptable transportation. The desire for sustainable transportation obliges managers and stakeholders to address the factors listed above, including planning initiatives to optimize the entire transportation process and its energy use. Achieving sustainable transportation also requires the progressive implementation of intermodal connections that allows the population to make use and be part of a single transport circuit, which can be efficient in terms of time and transportation data traceability. The most relevant indicators for sustainable transportation in cities need to be identified and monitored to ensure that the correct levels of control are being achieved where necessary, such as to limit air pollution levels. The development of new strategies to achieve sustainable transportation is a highly significant objective in most cities, especially since transportation-related problems in cities are only expected to become more prominent in the future as populations grow.

Keywords Transportation · Sustainable transportation · Air pollution · Cars · Train

A. Alvarez-Risco (✉)
Facultad de Ciencias Administrativas y Recursos Humanos, Universidad de San Martín de Porres, Lima, Peru
e-mail: aldo.alvarez@redsaf.org

S. Del-Aguila-Arcentales
Escuela Nacional de Marina Mercante "Almirante Miguel Grau", Callao, Peru

Universidad Nacional de la Amazonia Peruana, Iquitos, Peru
e-mail: sdelaguila@enamm.edu.pe

M. A. Rosen
Faculty of Engineering and Applied Science, University of Ontario Institute of Technology, Oshawa, ON, Canada
e-mail: marc.rosen@uoit.ca

A. Alvarez-Risco et al. (eds.), *Building Sustainable Cities*,
https://doi.org/10.1007/978-3-030-45533-0_12

12.1 Sustainability in Transportation

The concept of sustainability highlights the integration of human activities, making macro and micro planning necessary. In the case of transport in cities, planning clearly has an effect on a large number of people and organizations. This expected impact obliges transportation authorities to set objectives and define procedures that allow monitoring of established plans. In general, the goals for achieving sustainable transport have three dimensions: economic, social, and environmental, as shown in Fig. 12.1, along with some of the objectives within each dimension.

Currie and Hidalgo (2018) synthesized some papers presented at the *15th International Conference on Competition and Ownership in Land Passenger Transport* to discuss the integration of various types of transport in a city. In doing so, Currie and Hidalgo (2018) reviewed perspectives on intermodal integration and intermodal competition, user and operational views on bus rapid transit (BRT) system performance, BRT design and development challenges, and BRT implementation and future pathways. Transportation system diversity is one basis for citizens to experience the efficiency of transport. It is usually thought that traffic problems associated with an excess of private transport vehicles are the main hindrance to achieving sustainable transportation. These certainly need to be addressed; however, there are also several other actions that need to be implemented and standardized.

Environmental impact assessment (EIA) requirements are nowadays being carried out to plan changes and to support decision-making in transportation management. However, it is essential that EIAs include life cycle energy assessments of the different options currently available and also for future proposals. Different transport modes involve varying degrees of construction and maintenance activities. The various fuel options in particular necessitate a life cycle analysis. Figure 12.2 illustrates the general approach of a life cycle analysis (LCA).

Material and energy consumption at various stages of a transport project, i.e., construction, operation, and maintenance, needs to be examined in order to fully understand their impact on environment and to make informed decisions about providing an appropriately environmentally friendly transportation system to the people. The evaluation of a life cycle involves the following steps:

1. Developing a life cycle approach in the transportation sector.
2. Application of the life cycle approach for fuels for transportation.
3. Detailing of life cycle approach for fuel for transportation.
   (a) Fuel consumption expressed *per* passenger-*kilometers (pkm).*
   (b) CO_2 emissions in terms of passenger-*kilometers (pkm).*
   (c) Type and quantity of energy and materials utilized.
4. Calculation using the data in step 3 of the costs involved for each kind of fuel or each kind of transport mode (trains, buses, etc.).

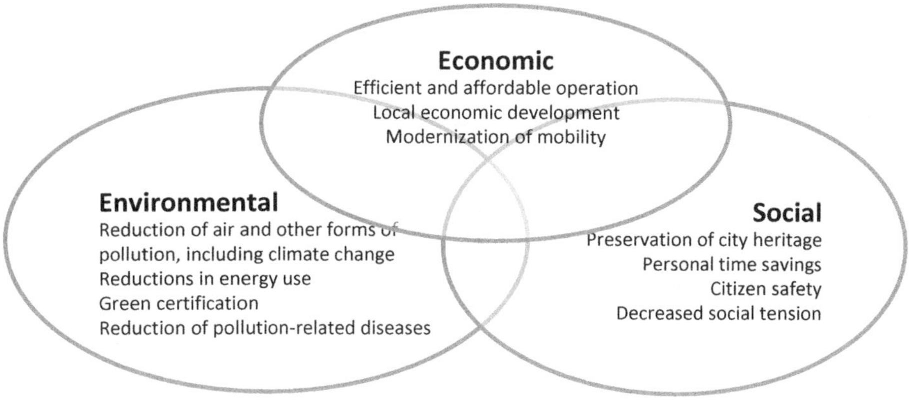

Fig. 12.1 Selected goals of sustainable transportation in cities

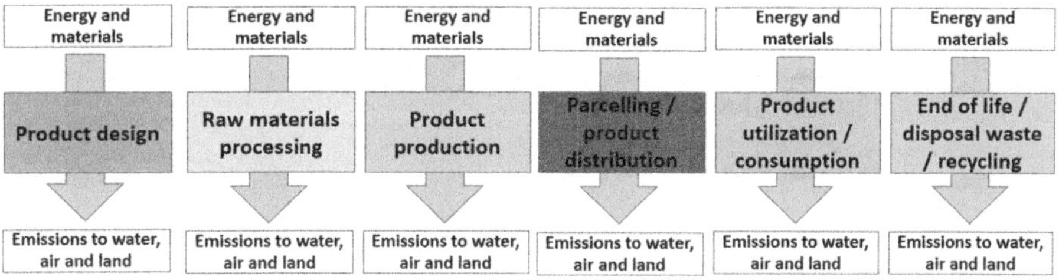

Fig. 12.2 General approach of life cycle analysis

Regarding step 2, mapping the processes of the different types of fuels used in transport buses in a city allows one to know in which phases each process can be optimized, since most if not all types of fuels have users. Thus, each process merits process optimization. Four phases are established: raw material acquisition, production, bus city service, and waste disposal. The different types of fuels for buses have different flows. The authors adopted the life cycle analysis approach in the ISO 14042 framework. It is necessary during city planning that mapping of fuel use is carried out so that the transition from currently available fuels towards eco-friendlier fuels can be managed, with costs accessible to users and with the corresponding certifications.

Life cycle analyses are illustrated for various types of transportation and their phases in Figs. 12.3–12.7. The life cycle analysis is shown in Fig. 12.3 for diesel bus transportation, in Fig. 12.4 for a liquefied natural gas (LNG) bus transportation, in Fig. 12.5 for liquefied petroleum gas (LPG) bus transportation, in Fig. 12.6 for hydrogen fuel cell (HFC) bus transportation, and in Fig. 12.7 for plug-in electric (PE) bus transportation.

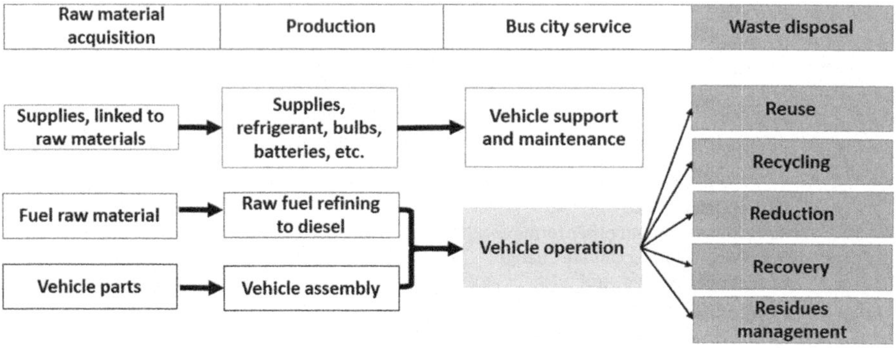

Fig. 12.3 Life cycle process of diesel bus transportation and its phases

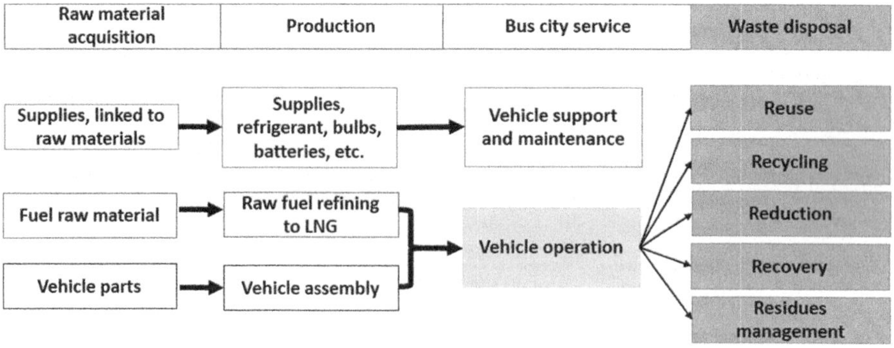

Fig. 12.4 Life cycle process of liquefied natural gas (LNG) bus transportation

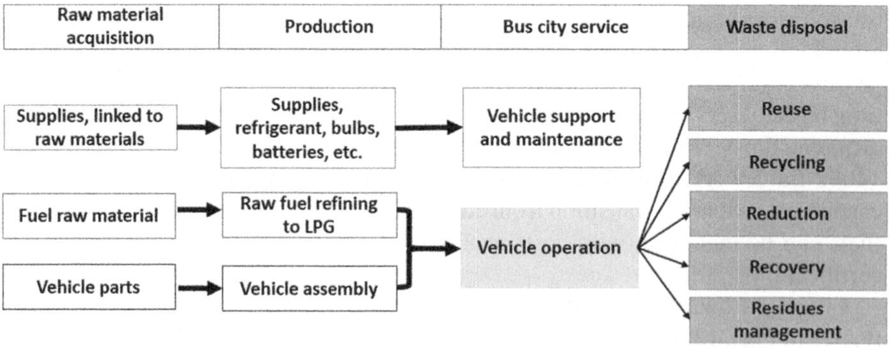

Fig. 12.5 Life cycle process of liquefied petroleum gas (LPG) bus transportation

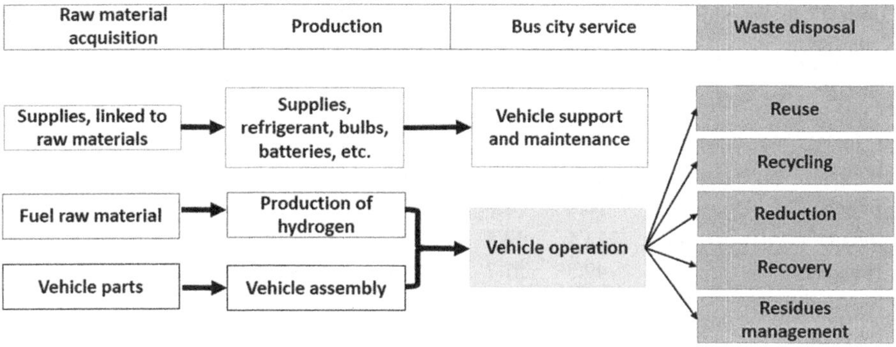

Fig. 12.6 Life cycle process of hydrogen fuel cell (HFC) bus transportation

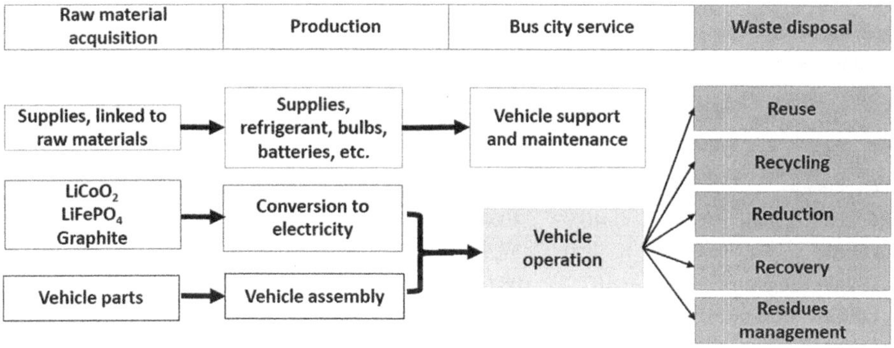

Fig. 12.7 Life cycle process of plug-in electric (PE) bus transportation

Recently, Google Maps has begun to offer a calculation of a city's carbon footprint. The Environmental Insights Explorer calculates emissions from buildings, car trips, and public transport to illustrate how a city's sustainability efforts are performing.

12.2 Statistics on Pollution from Transportation

Countries have taken varying efforts to reduce air pollution and other forms of environmental impact. Considering air pollution for countries that are part of the OECD, it is seen that in the 2013–2017 period several achieved a significant decrease while others have increased their pollution levels. Table 12.1 shows the variation of air pollution in selected countries. Significant variations were achieved by Chile (−65.1%), Estonia (−24.0%), Belgium (−20.1%), Lithuania (−19.5%), Ireland (−18.4%), and Hungary (−18.1%). On the other hand, Portugal (−1.8%), Spain

Table 12.1 Air pollution emissions in the form of PM 2.5 (in thousands of tonnes) in OECD countries

Country	2013	2014	2015	2016	2017	Variation from 2013 to 2017 (%)
Chile	377.6	274.1	121.1	128.0	131.6	−65.1
Estonia	12.1	8.9	9.6	7.8	9.2	−24.0
Belgium	28.8	22.3	24.4	25.0	23.0	−20.1
Lithuania	11.3	10.0	8.6	8.0	9.1	−19.5
Ireland	14.7	13.5	13.5	12.8	12.0	−18.4
Hungary	58.6	49.5	51.7	49.9	48.0	−18.1
Slovak Republic	21.6	16.7	18.5	18.8	18.0	−16.7
Sweden	23.7	20.7	19.7	19.8	20.1	−15.2
France	193.8	168.4	170.0	170.4	164.5	−15.1
Iceland	1.50	1.5	1.5	1.3	1.3	−13.3
Austria	18.0	16.1	16.1	15.7	15.5	−13.9
Luxembourg	1.6	1.6	1.4	1.5	1.4	−12.5
Switzerland	7.5	6.8	6.7	6.6	6.5	−13.3
Netherlands	15.4	15.0	14.5	13.7	13.6	−11.7
Slovenia	13.0	11.0	11.6	11.9	11.5	−11.5
Norway	31.1	27.9	28.1	27.2	27.9	−10.3
Finland	19.7	19.4	17.8	18.4	17.8	−9.6
Germany	109.1	104.4	103.6	100.8	99.1	−9.2
Czech Republic	43.5	40.9	40.4	39.1	39.9	−8.3
Greece	29.4	29.0	27.5	26.2	27.2	−7.5
United Kingdom	114.1	107.7	108.6	107.1	106.8	−6.4
OECD—Europe	1658.0	1555.3	1565.8	1556.3	1565.7	−5.6
Denmark	21.2	19.3	20.6	20.5	20.0	−5.7
United States	4229.3	4087.6	4071.1	4057.8	4042.9	−4.4
Italy	172.1	155.5	161.3	157.3	164.7	−4.3
Latvia	18.9	18.9	16.9	16.8	18.4	−2.6
Portugal	50.9	50.1	50.0	49.8	50.0	−1.8
Spain	106.0	104.7	105.1	103.2	105.1	−0.8
Canada	1677.9	1685.7	1674.4	1670.5	1673.5	−0.3
Poland	143.7	136.0	136.0	141.9	147.3	2.5
Turkey	376.5	379.3	382.12	385.0	387.8	3.0

Source: OECD (2019)
Pollutant: Particulates (PM 2.5). Variable: Total mobile sources

(−0.8%), and Canada (−0.3%) exhibited only minor reductions while other countries reported increases in air pollution emissions: Poland (+2.5%) and Turkey (+3.0%).

One important emission from transportation is fine particulate matter, which is particulate matter that is 2.5 microns in diameter and less. It is referred to as PM 2.5. Such particles are respirable in that they can easily travel in the lungs during breathing. The particles are often organic substances, dust, soot, metals, and chemicals.

PM 2.5 particles are produced from cars, as well as volcanoes and fires. PM 2.5 can cause health problems and can be lethal. Health specialists point out that people with respiratory or heart problems, or allergies, can find their symptoms aggravated by PM 2.5 particles. The elderly and children are more affected. Various articles describe PM 2.5 and its impacts (Deng, Deng, Miao, Guo, & Li, 2019; Thompson, 2018; Wu, Jin, & Carlsten, 2018; Yan et al., 2019).

Specifically, the most polluted cities, based on PM 2.5 air pollution in 2016, were reported (in micrograms in cubic meters) as (Business Insider, 2019):

Zabol (Iran): 217
Gwalior (India): 176
Allahabad (India): 170
Riyadh (Saudi Arabia): 156
Al Jubail (Saudi Arabia): 152
Patna (India): 149
Raipur (India): 144
Bamenda (Cameroon): 132
Xingtai (China): 128
Baoding (China): 128
Delhi (India): 122
Ludhiana (India): 122

2018 data, based on contamination from PM 2.5 per cubic meter, show the most polluted Latin American cities to be as follows (Greenpeace and Airvisual 2019):

Santiago (Chile): 29.4
Lima (Peru): 28.0
Toluca (Mexico): 26.4
Ecatepec de Morelos (Mexico): 24.9
Puerto Montt (Chile): 22.6
Guarulhos (Brazil): 21.6
León (Mexico): 20.8
Mexico City (Mexico): 19.7
Concepción (Chile): 19.6
Monterrey (Mexico): 17.2
Sao Paulo (Brazil): 16.2
Campinas (Brazil): 16.0
Santos (Brazil): 15.9
Bogota (Colombia): 13.9
Valparaiso (Chile): 13.4

Transportation is clearly an important air pollution factor. The above figures show that air pollution is a global problem. However, one may ask: Does this pollution have an effect on people's lives?

Some data to answer this question are given in Table 12.2, which presents the number of deaths attributable to ambient air pollution worldwide in 1990 and 2016, broken down by country, for selected countries. There are three countries that

Table 12.2 Number of deaths attributable to ambient air pollution worldwide in 1990 and 2016, for selected countries

Country	1990	2016	Variation from 1990 to 2016 (%)
Germany	56,444	36,938	−34.6
United States	103,945	93,376	−10.2
Myanmar	26,164	25,280	−3.4
Nigeria	65,736	68,887	4.8
Ethiopia	33,725	37,342	10.7
Tanzania	12,590	14,314	13.7
China	901,262	1,075,039	19.3
Brazil	41,088	49,724	21.0
Mexico	17,935	24,390	36.0
Bangladesh	79,902	108,687	36.0
South Africa	15,011	21,061	40.3
India	698,225	1,034,420	48.1
Indonesia	52,157	79,739	52.9
Pakistan	75,606	124,577	64.8
Thailand	15,368	25,432	65.5
Democratic Republic of the Congo	19,685	32,664	65.9
Philippines	24,423	44,389	81.8

Health Effects Institute and IHME (2016)

achieved a significant reduction: Germany (−34.6%), USA (−10.2%), and Myanmar (−3.4%). In the rest of the countries, the number of people that died due to air pollution increased.

12.3 Statistics on Public Transportation

The various types of transportation available currently face major challenges to becoming sustainable. An important aspect is the coverage of passengers who can travel comfortably and safely. Table 12.3 indicates the cities with the busiest bus rapid transit (BRT) networks worldwide in 2015, ranked by ridership.

Table 12.4 lists the cities with the busiest light rail worldwide in 2014, ranked by ridership.

Table 12.5 lists the cities with the busiest metro networks worldwide in 2017, ranked by number of trips.

Table 12.3 Cities with busiest bus rapid transit networks worldwide in 2019, ranked by ridership

City (country)	Ridership (millions of passengers)
Sao Paulo (Brasil)	990
Rio de Janeiro (Brasil)	953
Tehran (Iran)	600
Bogotá (Colombia)	657
Taipei (Taiwan)	626
Mexico City (Mexico)	372
Belo Horizonte (Brasil)	314
Guangzhou (China)	255
Buenos Aires (Argentina)	425
Istanbul (Turkey)	225

Source: Statista (2020). Public transportation

Table 12.4 Cities with busiest light rail worldwide in 2018, ranked by ridership

City (country)	Ridership (millions of passengers)
Budapest (Hungary)	430
Prague (Czech Republic)	370
Paris (France)	320
Vienna (Austria)	305
Warsaw (Poland)	280
Zurich (Switzerland)	210
Hong Kong (China)	205
Melbourne (Australia)	205
Berlin (Germany)	205
Zagreb (Croatia)	190

Source: Statista (2020). Public transportation

Table 12.5 Cities with busiest metro networks worldwide in 2017, ranked by number of trips

City (Country)	Number of journeys (in millions)
Tokyo Metropolitan Area (Japan)	3463
Moscow (Russia)	2369
Shanghai (China)	2044
Beijing (China)	1988
Seoul (South Korea)	1885
New York, metropolitan area (USA)	1806
New Delhi (India)	1789
Guangzhou	1730
Mexico City (Mexico)	1678
Hong Kong	1600

Source: Statista (2020). Public transportation

12.4 Indicators of Sustainable Transportation

There are several indicators of sustainable transportation. Some of the key ones are discussed here.

Indicator 1. Use of Space for Transport
Cities are growing based on their planning (Serrano-López, Linares-Unamunzaga, & San Emeterio, 2019; Sini, 2020) although many of them do not have a growth plan. Because of this, a city can grow and start needing new access routes (for example, more motorways) or even wider avenues for greater circulation, which in several cases involves the destruction of trees and the creation of roads. More space for pedestrians is also a growing need as there is an increasing need to find adequate transit spaces for pedestrians. In fact, it has been reported that this need is more critical in those countries that receive a significant number of immigrants (Lauby, 2019). Cities in which the number of vehicles is rapidly increasing have to assess whether spaces will have to be created so that more cars can pass and/or whether to decrease existing traffic. Planning of such works needs to take into account that when you plan to eliminate green spaces, it may be necessary to restore them in other areas where there are no green areas, so that oxygen production is maintained for the city. However, it should be noted that, viewed from an integral level, social sustainability is also linked to the landscape and that a city that is surrounded by green areas generally has greater attractiveness. This can in part directly compensate for surrounding pollution.

Indicator 2. Transportation Prices
Transportation prices in some cities are set by the state and in other cities depend on supply and demand. However, there are ways to establish more fair prices based on distances, as reported by Cosma, Pop, and Zelina (2019). Prices for intermodal connections can be predicted computationally, allowing the times and prices of the different connection possibilities to be known in advance (Hanbazazah, Abril, Erkoc, & Shaikh, 2019; Sun, Li, Xu, & Dai, 2019). With increasing corridors and greater complexity of transportation in cities (e.g., various types of transport with various forms of payment), the use of mobile applications that allow citizens to request various taxi and taxi-like services, which have similarities and differences, has appeared and expanded. The criteria to set prices. In this sense, different situations dictate the price that should be specified. In some countries, mobile applications allow one to choose the driver based on criteria of the driver's accumulated score, year of car, model of car, and internet availability, among other factors. Sometimes, applications do not offer this option, so a price is paid without knowing the characteristics of the service; then consumers end up blindly paying for services.

Indicator 3. Transportation Management
Governments and organizations in countries increasingly use crowdsourcing strategies to obtain the opinions of citizens to determine their needs and feelings towards proposals for solutions (Hong, Lee, Lee, & Kim, 2019; Marzano & Lubkina, 2019; Mouawi, Elhajj, Chehab, & Kayssi, 2019). This is important because it is the citizens who travel the city; they often know the problems in detail and have thoughts on potential solutions. Thus, it can be useful for cities to have think tanks focused on transportation so that they can investigate, develop, and propose conventional or disruptive measures not only to solve current transport problems in cities but also to anticipate future problems in the short, medium, and long terms. For example, what is the best way to optimize the route to and from the airport, since as the population increases, air travel also increases? Also, is the current type of transportation adequate for the next five or ten years? Ensuring the sustainability of transportation in a city needs to be focused on predictability by making the necessary projections and developing concrete actions that avoid complications in the future and provide lasting solutions.

Indicator 4. Transport Accidents
Each country has different speed limits on roads and highways, and there are differences in the effectiveness of traffic violation monitoring. A city that seeks to be sustainable needs transport to be safe and that leads to the lowest possible number of fatal accidents. However, the regulations in some cities do not provide harsh penalties for drivers, especially those who drive without authorization or those who drive under the influence or alcohol or drugs. National and city maps are required to follow in real time the number of car accidents, and to be able to monitor the effect of measures taken in a city to improve transportation and reduce accidents. An example of this is the Road Safety Atlas that shows data from Europe regarding accidents (European Commission, 2019). The impact of traffic accident deaths is very large, both socially, emotionally, and economically, especially for families who lose members. This is particularly true when the deceased was the economic earner in the family.

12.5 Strategies for Reducing Transportation Pollution

There is significant air pollution in most cities from transportation, with various impacts (Dons et al., 2019). In most jurisdictions, motor vehicles are the most widely used mode of transport, and more than 90% of transportation energy use is in the form of fossil fuels. An important aspect of planning by countries to combat environmental pollution, especially in transportation, is to ensure issues affecting poor populations are addressed. These populations sometimes are more subject to pollution. Furthermore, if individuals in such populations develop diseases associated with pollution, they often do not have the necessary resources to attend to their health. Thus, the elderly and children can be overly subject to the ills of air pollution. This is pointed out in the Sustainable Mobility for All initiative (SuM4All) (Sustainable Mobility for All, 2019), which reports that the poorest countries have higher concentrations of the fine particulate matter (PM 2.5).

More broadly, it is pointed out that two targets of the United Nations Sustainable Development Goals for 2015–2030, which were approved by the UN in 2015, are directly transportation-related:

(a) Target 3.6: By 2020, halve the number of global deaths and injuries from road traffic accidents.
(b) Target 11.2: By 2030, provide access to safe, affordable, accessible, and sustainable transport systems for all, improving road safety, notably by expanding public transport, with special attention to the needs of those in vulnerable situations—women, children, persons with disabilities, and older persons.

Various strategies can be implemented to contribute to make transport systems more sustainable, including the following.

12.5.1 Driving Restrictions

To contribute to the reduction of vehicular congestion and air pollution, some cities have chosen to prevent drivers from using cars during certain times of the day through higher traffic zones. Sometimes this is done by restricting according to the last digits of the license plate. This approach has been implemented in numerous countries:

Greece 1982
Chile 1986
Brazil 1997
Colombia 1998
Bolivia 2003
Costa Rica 2005
Honduras 2008
China 2008
Ecuador 2010
France 2014
Italy 2015
India 2016

In Quito, Ecuador, Carrillo, Malik, and Yoo (2016) showed that driving restrictions have achieved a 9–11% reduction in carbon oxide (CO) emissions during peak traffic hours. However, the effectiveness of this policy is dependent on the extent to which drivers can bypass restrictions by switching to other private travel modes.

In the city of Delhi, there is evidence showing the impact of the odd-even license plate number approach (Bernauer, Prakash, & Beiser-McGrath, 2018; Sharma et al., 2017; Tiwari et al., 2018). There are many old vehicles that emit significant quantities of pollution. However, in many cities they still circulate, contributing notably to environmental pollution and the damage associated with it. Therefore, it is important that emissions regulations be stringent and complied with. Another concern directly linked to the regulations is the importing to some countries of second-hand vehicles. It may make sense to restrict the import of second-hand cars to those not more than 5 years old, since these cars can contribute notably to pollution levels in a city, depending on the type and level of technology applied in the cars. Other approaches to dealing with such vehicles can be used. In Tanzania, for example, a car tax is charged for vehicles of an age greater than eight years.

12.5.2 Electric Cars

The increase in the development and use of electric cars is mainly based on environmental and ecological reasons, since this type of mobility does not directly emit exhaust gases and therefore does not directly pollute the environment. There may be indirect emissions, depending on the source of the electricity used by the vehicles. It is estimated that around 1.3 million vehicles travel the streets worldwide today (Bloomberg Businessweek, 2019, February 28). Hence the environmental pollution associated with car use is clearly a problem, especially in countries where cars use internal combustion engines and where they are not maintained properly and regularly. But addressing this issue through electric cars is not simple and immediate, as it requires specific infrastructure to support the operation of such vehicles, including charging stations, relevant maintenance services, and driver safety training. It is helpful to have in a country a government department or similar body that regulates, promotes, and supervises the use of electric cars. As an example, it is noted that such a body is currently in place in New Zealand (the Energy Efficiency and Conservation Authority). As of July 1, 2019, all electric cars in Europe must have equipment with sound emission technology to reduce the risk of danger to pedestrians who are accustomed to the sound of cars (mainly from the internal combustion engine) and the indication that gives of proximity. This acoustic vehicle warning system will make transport safer for pedestrians, even though this technology creates noise pollution.

To understand the factors that affect the acceptance of electric cars, it is necessary to carry out studies such as the one developed by Thøgersen and Ebsen (2019), who evaluate the motivational reasons for the low adoption of electric cars in Denmark. Thøgersen and Ebsen point out that Norway, Netherlands, UK, USA, France, Germany, Japan, and China are the leaders in the registered electric cars as of 2019. Despite the broad evidence of vehicle-related pollution and the need for improvements, there are still cities that do not implement transport-related measures to reduce their pollution. These cities continually have emissions and subject their

Table 12.6 Average life expectancy lost per person

Time (months)	Cause
22	Particulate pollution
19	Smoking
11	Alcohol and drug use
7	Unsafe water, sanitation, and handwashing
4.5	Road injuries
4	HIV/AIDS
4	Malaria
3.5	Tuberculosis
0.75	Conflict and terrorism

Source: Greenstone and Fan (2018)

citizens to related diseases. It is fair to ask in such circumstances: What are the consequences of inaction in a city to reduce environmental pollution?

This discussion leads to a question asked by Ortar and Ryghaug (2019): Should all cars must be electric in 2025? As shown in Table 12.6, the average life expectancy lost per person, due to particulate pollution, is compared with other causes of life expectancy loss. It is seen that the leading cause of loss of life expectancy is particulate pollution.

12.5.3 Bikes Use for Sustainable Transportation

Bike-share schemes have benefits that differ by city. In many cases, the user (including those residing in the city and internal and external tourists) values the experience of using bicycles, as they provide another transportation experience. The use of bikes can also be very beneficial in relation to travel times, especially in cities where vehicular traffic is a serious problem. This problem is exacerbated at specific times like weekday rush hours (6–8 am and 4–7 pm), and in those cases the bicycle can be a significantly beneficial alternative for certain distances and routes.

The health benefits from the use of bicycles for conventional mobility have been studied. For example, the benefits have been examined by Otero, Nieuwenhuijsen, and Rojas-Rueda (2018) for Europe and by Yu, Chen, Jiao, Zafari, and Muennig (2018) for the city of New York. However, from the point of view of health, Trivedi et al. (2019) detailed the damages that are generated through the use of bike-share electric scooters. Such vehicles are appearing with greater frequency in many cities, including those that do not yet have in place proper regulations for their safe use, for users, drivers, and other citizens. Thus, pedestrians walking the streets are exposed to potential collisions with these vehicles.

To ensure sustainable transport in the city, there must be easy pedestrian access, and ramps for the movement of people with disabilities with relatively easy access.

Another relevant aspect of sustainable transportation is that sustainable cities often have offer bicycles to ride around the city, as well as good interconnections between buses, taxis, and bikes.

Closing Remarks

Achieving sustainable transportation requires consideration of social, economic, and especially environmental dimensions. That is why increasing environmental impact assessments are being incorporated into the planning and development of transportation projects, taking into account factors ranging from product design to waste disposal and recycling.

It is necessary that in each city mapping is carried out, according to the type of energy source used, of the types of cars that travel so that processes can be optimized and appropriate measures can be taken to migrate to more ecological transport alternatives. The transition from the current model of transformation to an ecological one is urgent as the pollution emitted continues to cause human diseases and deaths around the world. Different types of transport need to be standardized to facilitate the creation of intermodal systems, facilitating traffic in the city, according to the different needs of citizens.

The use of space for transport, transportation prices, transportation management, and transport accidents are basic indicators that need to be monitored and incorporated into certifications of ecological transportation in cities. Finally, strategies that can be considered for implementation to support sustainability, such as driving restrictions, increased use of electric and hydrogen cars, and creation of facilities for significantly increased use of bicycles.

References

Bernauer, T., Prakash, A., & Beiser-McGrath, L. F. (2018). Do exemptions undermine environmental policy support? An experimental stress test on the odd-even road space rationing policy in India. *Regulation & Governance*. https://doi.org/10.1111/rego.12225

Bloomberg Businessweek. (2019, February 28). This is what peak car looks like. Retrieved February 5, 2020, from https://www.bloomberg.com/news/features/2019-02-28/this-is-what-peak-car-looks-like

Business Insider. (2019, September 1). Most polluted cities based on PM2.5 air pollution in 2016 (in micrograms per cubic meter) [Graph]. Retrieved February 5, 2020, from https://www.airvisual.com/world-most-polluted-cities

Carrillo, P. E., Malik, A. S., & Yoo, Y. (2016). Driving restrictions that work? Quito's Pico and Placa Program. *Canadian Journal of Economics/Revue Canadian d'économique, 49*(4), 1536–1568.

Cosma, O., Pop, P., & Zelina, I. (2019, May). An efficient soft computing approach for solving the two-stage transportation problem with fixed costs. In *International workshop on soft computing models in industrial and environmental applications* (pp. 523–532). Springer.

Currie, G., & Hidalgo, D. (2018). Workshop 1 report: Integrating rail and bus based modes (including BRT) into a user-relevant transport system. *Research in Transportation Economics, 69*, 35–38.

Deng, Q., Deng, L., Miao, Y., Guo, X., & Li, Y. (2019). Particle deposition in the human lung: Health implications of particulate matter from different sources. *Environmental Research, 169*, 237–245.

Dons, E., Laeremans, M., Orjuela, J. P., Avila-Palencia, I., de Nazelle, A., Nieuwenhuijsen, M., et al. (2019). Transport most likely to cause air pollution peak exposures in everyday life: Evidence from over 2000 days of personal monitoring. *Atmospheric Environment, 213*, 424–432.

European Commission. (2019, September 1). Mobility and transport road safety. Retrieved February 5, 2020, from https://ec.europa.eu/transport/road_safety/specialist/statistics_en

Greenpeace & Airvisual (IQAir). (2019, September 1). Air particulate matter concentration in selected Latin American cities in 2018 (PM2.5 per cubic meter) [Graph]. Retrieved February 5, 2020, from https://www.airvisual.com/world-most-polluted-cities?continent=59af929e3e7000 1c1bd78e50&country=&state=&page=1&perPage=50&cities=

Greenstone, M., & Fan, C. Q. (2018). *Introducing the Air Quality Life Index: Twelve facts about particulate air pollution, human health, and global policy*. Report by Energy Policy Institute at the University of Chicago, Chicago.

Hanbazazah, A. S., Abril, L., Erkoc, M., & Shaikh, N. (2019). Freight consolidation with divisible shipments, delivery time windows, and piecewise transportation costs. *European Journal of Operational Research, 276*(1), 187–201.

Health Effects Institute and IHME. (2016, February 12). Number of deaths attributable to ambient air pollution worldwide in 1990 and 2016, by key country∗ [Graph]. Retrieved February 5, 2020, from https://www.stateofglobalair.org/data/#/health/table

Hong, D. G., Lee, Y. C., Lee, J., & Kim, S. W. (2019). CrowdStart: Warming up cold-start items using crowdsourcing. *Expert Systems with Applications, 138*, 112813.

Lauby, F. (2019). Transportation and immigrant political incorporation. *Journal of Ethnic and Migration Studies*, 1–18. https://doi.org/10.1080/1369183X.2019.1635003

Marzano, G., & Lubkina, V. (2019, March). CityBook: A mobile crowdsourcing and crowdsensing platform. In *Future of information and communication conference* (pp. 420–431). Springer.

Mouawi, R., Elhajj, I. H., Chehab, A., & Kayssi, A. (2019). Crowdsourcing for click fraud detection. *EURASIP Journal on Information Security, 2019*(1), 11.

OECD. (2019, September 1). Emissions of air pollutants. Retrieved February 5, 2020, from https://stats.oecd.org/Index.aspx?DataSetCode=AIR_EMISSIONS

Ortar, N., & Ryghaug, M. (2019). Should all cars be electric by 2025? The electric car debate in Europe. *Sustainability, 11*(7), 1868.

Otero, I., Nieuwenhuijsen, M. J., & Rojas-Rueda, D. (2018). Health impacts of bike sharing systems in Europe. *Environment International, 115*, 387–394.

Serrano-López, R., Linares-Unamunzaga, A., & San Emeterio, C. M. (2019). Urban sustainable mobility and planning policies: A Spanish mid-sized city case. *Cities, 95*, 102356.

Sharma, S. K., Agarwal, P., Mandal, T. K., Karapurkar, S. G., Shenoy, D. M., Peshin, S. K., et al. (2017). Study on ambient air quality of megacity Delhi, India during odd–even strategy. *Mapan, 32*(2), 155–165.

Sini, R. (2020). Parks for the community: The city state's modernist planning model. In *Singapore's park system master planning* (pp. 87–130). Singapore: Springer.

Statista (2020). Tramway networks worldwide - leading cities by ridership 2018.

Statista (2020). Cities with the busiest bus rapid transit (BRT) networks worldwide as of 2019, ranked by ridership

Sun, J., Li, G., Xu, S. X., & Dai, W. (2019). Intermodal transportation service procurement with transaction costs under belt and road initiative. *Transportation Research Part E: Logistics and Transportation Review, 127*, 31–48.

Sustainable Mobility for All. (2019, September 1). 2017–2018 Annual report. Retrieved February 5, 2020, from http://www.sum4all.org/publications/sustainable-mobility-all-annual-report-2017-18

Thøgersen, J., & Ebsen, J. V. (2019). Perceptual and motivational reasons for the low adoption of electric cars in Denmark. *Transportation Research Part F: Traffic Psychology and Behavior, 65*, 89–106.

Thompson, J. E. (2018). Airborne particulate matter: Human exposure and health effects. *Journal of Occupational and Environmental Medicine, 60*(5), 392–423.

Tiwari, S., Thomas, A., Rao, P., Chate, D. M., Soni, V. K., Singh, S., et al. (2018). Pollution concentrations in Delhi India during winter 2015–16: A case study of an odd-even vehicle strategy. *Atmospheric Pollution Research, 9*(6), 1137–1145.

Trivedi, B., Kesterke, M. J., Bhattacharjee, R., Weber, W., Mynar, K., & Reddy, L. V. (2019). Craniofacial injuries seen with the introduction of bike-share electric scooters in an urban setting. *Journal of Oral and Maxillofacial Surgery*. https://doi.org/10.1016/j.joms.2019.07.014

Wu, W., Jin, Y., & Carlsten, C. (2018). Inflammatory health effects of indoor and outdoor particulate matter. *Journal of Allergy and Clinical Immunology, 141*(3), 833–844.

Yan, M., Wilson, A., Bell, M. L., Peng, R. D., Sun, Q., Pu, W., et al. (2019). The shape of the concentration—Response association between fine particulate matter pollution and human mortality in Beijing, China, and its implications for health impact assessment. *Environmental Health Perspectives, 127*(6), 067007.

Yu, W., Chen, C., Jiao, B., Zafari, Z., & Muennig, P. (2018). The cost-effectiveness of bike share expansion to low-income communities in New York city. *Journal of Urban Health, 95*(6), 888–898.

Chapter 13
Sustainability of Urban Infrastructure

Mori-Pelaez Demostenez, Aldo Alvarez-Risco, and Shyla Del-Aguila-Arcentales

Abstract As part of the efforts to materially build a sustainable city is the sustainability of urban infrastructure approach, which seeks to guarantee built physical spaces thinking of ensuring a harmonious life, mainly considering what the damage that climate change can cause to buildings can be and use technologies that allow to resist corrosion, humidity, and the various damages in the constructions, both in homes and in offices, schools, and mass call centers in the cities. When analyzing the literature, research trends focused on new strategies are shown.

Keywords Sustainability of urban infrastructure · Urban infrastructure · Sustainable urban infrastructure · Sustainability

13.1 Introduction

The concept of sustainability was forged in the second half of the twentieth century, motivated by the emerging problems related to the overexploitation of resources, environmental pollution, and industrialization at global levels. In 1987, the United Nations Organization (UN) in the Brundtland report of the World Commission on Environment and Development, said that sustainable development contemplated the use of the planet's resources to meet the needs of present generations, without compromising the ability to meet the needs of future generations. The term "sustainability," in recent decades, has become important and has become common use, even with some abuse and lightness in its use, sometimes with a clear interest in

M.-P. Demostenez
Pontificia Universidad Católica del Perú, Lima, Peru

A. Alvarez-Risco (✉)
Facultad de Ciencias Administrativas y Recursos Humanos, Universidad de San Martín de Porres, Lima, Peru
e-mail: aldo.alvarez@redsaf.org

S. Del-Aguila-Arcentales
Escuela Nacional de Marina Mercante "Almirante Miguel Grau", Callao, Peru

Universidad Nacional de la Amazonia Peruana, Iquitos, Peru
e-mail: sdelaguila@enamm.edu.pe

A. Alvarez-Risco et al. (eds.), *Building Sustainable Cities*,
https://doi.org/10.1007/978-3-030-45533-0_13

taking advantage of the concept economically. And it is that although the premise, in its conception, contains a responsible and ethical position, it is also somewhat ambiguous, it is difficult to clarify or delimit, for example, what would be the level of satisfaction of which it is spoken, or what are, exactly, the present needs, or even beyond, what the needs of future generations would be; it is a kind of premonition to try to establish such needs, since the needs change from generation to generation; and finally, who and with what scale or reference can measure all that. As a way of answering the previous questions, according to Ruano (2000), the notion of capital was introduced, which is transferred from generation to generation, which in turn is subdivided into three types: artificial capital (the built), capital human (science, knowledge, and technology), and natural capital (pure air or water, biodiversity, etc.). A critique of this scheme may be the fragility of one type of capital when weighting another; A very common example: artificial capital at the expense of natural capital. Today there is even talk of up to five types of capital: human capital, physical capital, financial capital, natural capital, and social capital. Each one with preservation and responsible use needs; the debate about the importance of whether capital can be squandered to the detriment of another remains open. However, the concept of sustainability is increasingly widespread, and part of the daily imagination, and although its foundation is not fully defined. it is molded enough to serve as a basis for reflections on sustainability and sustainable architecture. One of the considerations that are mistakenly about sustainability, especially when it comes to architecture, is that foreign problems and solutions are frequently imported, almost unreflexively, problems that although they also suffer in the Peruvian environment, are not the most pressing. For example, there is a generalized position on the use of electricity; however, the conditions of each country or each city show substantial differences; to point out one: much of the northern United States, even to this day, relies heavily on coal combustion to supply the necessary energy from its cities, a practice that emanates particles harmful to health and the environment in general, in so much so that the Peruvian territory is supplied with the energy produced by hydroelectric plants, that is, a renewable source.

The aforementioned may not necessarily have an unfavorable consequence in the Latin American or national environment, the inconveniences arise when the interest given to a problem hides more serious or more urgent ones. For example, in Peru, exists problems of vehicular overpopulation with its consequent problems of mobility and deterioration of the air due to emanations, pollution of rivers, lakes, of the coast and of the water that reaches the homes, depredation of the jungle, for reasons such as illegal logging or the uncontrolled extraction of gold, social conflicts of different nature scattered throughout much of the country, inequities, centralization, neglected populations, etc., without counting on the great infrastructure deficit even in the capitals. Globally, sustainability has been applied in many fields of human industry, finding more tangible results in engineering and science in which numerical indicators are established. In the case of architecture, unless it is viewed from a reduced point of view, as a product, based on its materiality, supplies, and waste, it is not so easy to establish the conditions for a building to be categorically referred to as sustainable; This is due to the nature and the multiple variables that influence the project. And it is that an architecture project, almost always has a client or a client profile, its impact always goes beyond what an architect can anticipate when designing it. With few exceptions, a building is a good that will remain in a fixed place for many years, has a very high cost compared to other goods, in addition to consuming a large amount of resources in its operation and in its manufacture; In fact, the construction sector consumes half of the planet's resources.

13.2 Beginnings of Sustainable Architecture

The concept of "sustainable architecture" comes from the aforementioned Brundtland report of 1987, but the principles on which it is based can be found practically since the architecture itself exists, it could even be said that an architecture that is considered appropriate and of quality, implicitly contemplates several sustainability principles. The relationship of man with nature is the beginning of architecture, and although the theory of the history of architecture stems sharply from European sources, American human settlements, even before colonization, include a rich architectural production, in fact, serve as an example on issues such as the use of resources, construction methods, or even urban planning. Buildings such as Luoyang's excavated homes in China, which protected their inhabitants from adverse surface conditions (Piedecausa, 2009, p. 172). Or the Alhambra in Granada, which with its set of courtyards regulates the temperature inside the complex, represents what are called passive design strategies, since they use natural processes to achieve thermal comfort.

The Rationalism was a architectural current that emerged in Europe in the first half of the twentieth century, both were linked and had a great impact on the global architecture landscape; not surprisingly, his influence also reached South America, and of course also Peru. In the postulates formulated within the framework of its international congresses, there is also a germ of what today composes the concept of sustainable architecture. The International Congresses of Modern Architecture (CIAM in Spanish) held, although irregularly, in different cities in Europe from 1928 to 1959, were attended by renowned European architects and urban planners, at least in their first versions. These congresses set out to respond to major problems that afflicted many European cities, problems that are rooted in the Industrial Revolution of the nineteenth century, including: poor urban planning, various levels of pollution, overcrowding, among others. The CIAM postulates propose a new urbanism based on human functions or activities, from the workplace to the configuration of housing, recreational areas; they also integrate the relatively new means of transport such as cars or airplanes.

While issues such as ventilation and sunlight were important as design qualities for CIAMs, other issues such as green areas, for example, were only seen as a support for one more function that the city should provide to the inhabitants. The congresses clearly proposed the arrangement of cities distributed in areas of different activities; unfortunately there is no reflection on the effects of industry on nature and even on populations (Ruano, 2000, p. 8), it can be said that The vision of con-

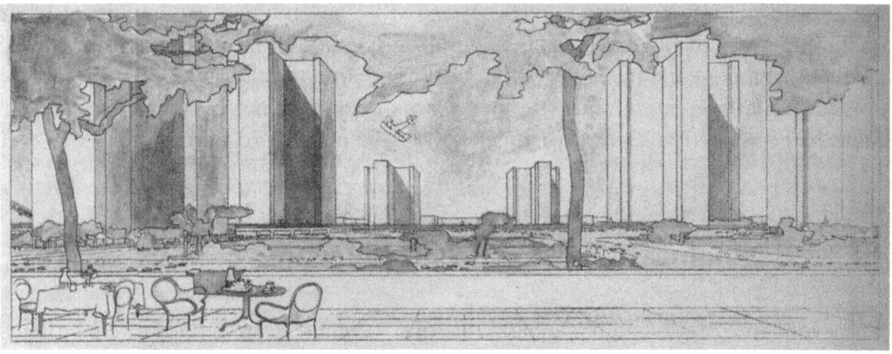

Fig. 13.1 Image of the "contemporary city," drawing by Le Corbusier (1922) for a city of three million inhabitants. Source: Le Corbusier Foundation

gresses is reduced practically to the treatment of cities, but it does not offer recognition and less solutions to more complex and global environmental problems. It is salvageable from the meetings of the CIAM, who worked hard to create images and physical models of the proposed urbanism, so that their approaches were clear and well represented, this observation is also made, because today, many ideas and proposals do not have a tangible image, especially if the urban or architectural proposals are based on economic motivations, landscape or urban issues are discussed and represented with numbers and prices but not in images (Fig. 13.1).

Today the effects of industry and the lifestyle of cities are more clearly known, however, this was not the case in the 1950s, in the period after World War II technological advances flourished with relative enthusiasm, and faith in science was also a source of such enthusiasm, nature was seen as a huge amount of resources that human beings could take advantage of. It is not until the 1970s that there was a recognition of the effects of the way of life of countries on nature, with this came the distrust that science could solve the energy and degradation problems of the biosphere. (Behling & Behling, 2002, p. 12). It is when a questioning and/or revaluation of the traditional modes of relationship with nature arises, this also recognized the need for a change in the way of exploitation of resources. The word "ecology," which refers to a lifestyle in tune with the rhythms of nature, began to be heard more and more among the environmentalist, protectionist, and alternative thinker's circles, with this we wanted to call for the fragility of our planet. Ruano (2000, p. 8), today more than ecology it is about sustainability, that is, the focus on the use of resources without neglecting development, technology, or the economy.

Another non-renewable resource, which produces a large amount of emissions, is the burning of fossil fuels, and although enormous progress has been made in the efficiency of its consumption, the global dependence on this resource has been known since the 1973 oil crisis. Today, car parks have grown considerably and faster in American and Asian countries, becoming one of the most pressing urban problems. This also confirms that although we are aware of the fragility of the planet, our habits, at least in the use of fuels, have not changed much. Since the eighties, the term to refer to sustainable architecture has undergone variants, being able to be called, eco-architecture, sustainable architecture, green architecture, etc., in addition other related branches such as bioclimatic architecture, sustainable urban planning, ecourbanism, and others more. This diversification of the concept is in turn due to the complexity of dealing with sustainability problems; today we are much more aware that the solutions are not always simple, as perhaps believed in the past decades.

13.3 Principles of sustainable architecture

An important antecedent that represents the will of the architects on the road to sustainability, and although it was not the first, had an impact on the urban planning of many cities on several continents. This is the Athens Charter, published in 1942 by the architects Josep Lluis Sert and Charles Édouard Jeanneret-Gris, known as Le Corbusier. The letter arises in the context of deterioration that many European cities suffered, essentially as a result of the industrial revolution. Also developed within the framework of the CIAM held in Europe, among other ideas, it states:

- The houses constitute neighborhoods and must be located in the best areas of the city, taking advantage of the climate, topography, green areas, and natural light.
- The location of the houses is based on hygiene criteria.
- There must be coherence on the density of housing with the characteristics of the house and the land to avoid slumming.
- For reasons of air safety and accident prevention, it was forbidden for homes to be aligned with the main roads.
- Promote the implementation of large green areas, on the edges of large and surrounding large housing towers.
- Pedestrian and car traffic were separated; also, proposed overpasses at crossroads of important roads.

As a city model, it was proposed that it be composed of four large areas: the living area, the circulation zone, the work zone, and the recreation area. Cities like Brasilia and others in Europe or Asia took these principles in their design, and although several of their rationalist considerations have left the current urban planning as an inheritance, especially those that allude directly to architectural design, the negative consequences of this manifesto, and in General of the postulates of the CIAM congresses on urban planning were also evident. One of the most ominous is probably the separation of uses by areas in the city, this postulate, apparently innocuous, fostered insecurity, since, at work hours, residents were not in their homes, which resulted in criminal acts. This problem was pointed out by activist and writer Jane Jacobs in the early sixties, she published: Death and life of the great cities of the United States; in his writing he points out the flaws of urban planning, but also lists solutions, one of them was to recognize that, the constant life on the streets and at all hours, avoided crime in the neighborhoods.

Today, there are certifying companies that establish the degree of sustainability of the architecture, they evaluate and score; Non-governmental organizations such as LEED, for example, categorize buildings after a review of their capacities or deficiencies on sustainability, all this also represents in some legislative frameworks a legal obligation, in addition to a commercial value. The criteria to evaluate are essentially guided by economic and environmental aspects; for example, they evaluate building materials, energy sources, or waste recycling among others; This way of assessing sustainable architecture finds its roots in the general concept of sustainable development, whose principles are declared in 1992 after the Earth Summit held in Rio de Janeiro, Gauzin-Müller (2002) points out:

- The study and analysis of the life cycle of materials.
- The development of raw materials and renewable energy.
- Recycling of material waste and reduction of the amounts of material and energy used in the extraction and exploitation of material resources.

In 1998, in a publication of the School of Architecture and Urban Planning at the University of Michigan, a new synthesis of the principles of sustainable architecture that were widely accepted by the academic world and by governmental institutions, such as:

- Special attention to climatic conditions, hydrography, and the different ecosystems that may constitute the environment of the buildings, seeks to obtain the highest possible yield with the least impact.
- Efficiency and moderation in the use of construction materials, it is recommended to use materials whose energy content, i.e., the energy resources used in their production and transport, is low.
- Reduction of energy consumption in buildings, whether in heating, ventilation, lighting, etc., covering, if possible, the demand from renewable energy sources.
- Minimize also the general balance of the building, based on factors such as its design, construction, use, and the end of its useful life.
- Obey the requirements of thermal comfort, ventilation, lighting, and habitability of the spaces that make up the building.

In the aforementioned principles there is obviously a preponderant concern for materials, energy use, or waste treatment, these issues give a mark to sustainable architecture. Today, architecture, and even more sustainable urbanism, broadens its spectrum of action, since they also include socio-economic considerations, impact of globalization, exploitation of resources in certain cultural contexts, lack of infrastructure and technology, degradation and humanization of landscape, etc.

It is also important to point out that sustainability establishes something that had not been taken into account in past generations, and that also often happens with contemporary architects: the building, although it generally has a useful life of many years, at some point must finish, the sustainable architecture must contemplate the steps to follow for an adequate disassembly of the building.

13.4 The Principles of Hannover

After the session of the United Nations of 1987, in which the Brundtland report was presented, and the celebration of the Earth Summit in Rio de Janeiro in 1992, after which the aforementioned principles of sustainable development were coined, in addition of the synthesis of the principles of sustainable architecture published in 1998; Another important document is the declaration of the nine principles of Hannover, which try to expand what is understood by sustainable design:

1. Defend the right of both humanity and nature to coexistence in healthy conditions, based on mutual, diverse, and sustainable support.
2. Recognizes the interdependence between human design and nature, also recognizes that the latter depends on the former and that the design has consequences at any of its scales, design considerations must even contemplate distant consequences.
3. Respect the relationships between spirit and matter, consider everything concerning human settlements, in their various scales and types, from the community, housing, industry, and commerce, and their changing relationships, between spiritual awareness and material.
4. Accepts responsibility for the consequences of the exercise of design on human well-being, the recognition of natural systems and the right to coexist.
5. Create safe and long-term valuables so as not to recharge future generations as a result of poor design, processes, or standards.
6. Eliminate the concept of waste by optimizing production systems until they are similar to the production of natural systems, in which there is no waste.
7. Trust natural energy flows, by deriving their design capabilities adapted to natural energies such as solar energy, using it efficiently and responsibly.
8. Know and understand the limitations of design, no human creation is eternal, and design cannot solve all problems. Those who create and plan must do so with humility in the face of nature, taking advantage of it as a model and mentor but not as an impediment or inconvenience to avoid.
9. Seeks constant improvement by sharing knowledge, open communication between colleagues, different professionals, or users is important to link long-term sustainability considerations, without neglecting ethics to repeatedly restore natural processes with human activity.

Those responsible for the elaboration of these principles are the architect William McDonough in cooperation with the chemist Michael Braungart, both worked on them since the 1990s, but published them within the framework of the Hannover world exhibition. The nine principles are valid for any process of human creation, reflect on past and present forms of production, so they are widely accepted in addressing the restoration between human creation and natural processes in a comprehensive and adequate manner. They are also guidelines open enough not to contradict or close possibilities such as the use of technology or the introduction of sociocultural considerations, provided they conform to an ethical, conscious, and responsible posture. However, this declaration of principles, which subsequently motivated different events with international participation, also had criticism, not necessarily for its content as it happened with the Athens Charter, but for the meetings themselves, as congresses such as the one in New York in 2000, Johannesburg in 2002, Rio de Janeiro 2012, etc., they do not represent, according to critics, tangibly an advance in the reduction of environmental problems, social conflicts, poverty reduction, or inequality.

13.5 Dimensions of Sustainable Architecture

It should be noted that sustainable architecture, beyond being a phrase that can serve to exalt the value of a particular building, company, or architect, is a position with which to face a design assignment, especially when the Sustainability legislation is weak, the designer must establish some complicity or at least the client's predisposition. This precision is made because the term "sustainability" is often used lightly and for purely commercial purposes, thus banalizing its foundations and objectives. There is a lot of architecture that is a good example of sustainability produced before the term even existed. Projects such as patio houses on the Peruvian coast, which provide ventilation and thermal control, thanks precisely to the yards that keep the air cool in the summer, or adobe houses in the mountains, which conserve heat from solar radiation during the day to regulate the temperature at night, or also the malocas in the jungles of Peru and Colombia, circular plant structures strongly linked to the transmission of knowledge, but also with a design closely related to its surroundings. Currently, what is understood by sustainable architecture contemplates a wide range of factors, which certainly should be reviewed, since every architecture project contemplates them, although probably with variation in the relevance of each one depending on the context of the project. In the following list, a series of elements to consider in the architectural design are listed, the list is based on a publication of the Universidad Diego Portales de Chile, prepared by Castillo and Castillo (2015), its postulates also cover non-domestic typologies such as factories, workshops, or other equipment whose energy demand and waste could represent a significant environmental impact.

Environmental Dimension It deals with natural resources in a large spectrum and in its interaction with architecture, Castillo and Castillo (2015) subdivides this dimension into the following themes and criteria, from which some appreciations are developed:

Land. In its topographical complexity, natural relief has to be considered as the location of the project, and in terms of its geological diversity, the type of soil must be recognized and worked on. Construction materials also come from the land, considering that they should not be indiscriminately exploited, for this purpose there are certifications that prove their legal origin. On the other hand, the possibility of

Fig. 13.2 Pools of Marés in Porto, Portugal, by architect Alvaro Siza (1966), takes advantage of the morphology of the beach to create pools with few materials and water from the sea itself. Its operation is subject to the rhythms of nature. Photography: Fernando Guerra and Sérgio Guerra

recycling can also be considered. Finally, the most natural and locally sourced materials are preferable (Fig. 13.2).

Water. Consider the appropriate means of collection and its economy, water saving devices can be mounted. Water can also be recycled for uses in which sewage can be used or even incorporated back into the hydrological cycle (Fig. 13.3).

Air. It is an important architectural element, its proper handling can provide cold or heat when necessary, it contributes to passive insulation (conservation of interior temperature). Ventilation is also crucial to renovate and maintain healthy enclosures, natural ventilation should be weighed if possible, and the incorporation of moisture if necessary. If the ventilation is of artificial origin, it should come from renewable sources of energy (Fig. 13.4).

Energy. Its treatment can have two approaches: the relationship of the project with solar radiation, and the consumption of electric energy. On the first, the way in which it is located on the ground and the spaces are distributed is essential to receive or not receive direct radiation at certain times of the day, this in consideration of the thermal comfort that is desired in each space. In this sense, it is also necessary to contemplate the use of natural light always, as well as the control of solar radiation by means of elements that produce shade. On the other hand, there is the use of energy, savings and/or reduction of energy consumption systems must be implemented, and if their origin is from renewable sources, the better (Fig. 13.5).

Fig. 13.3 Fletcher-Page House, by architect Glenn Murcutt (2000), among several other qualities about sustainability, the house also collects rainwater for later use. Photography: user Guess.Who / Flickr

Life. The architecture obeys the habitability needs, so it must basically provide thermal and acoustic comfort, it should be considered that buildings also produce waste, emit various types of substances into the environment, reduce their quantities, and ideally treat them great importance. The conservation of biodiversity is a variable that is neither new nor minor; An architectural project, if it is to make use of vegetation, consider those that are preferably local (Fig. 13.6).

Social Dimension Sustainability in a certain way deals with the relationship of culture with nature, in that sense, Castillo lists the following themes and criteria, on which some considerations are made:

Cohesion. It gathers identity issues based on culture and heritage; The purpose of working with identity is to reach levels of security, trust, and social integration. It is also important to recognize and work with the installed government structure, observing current legislation and political management. The participation of local agencies and actors sometimes turns out to be crucial depending on the architectural project to be designed (Fig. 13.7).

Mobility. Accessibility is an increasingly important and necessary resource, access not only to equipment but also to services may have greater or lesser relevance

Fig. 13.4 Eastgate shopping center, in Harare, Zimbabwe. Designed by the architect Mick Pearce (1996), it uses the Venturi effect to provide fresh air during the day to the stores, the concept of thermal design was inspired by the structure of a termite mound. Photography: www.arup.com

Fig. 13.5 Fab Lab House, designed in 2010 by the Advanced Architecture Institute of Catalonia (IAAC) in collaboration with MIT and the Fab Lab network. The house, by design, maximizes the collection of solar energy by having photovoltaic panels in its coverage depending on the movement of the sun. Photography: Fab Lab House

Fig. 13.6 Naturbad Riehen, built in the years 2010–2014 and located in Basel, Switzerland, the public pool designed by the architects Herzog and de Meuron, receives about 2000 bathers daily, cleaning does not depend on filters or chlorine, uses the vegetation that surrounds it, a process called phytodepuration. Photography: www.naturbadriehen.ch

depending on the project. Transport also retains relevance, especially at the urban level, public transport is prioritized with efficiency and travel by bicycle (Fig. 13.8).

Economic Dimension It is a sensitive dimension due to human relations around economic systems and pressures exerted by the market on architecture and the different means of production, Castillo states the following elements and criteria on which some considerations are extended:

Connectivity. Through infrastructure for means of transport, also giving importance to its approach as a network. Also contemplate the connectivity of services such as electricity, internet, telephony, etc. Connectivity can also be established by way of organization and cooperation between individuals (Fig. 13.9).

Efficiency. The economic activity, jobs, or appearance of new jobs should not represent a threat to ecology or sustainability. Always consider that local economies can commune with global ones. The idea of sustainability implies the demand for adaptation processes as well as innovative and creative solutions. Achieving efficiency can also be achieved with a useful, efficient, and low maintenance life (Fig. 13.10).

Table 13.1 shows the evaluation of the publications on "urban infrastructure" that have more citations. Google Scholar is used between 2010 and 2018.

Table 13.2 shows the academic journals that published more studies on "urban infrastructure." Scopus is used between 2010 and 2018.

Table 13.3 shows the list of institutions that have published scientific articles on "urban infrastructure." You can see universities of North America, Europe, and Asia.

The cocitation analysis allowed obtaining three conglomerates of authors who published studies on "urban infrastructure," which were indexed in Scopus (see Fig. 13.11).

Fig. 13.7 Located in Archidona, Ecuador, the Cocoa Interpretation Center was designed by the ENSUSITIO Arquitectura y Taller Con Lo Que hay 4 (2014) studies, it proposes spaces for the appreciation of traditional knowledge about culinary with cocoa, they also enable places for marketing in addition, the local construction tradition is recognizable as an important factor in architectural design. Photography: ENSUSITIO Architecture / Taller Con Lo Que hay 4

Fig. 13.8 Ambitious plan by architect Norman Foster presented in 2013, proposes to create 220Km of bike paths supported by pillars on the London railways. Image: Norman Foster study

Fig. 13.9 Elementary school in Gando (extension). Located in Burkina Faso and designed by the architect Francis Keré (2001), it has become a symbol of Gando. The project provides infrastructure for education, a valuable service especially in sensitive environments such as Burkina Faso. In its construction local materials were used and it had the organization and support of the population. Photography: GandoIT

Each conglomerate is identified by a characteristic color that groups the authors that are part of it from and that were obtained from the Scopus. Thus, in green color, Simon J. Marvin of University of Sheffield whose research is based on urban structure planning and transforming current designs. The yellow node figure Leonardo Dueñas-Osorio of Rice University with research oriented to the urban structure policies. The authors around the main author share similar contents and approach.

Closing Remarks

Through a slow but also sustained evolution, the concept of sustainability has been molded and disseminated, it is also a guide and standard for a large number of human activities. The emergence of the concept, the ideals of sustainability and its implementation are undeniably necessary, and although many detractors argue that international congresses and other activities do not have an impact on the behavior, customs, and concrete actions that involve human development, the truth is that, slowly, they have been incorporated into the particular laws of several nations, in addition to instilling sustainable awareness in the population. Undoubtedly, the road is long and will take a long time to achieve major progress towards a balance between production and degradation of the planet, but there is no other way, we are probably within the last generations that have been free to exploit nature and contaminate it almost without restrictions.

Fig. 13.10 Sanghat is the office of the architect Balkrishna Doshi (1981), in its design sustainability criteria were considered as the use of vaults that reflect solar radiation, the incorporation of vegetation for temperature control, the water path to humidify the facilities, among others. Photography: Nicholas Iyadurai

As for architecture and construction, in addition to other related activities, such as at no time in history, we can now have knowledge of everything that implies its exercise, although it is not easy to find solutions to all sustainability problems, it is it can just point out the problems. The advancement of research, technological development, and innovation should provide more knowledge about sustainable issues that perhaps today is still unsuspected, and in the best case, solutions. It is also important that the concept of sustainability is also shared domestically, its dissemination is necessary at full scale, in schools, institutes, universities, and any other context of dialogue. When it comes to architecture, and in cases where the regulations may not be very demanding on sustainability, the sustainable nature of the project must be contemplated from the beginning, this, in some cases, will require more knowledge and resources from professionals or of the client, will therefore imply an ethical commitment of all the actors involved.

Table 13.1 Publications on "urban infrastructure" most cited in Google Scholar

N°	Title	Author	Year	Cites
1.	A three-stage resilience analysis framework for urban infrastructure systems	Ouyang, Dueñas-Osorio, & Min	2012	310
2.	Developing sustainability criteria for urban infrastructure systems	Sahely, Kennedy, & Adams	2005	302
3.	Mitigating and adapting to climate change: Multi-functional and multi-scale assessment of green urban infrastructure	Demuzere et al.	2014	284
4.	Low-carbon transitions and the reconfiguration of urban infrastructure	Bulkeley, Castán Broto, & Maassen	2014	202
5.	Engineering novel habitats on urban infrastructure to increase intertidal biodiversity	Chapman & Blockley	2009	145
6.	"Cold spots" of urban infrastructure: "Shrinking" processes in eastern Germany and the modern infrastructural ideal	Moss	2008	126
7.	The evolution of the urban infrastructure in the nineteenth and twentieth centuries	Tarr	1984	121
8.	Time-dependent resilience assessment and improvement of urban infrastructure systems	Ouyang et al.	2012	118
9.	Maintaining climate change experiments: Urban political ecology and the everyday reconfiguration of urban infrastructure	Broto & Bulkeley	2013	101
10	Housing finance and urban infrastructure finance	Kim	1997	97

Source: Google Scholar, 07/28/2019
Search command: "urban infrastructure"

Table 13.2 Academic journals that published more studies on "urban infrastructure"

N°	Journal	Scopus
1.	Urban studies	13
2.	Journal of management in Engineering	12
3.	Procedia Engineering	8
4.	Urban policy and Research	8
5.	Cites	7
6.	Financing Cities Fiscal Responsibility And Urban Infrastructure In Brazil China India Poland And South Africa	6
7.	Habitat International	6
8.	International Journal Of Urban And Regional Research	6
9.	Journal Of Urban Technology	6
10.	Lecture Notes In Computer Science Including Subseries	4

Source: Scopus, 07/28/2019
Search command in Scopus: TITLE("urban infrastructure") OR (urban AND infrastructure)
Own elaboration

Table 13.3 Institutions that published more studies on "urban infrastructure"

N°	Institution	Scopus
1.	Georgia Institute of Technology	12
2.	Tianjin University	10
3.	Harbin Institute of Technology	10
4.	University of Toronto	8
5.	Newcastle University	7
6.	University of Durham	7
7.	University of Manchester	6
8.	Rice University	6
9.	Chinese Academy of Sciences	6
10.	Leibniz Institute for Research on Society and Space	6

Source: Scopus 07/28/2019. Search command in Scopus: TITLE ("Urban infrastructure")
Own elaboration

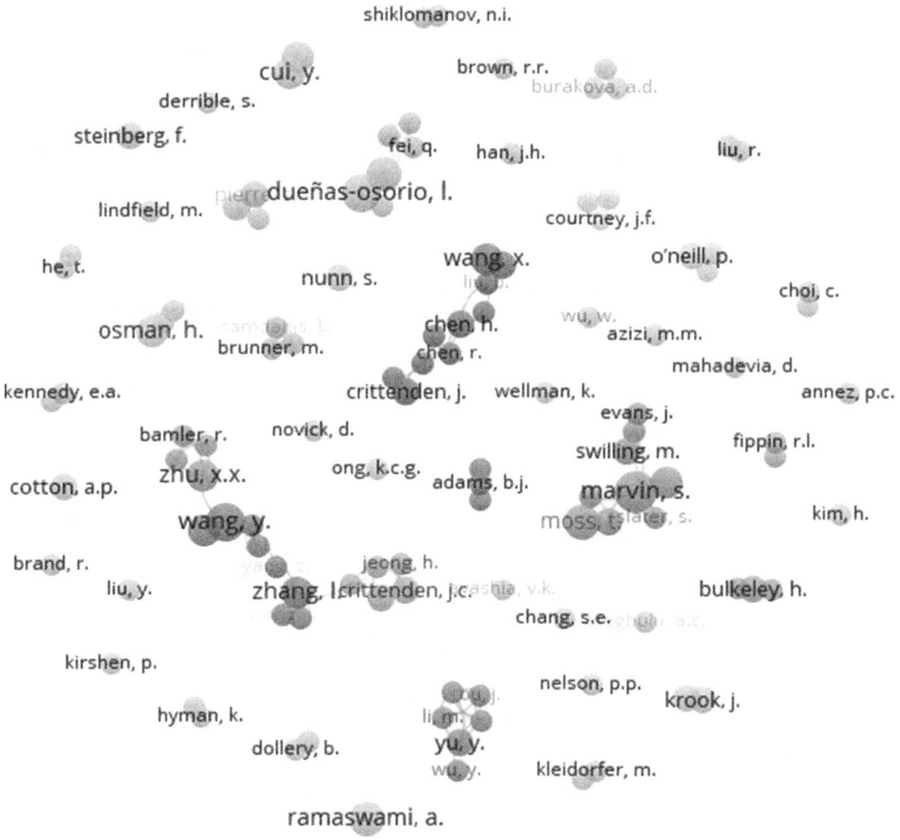

Fig. 13.11 Conglomerate of authors with greater cocitation in Scopus 2010–2018. Source: Scopus, 07/28/2019. Only authors who had at least 5 citations were considered

References

Behling, S., & Behling, S. (2002). *Sol Power*. Barcelona, España: Gustavo Gili.

Broto, V. C., & Bulkeley, H. (2013). Maintaining climate change experiments: Urban political ecology and the everyday reconfiguration of urban infrastructure. *International Journal of Urban and Regional Research, 37*(6), 1934–1948.

Bulkeley, H., Castán Broto, V., & Maassen, A. (2014). Low-carbon transitions and the reconfiguration of urban infrastructure. *Urban Studies, 51*(7), 1471–1486.

Castillo, C., & Del Castillo, M. (2015). Enseñanza, sustentabilidad, arquitectura. *Arquitecturas del Sur, 33*(48), 30–43.

Chapman, M. G., & Blockley, D. J. (2009). Engineering novel habitats on urban infrastructure to increase intertidal biodiversity. *Oecologia, 161*(3), 625–635.

Demuzere, M., Orru, K., Heidrich, O., Olazabal, E., Geneletti, D., Orru, H., et al. (2014). Mitigating and adapting to climate change: Multi-functional and multi-scale assessment of green urban infrastructure. *Journal of Environmental Management, 146*, 107–115.

Gauzin-Müller, D. (2002). *Ecological architecture. [Arquitectura ecológica]*. Barcelona, España: Gustavo Gili.

Kim, K. H. (1997). Housing finance and urban infrastructure finance. *Urban Studies, 34*(10), 1597–1620.

Moss, T. (2008). 'Cold spots' of urban infrastructure: 'Shrinking' processes in eastern Germany and the modern infrastructural ideal. *International Journal of Urban and Regional Research, 32*(2), 436–451.

Ouyang, M., Dueñas-Osorio, L., & Min, X. (2012). A three-stage resilience analysis framework for urban infrastructure systems. *Structural safety, 36*, 23–31.

Piedecausa, B. (2009). La vivienda enterrada: estudio de su evolución tipológica y adaptación geográfica. *Investigaciones Geográficas, 50*, 169–189.

Ruano, M. (2000). *Ecourbanismo*. Barcelona, España: Gustavo Gili.

Sahely, H. R., Kennedy, C. A., & Adams, B. J. (2005). Developing sustainability criteria for urban infrastructure systems. *Canadian Journal of Civil Engineering, 32*(1), 72–85.

Tarr, J. A. (1984). The evolution of the urban infrastructure in the nineteenth and twentieth centuries. In R. Hanson (Ed.), *Perspectives on Urban Infrastructure* (pp. 4–66). Washington, DC: National Academy Press.

Chapter 14
Sustainability of Urban Energy

Timothy Fraser and Andrew J. Chapman

Abstract The sustainability of urban energy generates a very big challenge for the countries, since it is necessary to ensure that the energy remains available for the activities of each city and at the same time it is necessary to invest in obtaining sustainable energy resources and the construction of the infrastructure corresponding for the use of this sustainable energy.

Keywords Sustainability of urban energy · Sustainable urban energy · Energy · Sustainability · Urban

14.1 Introduction

Cities face challenges in reducing their carbon footprint while maintaining and improving the social and economic conditions of life for their urban residents. However, recently, solar and wind have emerged as flexible, decentralized technologies well suited for cities. This chapter highlights the contributions of urban renewable energy to local economies and urban planning, using a case study of mega-solar in the city of Kagoshima, Japan. Despite conventional wisdom that large-scale renewables are best suited for rural areas, cities are employing innovative partnerships to build mega-solar farms on city harbors. These plans put underutilized land back to use, especially in small and medium sized cities. In this way, decentralized renewable power can boost social sustainability not only through tax revenue and some employment, but also through urban planning itself.

T. Fraser
Department of Political Science, Northeastern University, Boston, MA, USA
e-mail: fraser.ti@husky.neu.edu

A. J. Chapman (✉)
International Institute for Carbon Neutral Energy Research, Kyushu University, Fukuoka, Japan
e-mail: chapman@i2cner.kyushu-u.ac.jp

A. Alvarez-Risco et al. (eds.), *Building Sustainable Cities*,
https://doi.org/10.1007/978-3-030-45533-0_14

Cities are home to large numbers of people, often living close together. As a result, living in an urban setting offers advantages including employment opportunities, access to essential services, efficient transportation, and other life opportunities including education and social connectedness. On the other hand, due to the concentration of populations in a single area, the need for food, energy, and clean water often exceeds that which the city can provide for itself. One ideal of urban sustainability is that cities continue to sustain growth and innovation, without excessive reliance on imported resources, and the ability to power themselves through innovations such as renewable energy. Energy remains a key issue in urban sustainability, as, although cities only account for 3 percent of the earth's land, they also account for up to 80 percent of global energy consumption and 75 percent of energy related carbon emissions (UN, 2018). In line with the United Nation's (UN) sustainable development goals (SDGs), particularly SDG 11: sustainable cities and communities, in this chapter we focus on the need for cities to continue to innovate and to take advantage of their high density to achieve efficiency gains and technological advancement. In terms of sustainable energy, we focus on the deployment of specific renewable energy generation infrastructure which engenders multiple benefits including a reduction in energy related carbon emissions, associated health benefits, and a reduction in reliance on energy imports. Among the types of renewable energy available for deployment in cities, wind and solar power have proven to be the easiest to deploy, dominating all other types in terms of deployed capacity up until 2016, and from 2017 onward, it is estimated that a turning point will be reached where solar photovoltaics (PV) will become the largest contributor to new renewable energy generation (IEA, 2017).

The reasons for the emergence of this "new era" for solar PV include continuing cost reductions, underpinned by enabling policies around the world and rapid growth in deployment, particularly in Asia, led by China—responsible for over 40 percent of global renewables capacity growth (IEA, 2017). In terms of cities that have targeted a clean energy future and successfully ushered in a transition to renewables based, sustainable energy generation, there is evidence of over 100 global cities that now generate over 70% of their electricity from renewable sources, city-wide (CDP, 2018).

14.2 Sustainable Urban Energy Approaches

As sustainability has economic, environmental, and social aspects, the consideration of energy generation from renewable sources alone is insufficient to declare that cities are sustainable. Indeed, when considering appropriate energy types and deployment, city planners must also consider the impacts of these decisions. Ideally the deployment of renewable energy will not only alleviate the need for fossil fuels and reduce carbon emissions, engendering environmental benefits, but also improve life directly for cities. Good renewable energy policy stimulates city growth and offers employment opportunities, demonstrating economic benefits. It also reduces electricity bills, stimulates community development, and improves people's quality of life, demonstrating social benefits.

Each type of renewable energy offers different benefits, limitations, and deployment challenges in the urban setting. Wind and hydro power typically require large amounts of land, appropriate resources including large volumes of water or suitable wind speeds, and are difficult to scale to city-sized applications. Biomass-based generation is another option but is likely to be less desirable from an amenity, and resource collection point of view in urban areas. Solar power on the other hand offers an easy to deploy option for urban planners, both in a distributed (i.e., deployed on individual rooftops) or centralized (mega-solar) approach. More recently, the aggregation of individual rooftop power plants into "virtual" mega-solar plants in Japan and Australia has also occurred, in some cases incorporating battery storage to overcome intermittency issues (Fujitsu, 2015; Reuters, 2018). Based on these precedents, solar power appears to be a flexible and clean approach to urban energy generation, worthy of further investigation in terms of its contribution to urban energy sustainability. The next section of this chapter outlines a case study in Kagoshima City, Kagoshima Prefecture, Japan, where mega solar was deployed, investigating the capacity of cities to innovate on social sustainability and urban planning problems using large-scale renewables.

14.3 Case Study and Applications

Japan's sunny, southernmost mainland city of Kagoshima offers important lessons for how cities can reduce greenhouse gases while improving the social sustainability of urban areas. Cities often face more difficulty deploying large-scale renewables, as opposed to rooftop solar or energy efficiency improvements, with many mega-solar plants going to sparsely populated rural towns with cheaper land. Yet Kagoshima hosts Japan's fourth largest mega-solar plant, the Kagoshima Nanatsujima mega-solar power plant, despite being a dense city of over 600,000 people. This plant makes a major dent in Kagoshima's carbon footprint. It hosts an installed capacity of 70 MW (about 1/12 of a standard, 900 MW nuclear reactor), producing 78,800 kWh of electricity annually. It provides 8.2 percent of the city's power supply and 2.2 percent of the entire prefecture's—no small feat. To explain the surprising construction of such a large plant in an urban area, the authors conducted interviews with Kagoshima City Council member Eiichiro Noguchi, officials from the Kagoshima City Hall Department of Environment, and Nanatsujima plant representatives on January 25, 2017 (Fraser & Chapman, 2018). The way Kagoshima City came to host this facility is integrally related to mega-solar's potential to improve sustainability in the communities that host it.

The Nanatsujima plant resulted from a deal to put reclaimed land on the Kagoshima Bay harbor back to use. Kagoshima, one of the prefectures furthest from Japan's industrial corridor between Tokyo and Osaka, suffered especially from the country's lost decades of economic recession. Like many second and third tier cities that invested in port infrastructure to boost their economic profile in the late twentieth century, Kagoshima's efforts backfired. After Ishikawajima-Harima Heavy Industries (IHI) bought prime land on the outskirts of the harbor, the land sat unused for nearly 40 years. Lacking corporate interested in buying the land during the recession, IHI held onto the land in the event that it might become useful or profitable in the future. For every year the land sat idle, Kagoshima City and Prefecture lost potential employment gains, tax revenue, or tourism that the property could have brought to the region under different circumstances.

As a result, when the Japanese parliament passed the Feed-in Tariff policy in 2012 to subsidize renewable power through added costs on consumer electric bills, the Kagoshima Development Corporation saw an opportunity. This quasi-public development agency coordinates opportunities to promote industry siting in Kagoshima from its offices in Osaka. They established a twenty-year partnership enterprise between IHI and six other corporations and banks. These included Kyōcera, the electronics manufacturer, Kyūdenkō, the engineering firm linked to the regional power company, Takenaka Corp, a national construction firm, KDDI Corp, Japan's second-largest mobile phone company, Kyoto Bank, and Kagoshima Bank.

Over seven years, the joint enterprise would accrue a total value of 27.3 billion yen. Kyōcera, Kyūdenkō, and Takenaka Corp. managed the construction and manufacturing. KDDI, Kyoto Bank, and Kagoshima Bank provided the financing for the process. Finally, IHI agreed to rent the property for twenty years, at the end of which, KDDI would receive the land. Under this market-driven, state-coordinated investment, neither Kagoshima City, the prefecture or the Nanatsujima neighborhood government were financially responsible for redeveloping the land, nor has the city government been significantly involved in the management of the project. As a result of this arrangement, the plant has provided new employment and revenue in the city. 78,000 employees were involved in the project's construction from September 2012 to November 2013, with as many as 500 persons per day working on the project during peak construction. Post-construction, the plant still hosts 7 employees responsible for maintenance, tourism, and environmental education roles at the on-site solar energy education center.

The small number of these permanent jobs demonstrates a clear limitation of the employment potential of distributed power plant management. Best practices encourage local ownership of projects and locally based manufacturing of parts, but given the rise of less costly, Chinese-made parts, local manufacturing of panels has become more of a pipe dream in Japanese municipalities. Given that many mega-solar plants host *no* employees after construction finishes, these larger plants remain an important boon to the economy, especially because large plants like these are better able to also host multipurpose educational facilities due to higher revenue streams. Additionally, the plant contributes property tax revenue to the city, the other, direct form of financial benefits for residents. When the impacts of the Kagoshima plant and others like it were measured throughout Japan in terms of their effect on carbon emissions, pollution, electricity prices, community tax base, and employment, these mega-solar plants *all* improved the social equity of their host municipalities in the eyes of local officials (Chapman & Fraser, 2019).

A key prerequisite for this project and others like it was the Feed-in Tariff. This national level policy guarantees renewable power plant companies that utilities would buy a specific amount of electricity that they generate at a fixed rate of revenue for a 20-year period. This guarantee ensured access to the market for smaller companies, the 20-year subsidies added stability to the market, and the fixed rate of revenue created incentives for companies to deploy the largest plants possible in order to maximize their revenue. However, the most important aspect of these projects is coordination between regional government and deploying companies. Assuming that government coordination is present, a number of other renewable power subsidy programs could also provide the stability necessary for these projects.

Most importantly, the Nanatsujima plant provides a model for deploying large amounts of renewable power in cities while solving long-standing urban planning problems—in this case, redeveloping large amounts of underutilized land. Kagoshima is not alone in its success. Japanese cities large and small have turned to hosting solar on unsalable land plots, including reclaimed land like in Kagoshima, but also polluted, irradiated land like in Minamisōma, Fukushima Prefecture, as well as on numerous unprofitable golf courses and harvested quarries throughout Japan. In this way, the deployment of mega-solar can be a powerful tool for remedying long-standing urban planning problems caused by urban sprawl, pollution, or recession.

However, under the Kagoshima model, while city management or financing is possible, these private-sector led projects do not necessarily require the long-term attention of city officials, meaning that mega-solar can have a compounding effect. It can repurpose underutilized industrial or commercial sites, increasing the capacity for renewable power in a city, while private-sector management of these projects can free city officials to work on other means of boosting decentralized renewable power or increasing energy efficiency. Yet, this model does carry drawbacks. A lack of local government involvement and oversight can lead to companies ignoring local requests or landscape degradation. As a result, transparency and at least informal local government involvement to ensure procedural justice in the siting and operation of these plants is required.

While tax revenue, employment, and proactive urban planning are not the only components of social sustainability, they are three important ways that small and medium-sized cities can benefit from mega-solar while reducing greenhouse gas emissions. Smaller, peripheral cities often lack the financial resources of those larger cities party to the Cities Climate Leadership Group (C40), whose high tax bases enable innovation at city hall and who can deploy more rooftop solar due to their larger number of wealthy citizens. However, mega-solar allows small and medium-sized cities to take advantage of their size, of past urban sprawl, or of large-scale industrial development now extinct. This different set of resources can ensure that smaller cities like Kagoshima do not lose out on the economic and social benefits of the renewables boom.

As a result, Kagoshima's model has broad applicability to rustbelt cities in economic distress or recession, laden with former industrial land or polluted land that cities struggle to sell. Given the right national-level policy incentives, such as a

Feed-in Tariff, cities can use mega-solar to boost the social benefits of their renewable energy policies. Financially troubled Eurozone countries such as Greece or Portugal could look to this as an opportunity for development in peripheral cities. Similarly, as China's economy transitions from a boom economy, cities where rapid urban development resulted in urban or industrial sprawl should consider mega-solar as a means of repurposing this land should it fail to be parlayed into other, more profitable uses.

Closing Remarks
In conclusion, mega-solar appears to be a flexible and clean approach to urban energy generation, especially in small and medium-sized rust-belt cities that struggle to sell or repurpose underutilized land. Ideally, renewable power projects could contribute to the social sustainability of urban systems by boosting civic engagement and changing resident behaviors in energy consumption. However, these goals can be challenging for city officials to put forward at the beginning of an energy transition, especially when they lack significant financial backing from city hall. Instead, mega-solar development provides an important means by which private-sector development can boost employment and tax revenue from underutilized properties, helping reclaim economic value lost from years of abandonment. Finally, city and regional governments have an important role to play in identifying appropriate properties and connecting developers with financial backing, and especially in working with communities and residents to make sure that their interests and preferences are represented in these development processes. By making mega-solar development affordable and feasible, governments can give cities a powerful tool to remedy carbon emissions, land use, tax base, and employment issues in tandem.

References

Carbon Disclosure Project. (2018). The World's renewable energy cities. Retrieved February 5, 2020, from https://www.cdp.net/en/cities/world-renewable-energy-cities

Chapman, A., & Fraser, T. (2019). Japan's mega solar boom: Quantifying social equity expectations and realities at the local scale. *Sustainability Science, 14(2)*, 355-374.

Fraser, T., & Chapman, A. (2018). Social equity impacts in Japan's mega solar siting process. *Energy for Sustainable Development, 42*, 136–151. https://doi.org/10.1016/j.esd.2017.11.002

Fujitsu. (2015). Turning apartment rooftops into 'virtual power plants', green energy through mega-solar power generation. Retrieved February 5, 2020, from http://www.fujitsu.com/global/vision/customerstories/leopalace21

International Energy Agency. (2017). *Renewables 2017: Analysis and forecasts to 2017*. France: IEA Publications.

Reuters. (2018). South Australia promises world's largest virtual power plant. Retrieved February 5, 2020, from https://www.reuters.com/article/us-australia-power-tesla/south-australia-promises-worlds-largest-virtual-power-plant-idUSKBN1FO02G

United Nations. (2018). Sustainable development goals, goal 11: Make cities inclusive, safe, resilient and sustainable. Retrieved February 5, 2020, from https://www.un.org/sustainabledevelopment/cities

Chapter 15
Urban Waste Management

Eduardo De-La-Torre-Jave, Aldo Alvarez-Risco, Shyla Del-Aguila-Arcentales, and Alex Harras

Abstract Success waste management must use more than only the disposal of household waste due to both construction and demolition tasks increase and also the building of new homes and buildings entails large increases in the generation of construction waste, more in big cities. On the other hand, natural disasters also generate a large amount of waste and in an unplanned manner, which leads to having such waste out of the management budget. The literature shows the different efforts that are made to establish the best practices of urban waste management.

Keywords Urban waste management · Sustainable waste management · Waste management · Sustainability

15.1 Solid Waste and Its Context

15.1.1 Solid Waste and Sustainable Development Goals

Solid waste management is closely linked to several of the 17 sustainable development goals (SDG). To fulfill these goals by 2030, it will be necessary for governments at the national, regional, and local levels, as well as civil society and the private sector, to address those which are closely related with the environment and

E. De-La-Torre-Jave
Ciudad Saludable, Lima, Peru
e-mail: eduardo@ciudadsaludable.org

A. Alvarez-Risco (✉)
Facultad de Ciencias Administrativas y Recursos Humanos, Universidad de San Martín de Porres, Lima, Peru
e-mail: aldo.alvarez@redsaf.org

S. Del-Aguila-Arcentales
Escuela Nacional de Marina Mercante "Almirante Miguel Grau", Callao, Peru

Universidad Nacional de la Amazonia Peruana, Iquitos, Peru
e-mail: sdelaguila@enamm.edu.pe

A. Harras
Center for MSP Research, Lima, Peru

A. Alvarez-Risco et al. (eds.), *Building Sustainable Cities*,
https://doi.org/10.1007/978-3-030-45533-0_15

conservation, as well as complementary with the improvements required in solid waste management in particular. The first such complementary objective is SDG 8. It promotes *sustained, inclusive and sustainable economic growth, full and productive employment and decent work for all.* This is linked to the formalization and improvement of the work conditions for solid waste recyclers. These are the men and women involved with the segregation and selective collection of solid waste. Such programs in municipalities contribute not only to their improved incomes, but also to their personal and emotional well-being. Above all, they are recognized by their neighbors and receive better treatment from them.

SDG 11 promotes *sustainable cities and communities*. This is closely related to the integral management of solid waste, from its generation to the appropriate final disposal. It prioritizes the minimization of the waste and its recovery, whether in material or through energy conversion.

SDG 12 pertains to *responsible production and consumption*. Worldwide, one-third of all food produced goes to dumping grounds or landfills as organic waste. Because of the substantial volume, food waste must be taken into account by any management plan, as well as those activities or actions to promote responsible consumption and production that may be expected to have any positive effects on the reduction of waste.

SDG 14 requires consideration of *life below water*. This entails the conservation and sustainable use of the oceans, seas, and marine resources in general. For there to be any possibility of sustainable development, we must address the deterioration of these ecosystems due to the large accumulation and contamination by solid waste, mainly by plastics and microplastics (see Figs. 15.1 and 15.2).

Fig. 15.1 Marine garbage in Paracas, Pisco, Peru. (Source: Eduardo De La Torre-Ciudad Saludable)

Fig. 15.2 Plastic waste, San Andrés, Pisco, Peru. (Source: Eduardo De La Torre-Ciudad Saludable)

15.1.2 Latin America

Throughout Latin America and the Caribbean during the last 15 years, increases in both population and economic activity have led to corresponding increases in consumption. This in turn has led to a commensurate rise in the generation of solid waste. According to the Iniciativa Regional de Reciclaje (Regional Recycling Initiative) (2017), 430 thousand tons of waste is generated daily. Governments are implementing models of Integral Solid Waste Management. An important aspect within such models is recycling activity. It is estimated that there are 4 million people who earn their livelihoods by marketing recyclable waste. In 2009 and 2010, Peru and Brazil, respectively, were the first countries anywhere LAC to pass national solid waste laws, which recognized recyclers and protected their rights. The Iniciativa Regional de Reciclaje (2017) published a study entitled *Advances and challenges for inclusive recycling: Evaluation of 12 cities in Latin America*, which analyzed three dimensions including regulations, organizational structures, and markets. The results of this study showed that Sao Paulo, Buenos Aires, and Bogotá led in the three dimensions of inclusive recycling. Lima, Peru was in fourth place. On the other hand, Germany, the country with the most advanced Circular Economy, has a 63% annual recycled material rate. Some countries such as Chile, Argentina, and Colombia barely exceed 10%. And others such as Bolivia and Peru reach only slightly more than 3%.

In relation to the regulation on plastic, we have what described by Alvarez-Risco, Del-Aguila-Arcentales, and Rosen (2020), who describe how the historical evolution has been, Peru having the leadership in Latin America the contribution of plastic management through its recent regulation.

15.2 Case Peru

15.2.1 New Solid Waste Law

According to the Ministerio del Ambiente (Ministry of Environment) (2018), Peru generates 19,000 tons of solid waste daily with approximately 50% of this waste generated in the Lima metropolitan area, including Callao. Each resident there generates 870 grams of waste per day. Residents living in the interior of the country generate 800 grams per day. The physical composition of the waste includes organic waste (54%), recoverable waste (20%), non-recoverable waste (19%), and hazardous waste (7%). In the year 2000, Peru passed Law No. 27314: General Law of Solid Waste. In December 2016 it was updated and replaced by Legislative Decree No. 1278. This regulation is applicable to all productive sectors, including activities related to internment, storage, treatment, transport, and export of solid waste. The regulations changed the authority for the management of solid waste from the General Directorate of Environmental Health (DIGESA) under the Ministry of Health to the Ministry of Environment. Significant changes were made related to the rights, obligations, attributions, and responsibilities of society as a whole in the management of solid waste.

The law prioritizes the efficiency of materials, overall minimization, and waste recovery. Thus, it has promoted a change from a *linear economy model* where "(waste) is extracted, produced, consumed and discarded" to a *circular economy approach*, with final disposal being the last alternative. The new regulations apply valuation principles with respect to solid waste, thereby treating it as a valuable asset that can be reused, recycled, or composted. Additional value may be attributed based on possible energy recovery through biodegradation and co-processing processes. In addition, recyclers are integrated into the solid waste management system. Thus, the Source Segregation and Selective Collection Programs must be implemented taking into account an approach that includes recyclers operating in the formal economy. Also relevant, the law also simplifies procedures for the construction of infrastructures such as sanitary landfills. Under the previous law, approvals took an average of 5 years compared to the 1-year necessary today.

15.2.2 Degraded Areas Due to Accumulation of Waste

The most serious part of the waste problem is inadequate provision for solid waste final disposal. According to information from the Environmental Assessment and Inspection Agency (2018), 1585 degraded areas exist outside of Lima and Callao. From around the country, the greatest concentration of these areas is in the departments of Ancash (149 sites), Cajamarca (123 sites), and Puno (111 sites). Those that have the largest affected land surface areas are Lambayeque (438 hectares), Ica (276 hectares), and Piura (201 hectares). However, the largest and most notorious open pit dump, considered the third largest in Latin America according to the International Solid Waste Association (2014), is the *El Milagro* dump that receives about 800 tons of waste per day from all the districts in the province of Trujillo. However, according to Environmental Assessment and Inspection Agency (2018), that of the 1585 degraded areas, 27 such areas are slated for conversion to landfills, and the remaining 1558 degraded areas have been identified for cleanup and recovery. The current situation of informal recyclers remains worrisome since hundreds of people work in the open-air dumps under very hazardous conditions. They are routinely exposed to smoke from burning waste, to the gases generated from the decomposition of organic matter, and contact with hazardous waste, including waste from both hospital and industrial sources. If we consider that they do not have vaccines, do not use bodily protective equipment, and commonly take their children to accompany them, then the situation represents a much more complicated and dangerous public health situation.

Furthermore, according to the Environmental Ministry (2018) only 34 adequate sanitary landfills exist Peru. While 52% of the solid waste generated in our country is disposed of properly, 48% of the waste remains improperly disposed in open-air dumps. On the other hand, in Lima there are the four largest sanitary landfills in the country, where they dispose of more than 8000 tons per day.

In many cases, the breeding of hogs is still observed (Figs. 15.3 and 15.4).

On the other hand, according to the Ministerio del Ambiente (2018) there are only 34 sanitary landfills in Peru where 52% of the solid waste generated in our territory is properly disposed and in Lima there are the four largest landfills in the country, where they dispose of more than 8000 tons per day and 48% of the waste is improperly disposed in open-air dumps. On the other hand, in Lima there are the

Fig. 15.3 Recycler with her baby in her arms, Piura, Peru. (Source: Eduardo De La Torre-Ciudad Saludable)

Fig. 15.4 Recicladora informal, Chulucanas, Piura, Perú. (Source: Eduardo De La Torre-Ciudad Saludable)

four largest sanitary landfills in the country, where they dispose of more than 8000 tons per day.

15.2.3 Quality of Life of the Recyclers

Approved in 2009, Law 29419 regulates the activity of recyclers. It aims to protect, train, and promote social and labor development. In addition, it endorses labor formalization within the recycler sector and works to foster the formation of recycler labor associations. While it is estimated that only 12% of the recyclers are formalized, this greatly improves the chances that these laborers work in safer and cleaner conditions, and generate better incomes for their families. It helps make them visible to the municipal authorities, and the population generally, thereby elevating their status closer to that of other respectable occupations. Also in the Law of Integral Management of Solid Waste specifically mentions that selective collection can be carried out by the municipalities, operating companies, or associations of recyclers. While commercialization is permitted by operating companies and associations of recyclers, it disallowed by municipalities. Recyclers play a vitally important role in the solid waste value chain. They do the work of collecting, segregating the waste. And in some cases, they perform the additional task of conditioning it, which entails crushing and compacting the waste, thereby enhancing of the waste and preventing it from going to the open-air dumps or being buried in sanitary landfills. In some cases, they directly commercialize compacted PET plastic and deliver

Fig. 15.5 Formalized recyclers, Chincha, Ica, Peru. (Source: Eduardo De La Torre-Ciudad Saludable)

it to the company that transforms this material into new bottles. In this process, they take a major step to close the cycle and create a new virtuous one in compliance with the principle of Circular Economy (see Fig. 15.5).

As part of their formalization process, recyclers have taken training courses related to integral management of solid waste, occupational health and safety, business management and recycling, as well as social skills and personal development. They have also been vaccinated against hepatitis B and tetanus. Today they are better prepared for and able to play an effective role in segregation and selective collection programs on behalf of municipalities.

15.2.4 Municipal Programs

Since 2011, 250 municipalities have participated in the Incentive Plan for the Improvement of Municipal Management and Modernization (IP). Participants receive funding transfers on the condition they have fulfilled goals set for municipalities within an allotted period of time. These goals include those related to solid waste, including, improved municipal regulations for the formalization of recyclers, selective waste collection, waste management planning, waste segregation programs, various studies, among others. Meeting the goals set out ensures that the municipalities receive economic incentives while they participate and contribute to the improvement of the integral management of waste throughout the country. They include material recovery, including organic matter for compost or inorganic material through recycling, as well as a safe final disposal. Since its creation in 2008, the participation of the Ministry of Environment has been essential for the progress made in this sector. However, according to MINAM, as of 2017 only 3.4% of recyclable waste gets recovered.

Fig. 15.6 PET plastic for sale, El Carmen, Ica, Peru. (Source: Eduardo De La Torre-Ciudad Saludable)

Fig. 15.7 Selective collection, Chincha, Ica, Perú. (Source: Eduardo De La Torre-Ciudad Saludable)

Only 1.3% of the available organic waste gets processed by any of the municipal segregation and selective collection programs or through the participation of any of the 264 recyclers associations. Although these are steps in the right direction, greater participation among municipalities, along with greater citizen awareness, and visibility in the educational institutions remain necessary. More effective involvement companies that are large generators of recyclable waste must also be activated and involved in the municipal segregation programs (See Figs. 15.6 and 15.7).

Another important improvement must include Municipal Collection Centers. These are places designed to store non-hazardous solid waste recovered as a result of source segregation and selective collection programs. Because a large number of formalized recyclers do not have reliable access to such infrastructure, they can neither store large volumes of waste, nor can they make the finer waste separations. Not having such convenient aggregation facilities hinders recyclers and limits their ability to maximize their contribution. Furthermore, it limits their access to the premium prices paid for the highly segregated waste.

15.2.5 Less Plastic

Worldwide concern and information about the negative impacts of plastic, especially microplastics in the oceans, continues to grow. Plastic is considered as *marine litter*. It is causing a deterioration of these ecosystems, as well as the fauna and flora that inhabits them. In addition to the negative ecological consequences, they also bring economic impacts to sectors including fishing and tourism.

It is important to realize that the appropriate mitigative actions must be taken on land. It is estimated that 80% of marine litter is generated and enters the oceans from terrestrial sources. The United Nations Environment Program (2017) warns about the consequences of contaminating our oceans with plastics and microplastics. They note that 600 marine species are being damaged, and up to 15% of such species are ingesting or becoming entangled with garbage at sea. In some cases, such damage is severe enough to raise the possibility of extinction. They also predict that by 2050, 99% of seabirds will have ingested some quantity of plastic, and there will be more plastic by weight than fish in the sea. Given these considerations and background, the Ministry of Environment led a very strong effort, including much dialog with other ministries, legislators, and leaders of various political parties. Together they approved a law regulating single-use plastic containers and disposable containers in general. The passage of this law in December of 2018 has meant substantial changes in the reduction of waste generation, especially in both the food service sector, which has stopped using plastic straws, and the grocery sector, which has curtailed the use of plastic bags.

A noticeable change in behavior among consumers has also occurred. Indeed, many have stopped buying or using straws and plastic bags. In addition, social networks play an influential role helping make people more aware of the negative impacts of plastics on the environment. Such online interactions convey impactful messages. For example, the social acceptance of using both reusable shopping cloth bags, similar to the usage common from many years in the past, and the use of *reusable water bottles* as an alternative to the use of PET plastic bottles has started to spread. Companies have started to grow by selling biodegradable products as alternatives to plastic ones. For example, companies now sell straws made from a variety of non-plastic substances including grass fiber, wheat stalks, and bamboo. They sell kitchen dishes made from cellulose of sugarcane, leftover wheat fibers, and otherwise unused palm or banana leaves. Cutlery is being made with bamboo fiber. The most important result of these innovations is that all these products are being made from materials that were formerly treated as only surplus or waste from other industrial processes.

Civil society has joined in the effort and become involved. Beach cleaning campaigns are among the activities which have motivated participation of many volunteers. These efforts are important because they create individual awareness for the volunteers themselves, as well as create a larger social context when visitors go to these places and see these activities. The joint participation of the Ministry of Environment, the private sector and civil society becomes increasingly noticeable and relevant.

15.3 Waste Management: Global Approach

Society expects universities to carry out and produce knowledge across a wide range of possible topics, including solid waste recycling. Furthermore, universities are often the setting where such new knowledge is tested and applied first. However, such spinoffs from theory to practice do not always occur. Such is often not the case with waste management and recycling at university campuses according to Azevedo and Cruz (2019). They used task analysis and assessed the matter using the following criteria: destination of waste within the university, the amount and quality of waste generated, the waste trajectory outside the campus, and the requirements for selective collection. They examined the current system, with attention on people's mindset and behavior towards waste generation, and compared it with requirements for proper waste management. With respect to separation, they collected answers such as "Too much work," and "It's not my habit." More answers disparaged the system saying, "People don't respect the materials that must go into each waste bin," and "the cleaning staff often mix it all together in one bag that is later disposed of," and "There is no transparency as to where things go" and finally "Some waste bins have no indication of the kinds of material there should go there." The same useful methodology can be adapted by teachers and students at other universities to investigate and describe the current situation of waste management. The knowledge gathered can contribute to the optimization of the system.

Despite all the efforts made by cities, some wastes are illegally discharged into public or private areas, generating a variety of negative environmental, aesthetic, social, and economic impacts (Santos, Mendes, and Teixeira 2019). This clearly risks a public health danger for the entire population. These informal disposal sites require a large expenditure of money for ongoing management by municipalities, many of which do not have enough money or only inadequate capacity to oversee. The city streets become full of refuse. An article by Santos et al. (2019) proposes a model for participatory management relative to the efficient management of illegal solid waste. It provides the necessary guidelines for the efficient management of such waste, taking into account the environmental, economic, and social dimensions. This model is summarized by the following six managerial practices:

1. Understand the real situation of illegally deposited waste.
2. Integrate environmental, social, and economic objectives.
3. Welcome wide and open participation.
4. Define clear strategies.
5. Use life cycle principles that contribute to a circular economy.
6. Provide assistance to municipalities.

Does proper waste collection alone assure successful delivery to a landfill? As Slanina, Pokorny, and Dedek (2018) point out, landfilling itself remains a method that results in significant adverse effect on the environment. A large percentage of cities have the same inefficient waste management, such as in Argentina where Litter et al. (2019) reported arsenic in groundwater. Thus, more sustainable solutions are required, including the reuse of waste through the so-called *recycling of a utility product*, which requires technologies that process such products in an efficient, standard, and ultimately profitable way. The expected and continued growth of cities must lead to greater planning of sanitary waste management. This is an urgent requirement considering that there also exist cities with rapidly increasing populations due to immigration. Notably, this is the case in South America where immigrants are moving from Venezuela, and the case in Lebanon where immigrants are moving from Syria. To assess the response to waste disposal by municipal administrations, Boex, Malik, Brookins, Edwards, and Zaidi (2020) investigated 18 cities in six countries in Asia. They found deficiencies in waste management services. This points to a greater need for the reform of institutional structures to better align them with the essential responsibility for realizing cities' full economic potential and meeting targets set in the sustainable development goals related to waste management. Effective waste management must take into consideration more than just the disposal of household waste. As cities grow, both construction and demolition activities increase. The building of new homes and buildings entails large increases in the generation of construction waste. What laws are established to manage this waste? Although it is likely that such waste is going to landfills, it is also possible that it could become part of processes that allow the generation of new, economically valuable products. Addressing this possibility, Wu et al. (2019) evaluated certain indicators, finding that while it is true the environmental and economic aspects are moderately fulfilled, the social aspects have received very little attention. The article provides a research methodology to evaluate these aspects in other regions of the world. The research published by Jin, Yuan, and Chen (2019) shows a survey of construction and demolition waste management research published between 2009 and 2018. They found that growth in both housing and commercial construction is accompanied by an increasing pressure to "green." However, qualification systems are needed for such green constructions, specifically those that accompany construction waste management.

Lu, Chi, Bao, and Zetkulic (2019) evaluated the effects of three such green construction waste management systems, including Leadership in Energy and

Environmental Design (LEED), Mainland China's GB Evaluation Label (GBEL), and Hong Kong's Building Environmental Assessment Method (BEAM Plus). The researchers found that none of these systems greatly promoted superior performance of construction waste management. This should prompt us to make an in-depth analysis of construction waste management in order to ensure that the whole process is actually ecologically balanced. Hao, Yuan, Liu, Chin, and Lu (2019) described and developed a dynamic system model to investigate the economic performance of the reduction of construction waste. This brings a commercial point of view and describes how this type of management can generate economic gains for companies. They investigated the economic performance of the reduction of construction waste, based on the interrelationships of the main factors that affect such economic performance related to the reduction of construction residues. It involves three subsystems that cover the waste generation disposal, waste reduction, and economic performance. They found that four strategies effectively promote the economic performance of construction waste reduction:

1. Improve waste classification
2. Reduce illegal dumping behaviors
3. Promote the government financial subsidy for waste recycling
4. Increase the cost of landfills.

This research needs to be expanded through field tests in order to determine and confirm their applicability among other possible practical and profitable strategies for companies and society. Additionally, Ibáñez-Forés, Bovea, Coutinho-Nóbrega, and de Medeiros (2019) proposed a set of social impact categories, indicators, and metrics capable of assessing the socioeconomic and labor conditions of the different actors involved in the life cycle of a municipal waste management system (see Table 15.1). Specifically, 12 categories of social impact and 22 indicators with their corresponding metrics were raised.

Natural disasters create special requirements for waste management called disaster waste management (DWM). In that context, Zhang, Cao, Li, Liu, and Huisingh (2019) reviewed the existing DWM literature and found and classified different articles according to the type of disaster:

1. General (23 articles): Brown and Milke (2016), Takeda, Mori, Kubota, and Arai (2014), Trivedi, Singh, and Chauhan (2015), Brown, Milke, and Seville (2011a), Grzeda, Mazzuchi, and Sarkani (2014), Yusof, Zawawi, and Ismail (2016), Crowley and Flachsbart (2018), Kim and Hong (2017), Crowley (2017), Gabrielli et al. (2018), Regattieri, Gamberi, Bortolini, and Piana (2018), Boonmee, Arimura, and Asada (2018), Otsuka and Katsumi (2015), Wang, Hu, Guo, and Gong (2019), Ozdamar, Aksu, and Ergunes (2014), Lesperance, Stein, Upton, and Toomey (2011), Zawawi, Yusof, and Ismail (2018), Yoo, Lee, Chi, Hwang, and Kim (2017), Zawawi, Yusof, Kamaruzzaman, and Ismail (2015), Maryono, Nakayama, and Shimaoka (2015), Caniato, Tudor, and Vaccari (2016), Regattieri, Piana, Bortolini, Gamberi, and Ferrari (2016), and Baek, Kim, Choi, and Hong (2016).

Table 15.1 Social indicators for assessing the waste management system based in Ibáñez-Forés et al. (2019)

Working rights		
Social indicator	Metric	Stakeholder
1. Freedom for association and collective bargaining	Evidence for restrictions to the freedom of association and collective bargaining	W, PDS, SD
	Workers have access to meetings and the possibility to dispute resolution procedures	W, PDS
	Labor union presence	W, PDS, SD
Human rights		
2. Child/senior labor	Number of children working in the analyzed sector	W, PDS
Quality of job positions (working conditions)		
3A. Fair salary	Worker salary compared too minimum wage	W, PDS
3B. Working hours and/or weekly rest	Weekly hours actually worked by employees	W, PDS
Equal opportunities/discrimination		
4A. Gender discrimination	Number of women working in waste management	W, PDS, PDV
	Gender pay gaps	W, PDS
4B. Labor regulation	Number of undocumented workers in waste management	W, PDS, SD
4C. Workers from marginal classes	% of workers with no possibility of working in another sector	W, PDS
Health and safety		
5A. Security and safety of workers	% of workers who use personal protective equipment in their work	W, PDS, SD
5B. Long-term health	% de vaccinated workers	W, PDS, PDV
	% of workers with no problem health	W, PDS
Working benefits		
6A. Legal employment with social benefits/security	% of workers with information on the rights that correspond to the waste collector occupational code	W, PDS
6B. Workers and relatives with health insurance	% of workers with the possibility of paying the national health service	W, PDS, PDV
Socioeconomic conditions		

(continued)

Table 15.1 (continued)

Working rights		
Social indicator	Metric	Stakeholder
7A. Level of education: workers and their children	Workers' level of education	W, PDS
	Level of education of workers' families	W, PDS
7B. Social characteristics of population	Total monthly family income	W, PDS, PDV
7C. Safe and healthy living conditions	Quality of workers/costumers' houses	
Community satisfaction and participation		
8A. Customer/citizen satisfaction	Social welfare/satisfaction (quality of products/service)	U, PDS, PDV
8B. Customer/citizen participation	% of citizens with access to a reliable WM system	U, PDS
Value chain actors relationship		
9A. Transparency/corruption	Customer knowledge about the system	W, PDS
	Presence of periodical public company reports	MA, SD
10. Degree of environmental worker awareness	Workers' environmental education/awareness	W, PDS
Local development		
11A. Development of environmental awareness and responsibility	Custom environmental awareness	U, PDS
	% of users receiving environmental information on waste management	U, PDS, SD
11B. Local labor integration of formal workers from informal sector	% of formal workers from informal sector	W, PDS, SD
Governance		
12A. Public commitments to sustainability issues	% of actions made with public funds related to waste management	U, MA, PDS, SD
12 B. Maturity/existence of the informal WM system's regulation	Legislation on waste management	MA, SD

W workers, *PDS* primary data (survey), *PDV* primary data (visit), SD secondary data, *U* users, *MA* municipal authorities

2. Earthquake (33 articles): Sasao (2016), Asari et al. (2013), Tabata et al. (2017, 2016), Xiao, Xie, and Zhang (2012), García-Torres, Kahhat, and Santa-Cruz (2017), Berktas et al. (2016), Çelik, Ergun, and Keskinocak (2015), Sahin, Kara, and Karasan (2016), Domingo and Luo (2017), Karunasena and Amaratunga (2015, 2016), Shibata, Sologabriele, and Hata (2012), Kawamoto and Kim (2016, 2019), Faleschini, Zanini, Hofer, Zampieri, and Pellegrino (2017), Onan, Ulengin, and Sennaroglu (2015), Wakabayashi, Peii, Tabata, and Saeki (2017), Askarizadeh, Karbassi, Ghalibaf, and Nouri (2016), Saffarzadeh et al. (2017), Hu and Sheu (2013), Raila and Anderson (2017), Poudel, Hirai, Asari, and Sakai (2018), Cheng, Zhang, and Thompson (2018a), Karunasena, Amaratunga, and Haigh (2012), Pham, Apparicio, Gomez, Weber, and Mathon (2014), Askarizadeh, Karbassi, Ghalibaf, and Nouri (2017), Sakai, Poudel, Asari, and Kirikawa (2019), Pramudita and Taniguchi (2014), Memon (2015), Zhang, Wu, Tian, and Wang (2016), Hooper (2019), and Koyama, Gokon, Jimbo, Koshimura, and Sato (2016).
3. Tsunami (six articles): Asari et al. (2013), Prasetya, Black, Lange, Borrero, and Healy (2012), Tabata et al. (2017), Wakabayashi et al. (2017), Portugal-Pereira and Lee (2016), and Koyama et al. (2016).
4. Hurricane/Typhoon (11 articles): Fetter and Rakes (2011, 2012), Habib and Sarkar (2017), Habib et al. (2019), Kim, Deshmukh, and Hastak (2018), Lorca, Çelik, Ergun, and Keskinocak (2017), Fetter and Rakes (2013), Hu, Sheu, Wei, and Hu (2019), Jiang and Friedland (2016), Thompson, Escobedo, Staudhammer, Matyas, and Qiu (2011), and Szantoi et al. (2012).
5. Landslide (two articles): Tabata et al. (2017) and Wakabayashi et al. (2017).
6. Flood (seven articles): Leader, Gaustad, and Tomaszewski (2018), Pramudita and Taniguchi (2014), Phonphoton and Pharino (2019), Saat et al. (2016), Tabata, Morita, and Onishi (2018), Kim and Kim (2017), and Beraud, Barroca, and Hubert (2012).
7. Bushfire (three articles): Cheng and Thompson (2016), Cheng, Zhang, and Thompson (2019) and Brown, Milke, and Seville (2011b).
8. Thunderstorm (one article): Cheng, Zhang, and Thompson (2018b);
9. Man-made disaster (three articles): Aoki (2018), Noh et al. (2015), and Zhang et al. (2017)

Next, he proposed an analysis of the literature based on whether the different topics made contributions that were relevant to pre-disaster or post-disaster. Table 15.2 describes these articles.

Bibliometric Analysis
The authors have performed a bibliometric analysis to find out what containing the term "waste management" has been most cited in Google Scholar. Likewise, the journals that have published the more about "waste management" were specified. They also identified which institutions published more about "waste management" and performed an analysis based on articles indexed in Scopus with the term "waste management." The results of this work will help and guide researchers in their literature reviews by identifying the referential articles from the field of waste man-

Table 15.2 Literature review about waste management based in Zhang et al. (2019)

Topic	Authors
Pre-disaster	
Theoretical approach	
Outline of DWM	Brown et al. (2011a)
Topic: Strategies, plans and policies of DWM	Crowley and Flachsbart (2018), Yusof et al. (2016) and Zawawi et al. (2015, 2018)
Topic: Factors that impact the success of a DWM plan	Trivedi et al. (2015) and Crowley (2017)
Topic: Capacity building for DWM	Karunasena and Amaratunga (2015, 2016)
Model approach	
Topic: Quantities of disaster debris	Tabata, Zhang, Yamanaka, and Tsai (2016)
Topic: DWM and its effect	Wakabayashi et al. (2017)
Application approach	
Topic: Location of DWM site	Grzeda et al. (2014)
Post-disaster	
Theoretical approach	
Topic: Identification/separation of disaster debris	Shibata et al. (2012)
Topic: Estimation of disaster debris	Tabata et al. (2018)
Topic: DWM and its effect	Phonphoton and Pharino (2019), Aoki (2018), Raila and Anderson (2017), Noh et al. (2015), Caniato et al. (2016), Beraud et al. (2012) and Baek et al. (2016)
Topic: Treatment (option) of disaster waste	Brown and Milke (2016)
Topic: Organizations of DWM	Kawamoto and Kim (2016, 2019), Saat et al. (2016), Maryono et al. (2015) and Hooper (2019)
Topic: Experience of DWM (case study)	Sakai et al. (2019), Karunasena et al. (2012), Domingo and Luo (2017), Memon (2015), Lesperance et al. (2011), Kim and Kim (2017), Saffarzadeh et al. (2017), Gabrielli et al. (2018), Zhang et al. (2017) and Brown et al. (2011b)
Topic: Integrated issues	Xiao et al. (2012) and Asari et al. (2013)
Model approach	
Topic: Estimation of disaster debris	Poudel et al. (2018)
Topic: DWM and its effect	Cheng et al. (2018a, 2018b) and Tabata et al. (2017)
Topic: Treatment (option) of disaster waste	Sasao (2016)
Topic: Allocation of resources (e.g., facility)	Fetter and Rakes (2011, 2013)

Topic: Location of facility, temporary storage site and processing site	Wang et al. (2019)
Topic: Route planning of debris clearance	Çelik et al. (2015), Cheng et al. (2019), Takeda et al. (2014), Berktas et al. (2016), Pramudita and Taniguchi (2014), Hu et al. (2019) and Sahin et al. (2016)
Topic: Integrated issues	Lorca et al. (2017), Hu and Sheu (2013), Onan et al. (2015), Habib and Sarkar (2017), Boonmee et al. (2018) and Habib et al. (2019), Fetter and Rakes (2012), Kim et al. (2018) and Ozdamar et al. (2014)
Application approach	
Topic: Identification/separation of disaster debris	Jiang and Friedland (2016), Pham et al. (2014) and Zhang et al. (2016)
Topic: Estimation of disaster debris	Koyama et al. (2016), Szantoi et al. (2012), Thompson et al. (2011), García-Torres et al. (2017), Yoo et al. (2017) and Thompson et al. (2011);
Topic: Dispersal of disaster debris	Prasetya et al. (2012)
Topic: DWM and its effect	Faleschini et al. (2017)
Topic: Treatment (option) of disaster waste	Portugal-Pereira and Lee (2016), Regattieri et al. (2016, 2018) and Leader et al. (2018)
Topic: Location of facility, temporary storage site and processing site	Cheng and Thompson (2016)
Topic: Integrated issues	Otsuka and Katsumi (2015) and Askarizadeh et al. (2017)

Table 15.3 Publications on "waste management" most cited in Google Scholar

No.	Title	Author	Year	Cites
1.	Municipal solid waste management in Indian cities–A review	Sharholy, Ahmad, Mahmood, and Trivedi	2008	867
2.	Sustainable water and waste management in urban areas	Otterpohl, Grottker, and Lange	1997	289
3.	Urban solid waste management in low-income countries of Asia: How to cope with the garbage crisis	Zurbrugg	2002	228
4.	The regional urban solid waste management system: A modelling approach	Caruso, Colorni, and Paruccini	1993	222
5.	An environmentally sustainable decision model for urban solid waste management	Costi, Minciardi, Robba, Rovatti, and Sacile	2004	213
6.	Source control in urban sanitation and waste management: ten systems with reuse of resources	Otterpohl, Albold, and Oldenburg	1999	197
7.	Quality of life and alliances in solid waste management: contributions to urban sustainable development	Baud, Grafakos, Hordijk, and Post	2001	180
8.	A decision support system for urban waste management	Haastrup et al.	1998	123
9.	Solid waste management in urban areas: development and application of a decision support system	Fiorucci, Minciardi, Robba, and Sacile	2003	148
10.	Management of urban solid waste: Vermicomposting a sustainable option	Singh, Singh, Araujo, Ibrahim, and Sulaiman	2011	146

Search command: "waste management"
Source: Google Scholar, 08/01/2019

agement. In addition, it highlights those institutions that are working the most on this issue, and it enables future scholars to contact those institutions most likely interested in developing further research or technological advancements in the field of waste management.

Table 15.3 shows the evaluation of the publications on "urban infrastructure" that have more citations. Google Scholar is used between 2010 and 2018.

Table 15.4 shows the academic journals that published more studies on "waste management." Scopus is used between 2010 and 2018.

Table 15.5 shows the list of institutions that have published scientific articles on "waste management." You can see universities of Africa, North America, Europe, and Asia appear. Only one Latin American university is listed.

The cocitation analysis allowed obtaining three conglomerates of authors who published studies on "waste management," which were indexed in Scopus (see Fig. 15.8).

Each conglomerate is identified by a characteristic color that groups the authors. Thus, in yellow color, Thomas Christensen of Danmarks Tekniske Universitet, whose research is based on life cycle assessment of waste management. The light

Table 15.4 Academic journals that published more studies on "waste management"

No.	Journal	Scopus
1.	Journal of cleaner production	137
2.	Waste management and research	136
3.	Waste management	135
4.	Resources conservation and recycling	60
5.	Sustainability Switzerland	59
6.	Journal of material cycles and waste management	50
7.	IOP conference series earth and environmental science	36
8.	International journal of environment and waste management	35
9.	E3s web of conferences	31
10.	Environmental science and pollution research	30

Own elaboration
Source: Scopus, 08/01/2019. Search command in Scopus: TITLE("waste management") OR (waste AND management)

Table 15.5 Institutions that published more studies on "waste management"

No.	Institution	Scopus
1.	Universiti Teknologi Malaysia	35
2.	National Technical University of Athens	25
3.	Danmarks Tekniske Universitet	25
4.	University of Regina	24
5.	Chulalongkorn University	21
6.	University of Johannesburg	19
7.	North China Electric Power University	19
8.	Tsinghua University	19
9.	Shanghai Jiao Tong University	18
10.	Universidade Federal do Rio de Janeiro	17

Own elaboration
Source: Scopus 12/02/2019. Search command in Scopus: TITLE ("waste management")

blue node shows Chewtin Lee of the University of Technology in Malaysia, having research oriented to sustainable organic waste management.

Closing Remarks

Waste management is a necessity in activities related to municipal management, housing construction, and pre-disaster and post-disaster management. It is necessary to take into account that several indicators have been published that should be part of the planning at national and municipal institutions. However, the scope of such planning goes beyond only the correct management of waste, but also includes what is environmentally friendly, socially necessary, and economically sustainable so that the activities carried out by both companies and the state compatible with the needs of the population and the planet.

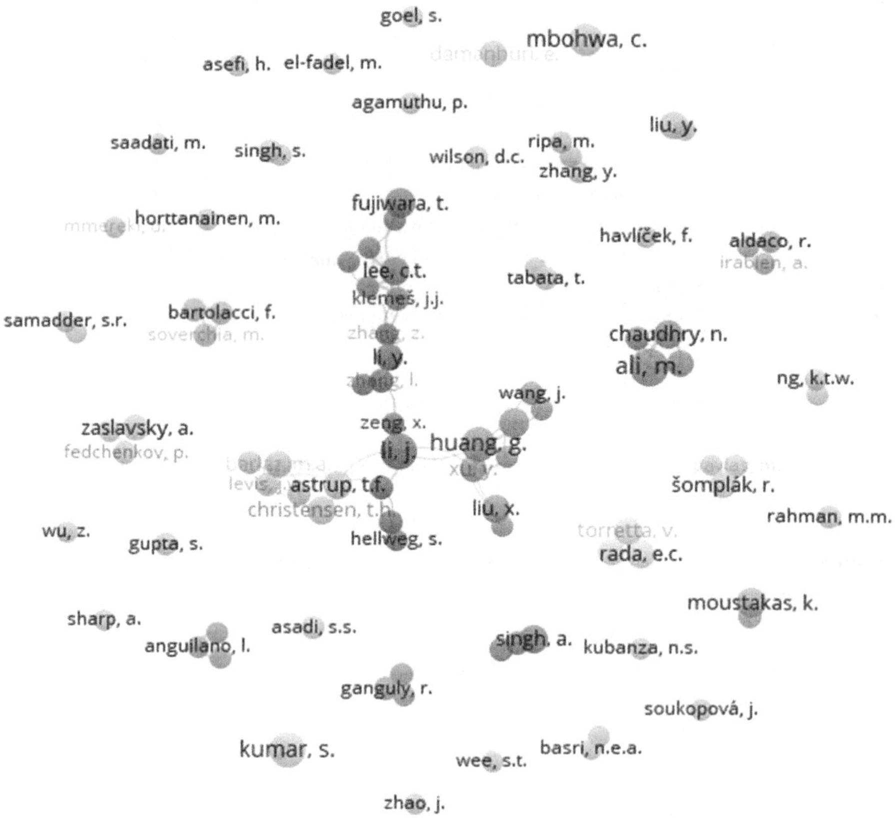

Fig. 15.8 Conglomerate of authors with greater cocitation in Scopus 2010–2018. (Source: Scopus, 07/28/2019. Only authors who had at least five citations were considered)

Following the COP meetings, the newly signed Paris Agreement entails taking concrete actions in business and state administrative activities. Both national and municipal government plans must include waste management. Not only must these plans provide the vehicles that collect the garbage and take it to a sanitary landfill, with the environmental contamination that this implies, but they must implement the necessary technology in the short term to generate new products from the treatment of this waste. The population needs to become a vigilant agent in the fulfillment of this waste management framework, and increasingly use social networks to monitor and pressure all those institutions that fail to manage cities in a sustainable way.

References

Alvarez-Risco, A., Del-Aguila-Arcentales, S., & Rosen, M. (2020). A new regulation for supporting a circular economy in the plastic industry: The case of Peru. *Journal of Landscape Ecology*. https://doi.org/10.2478/jlecol-2020-0004

Aoki, N. (2018). Who would be willing to accept disaster debris in their backyard? Investigating the determinants of public attitudes in post-Fukushima Japan. *Risk Analysis, 38*(3), 535–547.

Asari, M., Sakai, S. I., Yoshioka, T., Tojo, Y., Tasaki, T., Takigami, H., et al. (2013). Strategy for separation and treatment of disaster waste: A manual for earthquake and tsunami disaster waste management in Japan. *Journal of Material Cycles and Waste Management, 15*(3), 290–299.

Askarizadeh, L., Karbassi, A. R., Ghalibaf, M. B., & Nouri, J. (2016). Management of postearthquake construction debris in Tehran metropolitan. *International Journal of Environmental Science and Technology, 13*(2), 639–648.

Askarizadeh, L., Karbassi, A. R., Ghalibaf, M. B., & Nouri, J. (2017). Debris management after earthquake incidence in ancient City of Ray. *Global Journal of Environmental Science and Management, 3*(4), 447–456.

Azevedo, F., & Cruz, L. (2019). *Service mapping: Case study of university waste management system*. International Conference on Applied Human Factors and Ergonomics. Springer, pp. 742–752.

Baek, S. C., Kim, Y. C., Choi, J. H., & Hong, W. H. (2016). Determination of the essential activity elements of an asbestos management system in the event of a disaster and their prioritization. *Journal of Cleaner Production, 137*, 414–426.

Baud, I. S. A., Grafakos, S., Hordijk, M., & Post, J. (2001). Quality of life and alliances in solid waste management: Contributions to urban sustainable development. *Cities, 18*(1), 3–12.

Berktaş, N., Kara, B. Y., & Karaşan, O. E. (2016). Solution methodologies for debris removal in disaster response. EURO Journal on Computational Optimization, 4(3-4), 403–445.

Beraud, H., Barroca, B., & Hubert, G. (2012). Functional analysis, a resilience improvement tool applied to a waste management system–application to the "household waste management chain". *Natural Hazards and Earth System Sciences, 12*(12), 3671–3682.

Boex, J., Malik, A. A., Brookins, D., Edwards, B., & Zaidi, H. (2020). *The political economy of urban governance in Asian cities: Delivering water, sanitation and solid waste management services* (In New Urban Agenda in Asia-Pacific) (pp. 301–329). Singapore: Springer.

Boonmee, C., Arimura, M., & Asada, T. (2018). Location and allocation optimization for integrated decisions on post-disaster waste supply chain management: On-site and off-site separation for recyclable materials. *International Journal of Disaster Risk Reduction, 31*, 902–917.

Brown, C., & Milke, M. (2016). Recycling disaster waste: Feasibility, method and effectiveness. *Resources, Conservation and Recycling, 106*, 21–32.

Brown, C., Milke, M., & Seville, E. (2011a). Disaster waste management following the 2009 Victorian bushfires. *The Australian Journal of Emergency Management, 26*(2), 17–22.

Brown, C., Milke, M., & Seville, E. (2011b). Disaster waste management: A review article. *Waste Management, 31*(6), 1085–1098.

Caniato, M., Tudor, T. L., & Vaccari, M. (2016). Assessment of health-care waste management in a humanitarian crisis: A case study of the Gaza Strip. *Waste Management, 58*, 386–396.

Caruso, C., Colorni, A., & Paruccini, M. (1993). The regional urban solid waste management system: A modelling approach. *European Journal of Operational Research, 70*(1), 16–30.

Çelik, M., Ergun, Ö., & Keskinocak, P. (2015). The post-disaster debris clearance problem under incomplete information. *Operations Research, 63*(1), 65–85.

Cheng, C., & Thompson, R. G. (2016). Application of boolean logic and GIS for determining suitable locations for temporary disaster waste management sites. *International Journal of Disaster Risk Reduction, 20*, 78–92.

Cheng, C., Zhang, L., & Thompson, R. G. (2018a). Reliability analysis for disaster waste management systems. *Waste Management, 78*, 31–42.

Cheng, C., Zhang, L. H., & Thompson, R. G. (2018b). Disaster waste clean-up system performance subject to time-dependent disaster waste accumulation. *Natural Hazards, 91*(2), 717–734.

Cheng, C., Zhang, L. H., & Thompson, R. G. (2019). Reliability analysis of road networks in disaster waste management. *Waste Management, 84*, 383–393. https://doi.org/10.1016/j.wasman.2018.11.027

Costi, P., Minciardi, R., Robba, M., Rovatti, M., & Sacile, R. (2004). An environmentally sustainable decision model for urban solid waste management. *Waste Management, 24*(3), 277–295.

Crowley, J. (2017). A measurement of the effectiveness and efficiency of pre-disaster debris management plans. *Waste Management, 62*, 262–273.

Crowley, J., & Flachsbart, J. (2018). Local debris management planning and FEMA policies on disaster recovery in the United States. *International Journal of Disaster Risk Reduction, 27*, 373–379.

Domingo, N., & Luo, H. (2017). Canterbury earthquake construction and demolition waste management: Issues and improvement suggestions. *International Journal of Disaster Risk Reduction, 22*, 130–138.

Environmental Assessment and Inspection Agency (2018). OEFA identifies 1,585 informal dumps nationwide [OEFA identifica 1,585 botaderos informales a nivel nacional]. Retrieved February 5, 2020, from https://www.oefa.gob.pe/noticias-institucionales/oefa-identifica-1585-botaderos-informales-nivel-nacional.

Environmental Ministry. (2018). Current situation of solid waste management in Peru [Situación actual de la gestión de los residuos sólidos en el Perú]. [PowerPoint Slides].

Faleschini, F., Zanini, M. A., Hofer, L., Zampieri, P., & Pellegrino, C. (2017). Sustainable management of demolition waste in post-quake recovery processes: The Italian experience. *International Journal of Disaster Risk Reduction, 24*, 172–182.

Fetter, G., & Rakes, T. R. (2011). A self-balancing CUSUM approach for the efficient allocation of resources during post-disaster debris disposal operations. *Operations Management Research, 4*(1–2), 51–60.

Fetter, G., & Rakes, T. R. (2012). Incorporating recycling into post-disaster debris disposal. *Socio-Economic Planning Sciences, 46*(1), 14–22.

Fetter, G., & Rakes, T. R. (2013). An equity approach to contractor assignment in postdisaster debris disposal operations. *International Journal of Emergency Management, 9*(2), 170–186.

Fiorucci, P., Minciardi, R., Robba, M., & Sacile, R. (2003). Solid waste management in urban areas: Development and application of a decision support system. *Resources, Conservation and Recycling, 37*(4), 301–328.

Gabrielli, F., Amato, A., Balducci, S., Galluzzi, L. M., & Beolchini, F. (2018). Disaster waste management in Italy: Analysis of recent case studies. *Waste Management, 71*, 542–555.

García-Torres, S., Kahhat, R., & Santa-Cruz, S. (2017). Methodology to characterize and quantify debris generation in residential buildings after seismic events. *Resources, Conservation and Recycling, 117*, 151–159.

Grzeda, S., Mazzuchi, T. A., & Sarkani, S. (2014). Temporary disaster debris management site identification using binomial cluster analysis and GIS. *Disasters, 38*(2), 398–419.

Haastrup, P., Maniezzo, V., Mattarelli, M., Rinaldi, F. M., Mendes, I., & Paruccini, M. (1998). A decision support system for urban waste management. *European Journal of Operational Research, 109*(2), 330–341.

Habib, M. S., & Sarkar, B. (2017). An integrated location-allocation model for temporary disaster debris management under an uncertain environment. *Sustainability, 9*(5). https://doi.org/10.3390/su9050716

Habib, M. S., Sarkar, B., Tayyab, M., Saleem, M. W., Hussain, A., Ullah, M., et al. (2019). Large-scale disaster waste management under uncertain environment. *Journal of Cleaner Production, 212*, 200–222.

Hao, J., Yuan, H., Liu, J., Chin, C. S., & Lu, W. (2019). A model for assessing the economic performance of construction waste reduction. *Journal of Cleaner Production, 232*, 427–440.

Hooper, M. (2019). When diverse norms meet weak plans: The organizational dynamics of urban rubble clearance in post-earthquake Haiti. *International Journal of Urban and Regional Research, 43*(2), 292–312.

Hu, Z. H., & Sheu, J. B. (2013). Post-disaster debris reverse logistics management under psychological cost minimization. *Transportation Research Part B: Methodological, 55*, 118–141.

Hu, Z. H., Sheu, J. B., Wei, C., & Hu, S. L. (2019). Post-storm debris removal considering traffic and psychological impacts. *Transportmetrica A: Transport Science, 15*(2), 1145–1174. https://doi.org/10.1080/23249935.2019.1567618

Ibáñez-Forés, V., Bovea, M. D., Coutinho-Nóbrega, C., & de Medeiros, H. R. (2019). Assessing the social performance of municipal solid waste management systems in developing countries: Proposal of indicators and a case study. *Ecological Indicators, 98*, 164–178.

Iniciativa Regional de Reciclaje. (2017). Advances and challenges for inclusive recycling: Evaluation of 12 cities in Latin America [Avances y desafíos para el reciclaje inclusivo: Evaluación de 12 ciudades de América Latina]. Retrieved February 5, 2020, from https://reciclajeinclusivo.org/avances-y-desafios-para-el-reciclaje-inclusivo-evaluacion-de-12-ciudades-de-america-latina-y-el-caribe/

International Solid Waste Association. (2014). Waste Atlas. Retrieved February 5, 2020, from http://www.atlas.d-waste.com

Jiang, S. S., & Friedland, C. J. (2016). Automatic urban debris zone extraction from posthurricane very high-resolution satellite and aerial imagery. *Geomatics, Natural Hazards and Risk, 7*(3), 933–952.

Jin, R., Yuan, H., & Chen, Q. (2019). Science mapping approach to assisting the review of construction and demolition waste management research published between 2009 and 2018. *Resources, Conservation and Recycling, 140*, 175–188.

Karunasena, G., & Amaratunga, D. (2015). Capacity gaps in post disaster construction & demolition waste management. *Engineering, Construction and Architectural Management, 22*(4), 446–466.

Karunasena, G., & Amaratunga, D. (2016). Capacity building for post disaster construction and demolition waste management: A case of Sri Lanka. *Disaster Prevention and Management, 25*(2), 137–153.

Karunasena, G., Amaratunga, D., & Haigh, R. (2012). Post-disaster construction & demolition debris management: A Sri Lanka case study. *Journal of Civil Engineering and Management, 18*(4), 457–468.

Kawamoto, K., & Kim, K. (2016). Social capital and efficiency of earthquake waste management in Japan. *International Journal of Disaster Risk Reduction, 18*, 256–266.

Kawamoto, K., & Kim, K. (2019). Efficiencies of bonding, bridging and linking social capital: Cleaning up after disasters in Japan. *International Journal of Disaster Risk Reduction, 33*, 64–73.

Kim, J., Deshmukh, A., & Hastak, M. A. (2018). A framework for assessing the resilience of a disaster debris management system. *International Journal of Disaster Risk Reduction, 28*, 674–687.

Kim, T., & Kim, T. (2017). Smart and resilient urban disaster debris cleanup using network analysis. *Spatial Information Research, 25*(2), 239–248.

Kim, Y. C., & Hong, W. H. (2017). Optimal management program for asbestos containing building materials to be available in the event of a disaster. *Waste Management, 64*, 272–285.

Koyama, C. N., Gokon, H., Jimbo, M., Koshimura, S., & Sato, M. (2016). Disaster debris estimation using high-resolution polarimetric stereo-SAR. *ISPRS Journal of Photogrammetry and Remote Sensing, 120*, 84–98.

Leader, A., Gaustad, G., & Tomaszewski, B. (2018). The consequences of electronic waste post-disaster: A case study of flooding in Bonn, Germany. *Sustainability, 10*(11). https://doi.org/10.3390/su10114193

Lesperance, A. M., Stein, S., Upton, J. F., & Toomey, C. (2011). Challenges in disposing of anthrax waste. *Biosecurity and Bioterrorism: Biodefense Strategy, Practice, and Science, 9*(3), 310–314.

Litter, M. I., Ingallinella, A. M., Olmos, V., Savio, M., Difeo, G., Botto, L., et al. (2019). Arsenic in Argentina: Technologies for arsenic removal from groundwater sources, investment costs and waste management practices. *Science of the Total Environment, 690*, 778–789.

Lorca, Á., Çelik, M., Ergun, Ö., & Keskinocak, P. (2017). An optimization-based decision-support tool for post-disaster debris operations. *Production and Operations Management, 26*(6), 1076–1091.

Lu, W., Chi, B., Bao, Z., & Zetkulic, A. (2019). Evaluating the effects of green building on construction waste management: A comparative study of three green building rating systems. *Building and Environment, 155*, 247–256.

Maryono, Nakayama, H., & Shimaoka, T. (2015). Identification of factors affecting stakeholders' intentions to promote preparedness in disaster waste management: A structural equation modeling approach. *Memoirs of the Faculty of Engineering, Kyushu University, 74*(3), 79–98.

Memon, M. A. (2015). *Disaster waste recovery and utilization in developing countries learning from earthquakes in Nepal*. The 15th Asian Regional Conference on Soil Mechanics and Geotechnical Engineering. Japan, pp. 143–147.

Noh, S. R., Cheong, H. K., Ha, M., Eom, S. Y., Kim, H., Choi, Y. H., et al. (2015). Oxidative stress biomarkers in long-term participants in clean-up work after the Hebei Spirit oil spill. *Science of the Total Environment, 515*, 207–214.

Organismo de Evaluación y Fiscalización Ambiental. (2018). OEFA identifies 1,585 informal dumps nationwide [OEFA identifica 1,585 botaderos informales a nivel nacional]. Retrieved February 5, 2020, from https://www.oefa.gob.pe/noticias-institucionales/oefa-identifica-1585-botaderos-informales-nivel-nacional

Otsuka, Y., & Katsumi, T. (2015). *Analysis of the integrated data on disaster debris treatment in Yamada town, Iwate prefecture*. The 15th Asian Regional Conference on Soil Mechanics and Geotechnical Engineering. Japan, pp. 154–157.

Otterpohl, R., Albold, A., & Oldenburg, M. (1999). Source control in urban sanitation and waste management: Ten systems with reuse of resources. *Water Science and Technology, 39*(5), 153.

Otterpohl, R., Grottker, M., & Lange, J. (1997). Sustainable water and waste management in urban areas. *Water Science and Technology, 35*(9), 121–133.

Ozdamar, L., Aksu, D. T., & Ergunes, B. (2014). Coordinating debris cleanup operations in post disaster road networks. *Socio-Economic Planning Sciences, 48*(4), 249–262.

Pham, T. T. H., Apparicio, P., Gomez, C., Weber, C., & Mathon, D. (2014). Towards a rapid automatic detection of building damage using remote sensing for disaster management: The 2010 Haiti earthquake. *Disaster Prevention and Management, 23*(1), 53–66.

Phonphoton, N., & Pharino, C. (2019). A system dynamics modeling to evaluate flooding impacts on municipal solid waste management services. *Waste Management, 87*, 525–536.

Portugal-Pereira, J., & Lee, L. (2016). Economic and environmental benefits of waste-to-energy technologies for debris recovery in disaster-hit Northeast Japan. *Journal of Cleaner Production, 112*, 4419–4429.

Poudel, R., Hirai, Y., Asari, M., & Sakai, S. I. (2018). Establishment of unit generation rates of building debris in Kathmandu Valley, Nepal, after the Gorkha earthquake. *Journal of Material Cycles and Waste Management, 20*(3), 1663–1675.

Pramudita, A., & Taniguchi, E. (2014). Model of debris collection operation after disasters and its application in urban area. *International Journal of Urban Sciences, 18*(2), 218–243.

Prasetya, G., Black, K., Lange, W. D., Borrero, J., & Healy, T. (2012). Debris dispersal modeling for the great Sumatra Tsunamis on Banda Aceh and surrounding waters. *Natural Hazards, 60*(3), 1167–1188.

Raila, E. M., & Anderson, D. O. (2017). Healthcare waste management during disasters and its effects on climate change: Lessons from 2010 earthquake and cholera tragedies in Haiti. *Waste Management & Research, 35*(3), 236–245.

Regattieri, A., Gamberi, M., Bortolini, M., & Piana, F. (2018). Innovative solutions for reusing packaging waste materials in humanitarian logistics. *Sustainability, 10*(5). https://doi.org/10.3390/su10051587

Regattieri, A., Piana, F., Bortolini, M., Gamberi, M., & Ferrari, E. (2016). Innovative portable solar cooker using the packaging waste of humanitarian supplies. *Renewable and Sustainable Energy Reviews, 57*, 319–326.

Saat, N. Z. M., Malia, W. N., Ikram, A. M., Aishah, H. S., Zawaha, I., Ashikin, M. N., et al. (2016). Perception of flood waste management among stakeholders in Kelantan. *Journal of Environmental Science and Technology, 9*(4), 317.

Saffarzadeh, A., Shimaoka, T., Nakayama, H., Hanashima, T., Yamaguchi, K., & Manabe, K. (2017). Tasks and problems involved in the handling of disaster waste upon April 2016 Kumamoto earthquake, Japan. *Natural Hazards, 89*(3), 1273–1290.

Sahin, H., Kara, B. Y., & Karasan, O. E. (2016). Debris removal during disaster response: A case for Turkey. *Socio-Economic Planning Sciences, 53*, 49–59.

Sakai, S., Poudel, R., Asari, M., & Kirikawa, T. (2019). Disaster waste management after the 2016 Kumamoto earthquake: A mini-review of earthquake waste management and the Kumamoto experience. *Waste Management & Research, 37*(3), 247–260.

Santos, A. C., Mendes, P., & Teixeira, M. R. (2019). Social life cycle analysis as a tool for sustainable management of illegal waste dumping in municipal services. *Journal of Cleaner Production, 210*, 1141–1149.

Sasao, T. (2016). Cost and efficiency of disaster waste disposal: A case study of the great East Japan earthquake. *Waste Management, 58*, 3–13.

Sharholy, M., Ahmad, K., Mahmood, G., & Trivedi, R. C. (2008). Municipal solid waste management in Indian cities—A review. *Waste Management, 28*(2), 459–467.

Shibata, T., Sologabriele, H., & Hata, T. (2012). Disaster waste characteristics and radiation distribution as a result of the great East Japan earthquake. *Environmental Science & Technology, 46*(7), 3618–3624.

Singh, R. P., Singh, P., Araujo, A. S., Ibrahim, M. H., & Sulaiman, O. (2011). Management of urban solid waste: Vermicomposting a sustainable option. *Resources, Conservation and Recycling, 55*(7), 719–729.

Slanina, Z., Pokorny, R., & Dedek, J. (2018). *Waste management-weighing-machine automation*. International Conference on Advanced Engineering Theory and Applications. Springer, Cham, pp. 747–757.

Szantoi, Z., Malone, S., Escobedo, F., Misas, O., Smith, S., & Dewitt, B. (2012). A tool for rapid post-hurricane urban tree debris estimates using high resolution aerial imagery. *International Journal of Applied Earth Observation and Geoinformation, 18*, 548–556.

Tabata, T., Morita, H., & Onishi, A. (2018). What is the quantity of consumer goods stocked in a Japanese household? Estimating potential disaster waste generation during floods. *Resources, Conservation and Recycling, 133*, 86–98.

Tabata, T., Wakabayashi, Y., Tsai, P., & Saeki, T. (2017). Environmental and economic evaluation of pre-disaster plans for disaster waste management: Case study of Minami-Ise, Japan. *Waste Management, 61*, 386–396.

Tabata, T., Zhang, O., Yamanaka, Y., & Tsai, P. (2016). Estimating potential disaster waste generation for pre-disaster waste management. *Clean Technologies and Environmental Policy, 18*(6), 1735–1744.

Takeda, T., Mori, Y., Kubota, N., & Arai, Y. (2014). *A route planning for disaster waste disposal based on robot technology*. IEEE Symposium on Robotic Intelligence in Informationally Structured Space. Orlando, FL, pp. 74–79.

Thompson, B. K., Escobedo, F. J., Staudhammer, C. L., Matyas, C. J., & Qiu, Y. (2011). Modeling hurricane caused urban forest debris in Houston, Texas. *Landscape and Urban Planning, 101*(3), 286–297.

Trivedi, A., Singh, A., & Chauhan, A. (2015). Analysis of key factors for waste management in humanitarian response: An interpretive structural modelling approach. *International Journal of Disaster Risk Reduction, 14*, 527–535.

United Nations Environment Program. (2017). Retrieved February 5, 2020, from https://news.un.org/es/story/2017/05/1378771

Wakabayashi, Y., Peii, T., Tabata, T., & Saeki, T. (2017). Life cycle assessment and life cycle costs for pre-disaster waste management systems. *Waste Management, 68*, 688–700.

Wang, Z., Hu, H., Guo, M., & Gong, J. (2019). Optimization of temporary debris management site selection and site service regions for enhancing post disaster debris removal operations. *Computer-Aided Civil and Infrastructure Engineering, 34*(3), 230–247.

Wu, Z., Ann, T. W., & Poon, C. S. (2019). An off-site snapshot methodology for estimating building construction waste composition-a case study of Hong Kong. Environmental Impact Assessment Review, 77, 128–135.

Xiao, J. Z., Xie, H., & Zhang, C. (2012). Investigation on building waste and reclaim in Wenchuan earthquake disaster area. *Resources, Conservation and Recycling, 61*, 109–117.

Yoo, H. T., Lee, H., Chi, S., Hwang, B. G., & Kim, J. (2017). *A preliminary study on disaster waste detection and volume estimation based on 3D spatial information*. ASCE International Workshop on Computing in Civil Engineering. Washington, DC, pp. 428–435.

Yusof, N. S., Zawawi, E. M. A., & Ismail, Z. (2016). *Disaster waste management in Malaysia: significant issues, policies & strategies*. 4th International Building Control Conference. Kuala Lumpur, Malaysia, pp. 1–7.

Zawawi, E. M. A., Yusof, N. S., & Ismail, Z. (2018). Adoption of post-disaster waste management plan into disaster management guidelines for Malaysia. *Journal of Material Cycles and Waste Management, 20*(1), 223–236.

Zawawi, E. M. A., Yusof, N. S., Kamaruzzaman, S. N., & Ismail, Z. (2015). Important criteria for managing disaster waste in Malaysia. *Jurnal Teknologi, 75*(9), 89–93.

Zhang, F., Cao, C., Li, C., Liu, Y., & Huisingh, D. (2019). A systematic review of recent developments in disaster waste management. *Journal of Cleaner Production*. https://doi.org/10.1016/j.jclepro.2019.06.229

Zhang, H., Duan, H., Zuo, J., Song, M., Zhang, Y., Yang, B., et al. (2017). Characterization of post-disaster environmental management for hazardous materials incidents: Lessons learnt from the Tianjin warehouse explosion, China. *Journal of Environmental Management, 199*, 21–30.

Zhang, L., Wu, L., Tian, F., & Wang, Z. (2016). Retrospection-simulation-revision: Approach to the analysis of the composition and characteristics of medical waste at a disaster relief site. *PLoS One, 11*(7), e0159261.

Zurbrugg, C. (2002). *Urban solid waste management in low-income countries of Asia: How to cope with the garbage crisis*. Presented for: Scientific Committee on Problems of the Environment (SCOPE) Urban Solid Waste Management Review Session, Durban, South Africa, pp. 1–13.

Chapter 16
Management of Water

Aldo Alvarez-Risco, Shyla Del-Aguila-Arcentales, and Marc A. Rosen

Abstract The information available to international organizations on water management suggest that some results are being achieved that move toward optimal water management and sustainability in this area. But there also are projections for the next 20–30 years that indicate that various places in the world will have serious problems with water supply. More consolidated information is required to understand and improve the current institutional behavior of water management as well as its multiple intervening factors. To achieve better understanding and planning and to facilitate cost-efficient results, the use of artificial intelligence as a key tool for handling large volumes of data will likely prove necessary.

Keywords Water · Sustainability · Water footprint · Water management · Climate change

16.1 Water Use and Statistics on Water Problems

The global demand for water continues to increase steadily, and by the year 2050 it is expected that water use will rise to 20–30% above the current level (Burek et al., 2016). The main water demands are expected to come from the industrial and domestic sectors (OECD, 2012).

A. Alvarez-Risco (✉)
Facultad de Ciencias Administrativas y Recursos Humanos, Universidad de San Martín de Porres, Lima, Peru
e-mail: aldo.alvarez@redsaf.org

S. Del-Aguila-Arcentales
Escuela Nacional de Marina Mercante "Almirante Miguel Grau", Callao, Peru

Universidad Nacional de la Amazonia Peruana, Iquitos, Peru
e-mail: sdelaguila@enamm.edu.pe

M. A. Rosen
Faculty of Engineering and Applied Science, University of Ontario Institute of Technology, Oshawa, ON, Canada
e-mail: marc.rosen@uoit.ca

A. Alvarez-Risco et al. (eds.), *Building Sustainable Cities*,
https://doi.org/10.1007/978-3-030-45533-0_16

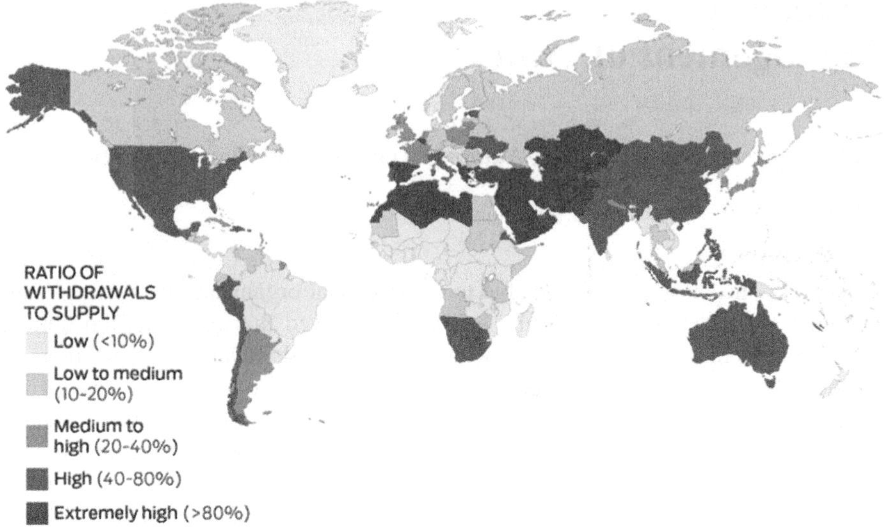

Fig. 16.1 Projection water stress across the globe by 2040

Concerns relating to providing adequate water supplies often relate to poor management. According to the World Water Council (2019), "There is a water crisis today. But the crisis is not about having too little water to satisfy our needs. It is a crisis of managing water so badly that billions of people—and the environment—suffer badly."

The World Resources Institute (WRI, 2015) carried out a projection of water stress by 2040, and the results are shown in Fig. 16.1.

The countries listed in red are of particular relevance in Fig. 16.1.

However, the data presented in projections like those in Fig. 16.1 and others only show averages that can mask issues and suggest that only some countries have water problems. But as indicated by Mekonnen and Hoekstra (2016) about four billion people (approximately two-thirds of the world's population) experience a severe water shortage for at least one month of the year. Such shortages generate numerous serious impacts in the social and economic dynamics of a population, such as the following:

(a) Impacts on agriculture, which leads to greater food insecurity.
(b) Impact on sanitation conditions, which leads to diarrheal diseases and consequent effects on health as well as school and work absenteeism.

Sometimes, countries that appear on the map in Fig. 16.1 to have minimal water stresses internally have stress areas throughout the year but on average they become hard to detect. A similar can occur in cities where there is water availability but there is an absence of sufficient infrastructure for the collection, transport, and treatment of water for human use. This more detailed interpretation of the data needs to be considered to generate more precise and efficient initiatives to improve the management of water.

Many other interesting and significant observations have been made regarding water supply and management, and water-related issues. The following points are made in the document "WASH in Health Care Facilities: Global Baseline Report" (JMP, 2019). Note that the World Health Organization (WHO)/UNICEF Joint Monitoring Programme for Water Supply, Sanitation and Hygiene (JMP) has reported country, regional, and global estimates of progress on drinking water, sanitation, and hygiene (WASH) since 1990.

Water

On a global basis, 14% of health care facilities globally had limited water services, 12% of health care facilities had no water service, 4% of hospitals and 11% of other health care facilities had no water service, 12% of government health care facilities and 6% of non-government health care facilities had no water service, and 896 million people had no water service at their health care facility.

Sanitation

Globally, 21% of health care facilities had no sanitation service, 9% of hospitals and 20% of other health care facilities had no sanitation service, 16% of government health care facilities and 36% of non-government health care facilities had no sanitation service, and more than 1.5 billion people had no sanitation service at their health care facility.

Hygiene

Again on a global basis, 16% of health care facilities had no hygiene service. In sub-Saharan Africa, half of health care facilities (51%) had alcohol-based hand rub at points of care, and 84% of hospitals had hand hygiene facilities at points of care, compared to 64% of other health care facilities.

Several additional important observations can be found in the JMP, 2018 global report on "WASH in Schools" (JMP, 2018).

Drinking Water in Schools

12% of schools had limited drinking water service and 19% of schools had no drinking water service.

Nearly 570 million children lacked a basic drinking water service at their school, less than half of schools in Oceania and only two-thirds of schools in central and southern Asia had a basic drinking water service. Furthermore, nearly half of schools in sub-Saharan Africa and over a third of schools in Small Island Developing States had no drinking water service, and one in four primary schools and one in six secondary schools had no drinking water service.

Sanitation in Schools

12% of schools had limited sanitation service and 23% of schools had no sanitation service. In addition, over 620 million children worldwide lacked a basic sanitation service at their school, and a third of schools in sub-Saharan Africa and in eastern and southeastern Asia had no sanitation service.

Hygiene in Schools

11% of schools had limited hygiene service, defined as a handwashing facility with water but no soap available. Specifically, 36% of schools had no hygiene service, defined as no facility or no water available, nearly 900 million children worldwide lacked a basic hygiene service at their school, and more than one in three primary schools and a quarter of secondary schools had no hygiene service.

According to the World Water Development Report 2019 (WWAP, 2019), water use has increased 1% since the 1980s due to population growth, socioeconomic development, and change in consumption models, with an increase expected between 20 and 30% above the current level of water use. More than two billion people live in countries that suffer from severe water shortages, and approximately four billion people experience severe water shortages for at least one month a year, which is expected to worsen in the short term. Also, 30% of people do not have access to safe drinking water. Almost 50% of people who drink water from unprotected sources live in sub-Saharan Africa. Approximately 60% of people do not have access to safe sanitation services, and one in nine practices open defecation although these data mask the real inequalities between and within regions, countries, communities, and even neighborhoods.

Water availability is contingent on the amount of physically available water and how it is managed and allotted to users. This includes aspects related to surface and groundwater management, as well as water recycling and reuse. In places where pipe networks are not available, people depend mainly on community water wells or systems, and have to pay several times higher prices for lower quality water, which further accentuates inequalities between the wealthy and the disadvantaged.

16.2 Migration, Climate Change, and Water

Water scarcity, along with other drivers of mobility, contributes to causing human migration. Human migration in response to ecological change has occurred since the origin of humanity. However, anthropogenic climate change has begun to intensify it, and that impact is predicted by many to likely become much worse in the future. That is in part because climate change can act as a threat multiplier and can worsen sociopolitical and economic vulnerabilities. Despite the wide range of research into climate change, little analysis has been reported on its impact on migration and on the health of migrant populations. There are numerous direct and indirect effects of climate change on human health. Many of these are related to the availability of water for the population, to food security, and to changes in ecosystems of infectious microorganisms and transmission vectors.

Flörke, Schneider, and McDonald (2018) project an urban surface-water deficit of 1.4–6.8 billion m^3. More than 27% of cities studied, accounting for 233 million people, are anticipated to have water demands that exceed surface-water availability. The authors list the 20 cities that are expected to have the greatest urban water deficit by 2050 as follows:

Los Angeles (USA)
Jaipur (India)
Dar es Salaam (Tanzania)
Dalian (China)
San Diego (USA)
Karachi (Pakistan)
Harbin (China)
Phoenix (USA)
Pôrto Alegre (Brazil)
Monterrey (Mexico)
Lima (Peru)
Salvador (El Salvador)
Santa Ana (USA)
Jodhpur (India)
Luanda (Angola)
Caracas (Venezuela)
Long Beach (USA)
Quetta (Pakistan)
Kabul (Afghanistan)
Chennai (India)

Table 16.1 lists the number of people living in cities with a surface-water deficit, at present and projected to 2050, for various global regions. Reference ranges of the estimates are presented.

The problems associated with water occur at all socioeconomic levels, with the greatest impact on vulnerable populations such as those with low incomes and those who are migrants due to climatic conditions. Responsible water management by organizations is an urgent task and needs planning. The concerns related not only to numbers of deaths due to water scarcity, but also to the quality of life that survivors have due to very limited water consumption.

16.3 Monitoring for Water

The United Nations Sustainable Development Goals for 2015–2030 speak directly to the importance of water through Sustainable Development Goal number 6 (SDG 6): Clean Water and Sanitation. Monitoring SDG 6 requires a clear and efficient methodology to ensure that countries and institutions can plan their activities with this objective in mind. For this, several guides have been designed which require step-by-step monitoring of SDG 6 and its targets. Table 16.2 shows the SDG 6 integrated monitoring proposed by GEMI (2017).

Countries can start their monitoring activities with existing data on basic services and gradually incorporate more information about water quality and availability. Tables 16.3 and 16.4 list the steps for progressive monitoring for indicators of SDG 6 (GEMI, 2017).

Table 16.1 Number of people (millions) living in cities with a surface-water deficit, by region

Geographic zone	Baseline value	2050 projection
Australia and New Zealand	1.1–1.6	1.7–2.4
Central Asia	0.0–2.6	0.0–6.1
Central Europe	7.0–12.8	11.3–18.2
Eastern Africa	0.0–0.0	0.0–3.7
Eastern Europe	4.9–7.9	8.2–9.8
Mashriq	1.7–5.2	4.2–16.2
Mesoamerica	3.0–4.2	5.8–5.9
North America	21.3–35.5	37.7–61.3
Northwest Pacific and East Asia	44.0–81.3	71.5–136.2
Northern Africa	3.9–5.3	17.0–17.1
South America	21.9–28.9	41.4–48.6
South Asia	13.7–90.5	146.8–242.3
Southeast Asia	3.2–3.3	11.9–19.2
Southern Africa	0.5–2.1	30.9–31.0
Western Africa	2.6–4.3	29.5–29.6
Western Europe	12.9–15.2	21.0–24.5

Source: Flörke et al. (2018)

Table 16.2 SDG 6 integrated monitoring

Indicator	Organization in charge
6.1.1 On drinking water	WHO, UNICEF
6.2.1 On sanitation and hygiene	WHO, UNICEF
6.3.1 On wastewater treatment	WHO, UN—Habitat, UNSD
6.3.2 On ambient water quality	UN Environment
6.4.1 On water use efficiency	FAO
6.4.2 On water stress	FAO
6.5.1 On integrated water resources management	UN Environment
6.5.2 On transboundary cooperation	UNESCO, UNECE
6.6.1 On water-related ecosystems	UN Environment
6.a.1 On international cooperation	WHO; UN Environment, OECD
6.b.1 On stakeholder participation	WHO; UN Environment, OECD

Source: GEMI (2017)

Table 16.3 Data and progressive monitoring of SDG 6 indicators (indicators 6.1.1/6.2.1/6.3.1/6.3.2/6.4.1)

Data sources and compilation	First step of progressive monitoring
Indicator 6.1.1	
National sources: Household surveys and institutional/utility records	Household surveys combined with population records for information on access and type of services
Global databases: WHO/UNICEF JMP	No information on water quality, reporting to the basic services level but not to the safely managed services level
Global compilation: WHO/UNICEF	Disaggregation of household data by place of residence, subnational region, and wealth
Indicator 6.2.1	
National sources: Household surveys, institutional/utility records, and licensed emptying service providers	Household surveys combined with population records for information on access and type of services
Global databases: WHO/UNICEF JMP	Estimation of proportion of total population using basic sanitation services, but no national data on management of fecal waste
Global compilation: WHO/UNICEF	Disaggregation of household data by place of residence, subnational region, and wealth
Indicator 6.3.1	
National sources: Line ministries and institutions (e.g., for water, sanitation, environment, health, public services, planning, housing, infrastructure, production), utilities and on-site service providers, National Statistical Office (NSO) for household surveys and registers of economic activities	Estimation of total wastewater generation by households from household surveys and population records
Global databases: FAO AQUASTAT, IBNET, WHO/UNICEF JMP, UNSD/UN environment water questionnaire for non-OECD/Eurostat countries, OECD/Eurostat questionnaire for OECD countries, and UNIDO statistics data portal	Estimation of total wastewater generation by economic activities from industry inventories, focusing on a few economic activities
Global compilation: WHO/UN Habitat on behalf of UN-water	Estimation of proportion of wastewater received and treated from institutional/utility records
Indicator 6.3.2	
National sources: Line ministries and institutions (e.g., for water, environment, natural resources), universities and research institutions, non-governmental organizations (NGOs), and citizens' science initiatives	Household surveys combined with population records for information on access and type of services
Global/regional databases: UN Environment GEMStat, OECD Lake and river quality and earth observations	Estimation of proportion of total population using basic sanitation services, but no national data on management of fecal waste
Global compilation: UN environment On behalf of UN-water	Disaggregation of household data by place of residence, subnational region, and wealth

(continued)

Table 16.3 (continued)

Data sources and compilation	First step of progressive monitoring
Indicator 6.4.1	
National sources: NSO, line ministries, and institutions (e.g., for water, agriculture, and environment)	Estimations based on national data. If needed, data can be retrieved from international databases. The agricultural rain-fed production factor Cr is calculated following the default coefficient provided in the indicator's guidelines
Global databases: FAO AQUASTAT, UNSD/UN environment water questionnaire for non-OECD/Eurostat countries, OECD/Eurostat questionnaire for OECD countries, FAO FAOSTAT, World Bank, UNSD National Accounts Estimates of Main aggregates, world energy outlook, and IBNET	

Source: GEMI (2017)

Table 16.4 Steps for progressive monitoring for indicators 6.4.2/6.5.1/6.5.2/6.6.1/6.a/6.b

Data sources and compilation	First step of progressive monitoring
Indicator 6.4.2	
National sources: NSO, line ministries and institutions (e.g., for water, agriculture, and environment)	Estimations based on internationally available data on water availability and withdrawals by different sectors, including data available through modeling
Global databases: FAO AQUASTAT, UNSD/UN environment water questionnaire for non-OECD/ Eurostat countries, OECD/Eurostat questionnaire for OECD countries and WMO WHOS	
Global compilation: FAO on behalf of UN-water	
Indicator 6.5.1	
National sources: Line ministries and institutions, NGOs, academia, and business. Global database: IWRM data portal	Survey response consolidated by ad hoc consultations between stakeholders
Global compilation: UN environment on behalf of UN-water	
Indicator 6.5.2	
National sources: Line ministries and institutions (e.g., for water, environment, natural resources, hydrology, geology)	Registration of data
In the absence of national data, sources such as the following could be referred to: UNESCO ISARM (for aquifers), GEF TWAP (for surface-water basins), OSU atlas of international freshwater agreements (for agreements and joint bodies)	
Global compilation: UNECE/UNESCO on behalf of UN-water	
Indicator 6.6.1	
National sources: Line ministries and institutions (e.g., for environment, water, natural resources), universities and research institutions, NGOs and citizens' science initiatives (ground-based surveys), space agencies (earth observations)	Monitoring of surface water-related ecosystems (e.g., vegetated wetlands, rivers, and open water bodies)
Global databases (selection): Ramsar convention on wetlands, global runoff database at GRDC, global groundwater information system of IGRAC, GlobWetland II, Hydroweb from LEGOS	
Global compilation: UN environment On behalf of UN-water	
Indicator 6.a.1	
National sources: Line ministries and institutions (e.g., for water, sanitation, environment, health, Public services, planning, finance), NSO, TrackFin Global databases: OECD creditor Reporting system and UN-water GLAAS	Information on the amount of official development assistance (ODA) received
Global compilation: WHO/UN environment/OECD on behalf of UN-water	
Indicator 6.b.1	

(continued)

Table 16.4 (continued)

Data sources and compilation	First step of progressive monitoring
National sources: Line ministries and institutions (e.g., for water, sanitation, environment, health, public services, planning, finance) global databases: UN-water GLAAS, IWRM Data Portal, OECD WGI	Qualitative estimation of degree of stakeholder participation at the national level
Global compilation: WHO/UN Environment/OECD on behalf of UN-water	

Source: GEMI (2017)

16.4 Strategies for Water Management

The Global Environmental Management Initiative (GEMI) is an inter-agency initiative composed of the World Meteorological Organization (WMO), the World Health Organization (WHO), the United Nations Educational, Scientific and Cultural Organization (UNESCO), the Food and Agriculture Organization (FAO), the United Nations International Children's Emergency Fund (UNICEF), the United Nations Human Settlements Program (UN-Habitat), the United Nations Environment Program (UNEP), and the United Nations Economic Commission for Europe (UNECE). GEMI was created to monitor various specific initiatives, many related to water, such as the case of wastewater treatment and water quality (6.3.2), water use (6.4.1), and scarcity (6.4.2), integrated water resources management (6.5.1), transboundary cooperation (6.5.2), and water-related ecosystems. On its website, GEMI provides the Water Sustainability Tool which aims to support organizations to improve understanding of water issues and at the same time promotes the development of water-based business strategies by companies. Figures 16.2–16.5 show the steps that companies are recommended to follow to achieve a successful business based on efficient water management (GEMI, 2018).

Water use for agriculture purposes is also significant, as it directly affects food supplies and human health. Based on the quantities of water use for agriculture and the needs for limitations on it, various initiatives have been launched to develop new or modified plants that use water efficiently for their growth. Plants involved in such initiatives include banana (Kissel, Eyland, Lawson, Swennen, & Carpentier, 2019), rice (Darbyshire, Crean, Dunn, & Dunn, 2019; Mahajan, Singh, Singh, Kaur, & Chauhan, 2018), and corn (Edge, Oikeh, Kyetere, Mugo, & Mashingaidze, 2018). It is important to monitor progress on these and other water initiatives to determine when they are successful and worthy of implementation. Investments in technologies are needed to achieve greater efficiency in water use to achieve the sustainability in agricultural activity (Malek, Adam, Stockle, Brady, & Rajagopalan, 2018).

The implementation and use of green infrastructure for water supply systems is another important issue (Jeong, Broesicke, Drew, & Crittenden, 2018; Tavakol-

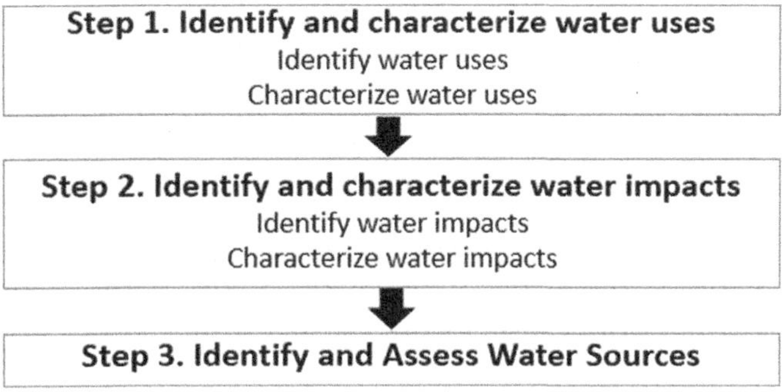

Fig. 16.2 Water use, impact, and source assessment. (Source: GEMI, 2018)

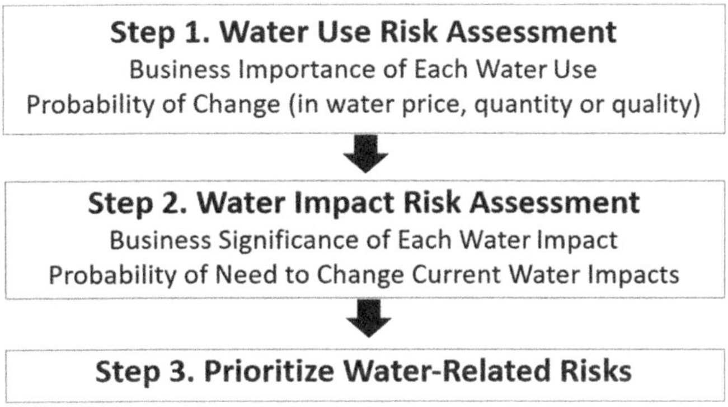

Fig. 16.3 Business risk assessment. (Source: GEMI, 2018)

Fig. 16.4 Business opportunity assessment. (Source: GEMI, 2018)

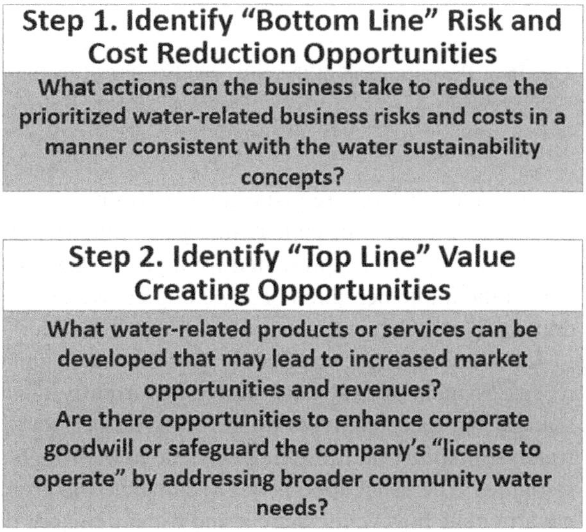

Fig. 16.5 Strategy development and implementation. (Source: GEMI, 2018)

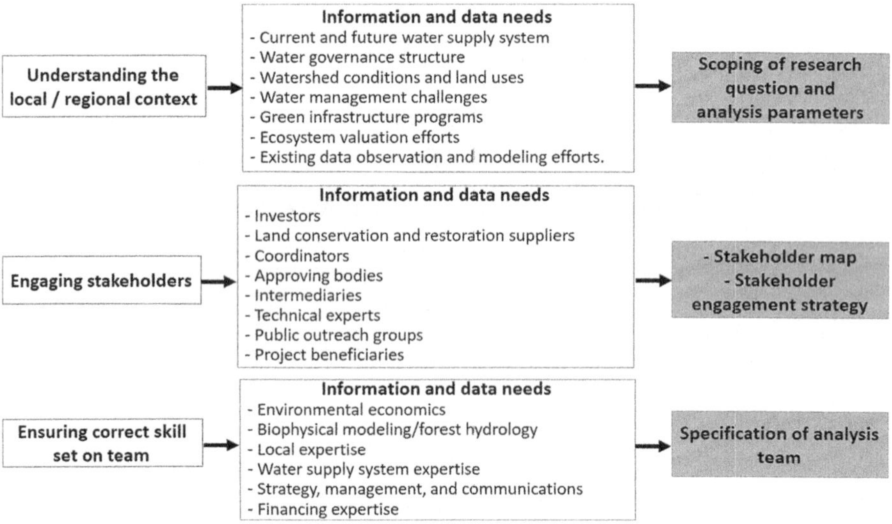

Fig. 16.6 Preassessment of the costs and benefits of the implementation of an ecological water supply system

Davani et al., 2018), largely because cost-efficient alternatives are needed to implement systems that contribute to the sustainable use of water for various purposes. In Fig. 16.6, a scheme is presented for a preassessment to evaluate the costs and benefits of the implementation of an ecological water supply system, based on the work done by Gray, Ozment, Altamirano, Feltran-Barbieri, and Morales (2019).

Lin, Petway, Lien, and Settele (2018) raise a significant issue regarding the correct assessment of freshwater scarcity. Currently, analyses may underestimate water scarcity but these analyses do not fully reflect how climate impacts the change in water availability at different levels due to the irregular spatial distribution of water resources. The usual approaches to calculate the problem of water scarcity often do not articulate how water sources and water demands through river network flows are geographically related. Thus, a distributed network approach is required. The use of blockchain technology may be beneficial to record public water transactions globally, in addition to ensuring trust and participation in data fidelity, data security, and data verification. With remote sensors, water use information can be obtained, and it can ultimately become "big data" given the massive amount of information to be collected; we are only in the initial stages of such efforts and the necessary technology and infrastructure must be developed and implemented to bring it to fruition. Lin et al. (2018) suggest that artificial intelligence can accelerate the identification of complex patterns in large data on the changes in water distributions, through the constant execution of small-scale scientific processes. This technological integration can advantageously and efficiently aid in the identification of patterns of water abundance and scarcity, and assist in the equitable management of water resources at multiple scales under climate change conditions. It is noted that the latest research corroborates the growing interest in more comprehensive mechanisms that allow a complete analysis of the water data that is currently collected by various countries and regions (Al Aani, Bonny, Hasan, & Hilal et al., 2019; Perea, Poyato, Montesinos, & Díaz, 2019; Zaji & Bonakdari, 2019).

Closing Remarks

Much information exists at the global level about problems associated with water supplies, but there are a relatively limited number of initiatives to mitigate the water concerns. Many of these concerns are serious and constitute crises. There are certainly needs for government efforts in these areas, but initiatives are also needed by companies, so that they can generate business models based on sustainable water management. The active involvement of companies in such initiatives is more likely to lead to the success and sustainability of efforts. Going forward, the handling of large amounts of water supply and use data suggests that mechanisms are needed that allow comprehensive and real-time analysis, utilizing artificial intelligence where suitable. Such initiatives are anticipated in the short and long terms to be promising for enhancing water analysis and planning as well as for preventing water crises and attaining more sustainable cities.

References

Al Aani, S., Bonny, T., Hasan, S. W., & Hilal, N. (2019). Can machine language and artificial intelligence revolutionize process automation for water treatment and desalination? *Desalination, 458*, 84–96.

Burek, P., Satoh, Y., Fischer, G., Kahil, M. T., Scherzer, A., Tramberend, S., et al. (2016). Water futures and solution-fast track initiative. Retrieved February 5, 2020, from http://pure.iiasa.ac.at/id/eprint/13008/1/WP-16-006.pdf

Darbyshire, R., Crean, E., Dunn, T., & Dunn, B. (2019). Predicting panicle initiation timing in rice grown using water efficient systems. *Field Crops Research, 239*, 159–164.

Edge, M., Oikeh, S. O., Kyetere, D., Mugo, S., & Mashingaidze, K. (2018). Water efficient maize for Africa: A public-private partnership in technology transfer to smallholder farmers in sub-Saharan Africa. In *From agriscience to agribusiness* (pp. 391–412). Cham: Springer. https://doi.org/10.1007/978-3-319-67958-7_19.

Flörke, M., Schneider, C., & McDonald, R. I. (2018). Water competition between cities and agriculture driven by climate change and urban growth. *Nature Sustainability, 1*(1), 51.

GEMI. (2017). Integrated monitoring guide for sustainable development goal 6 on water and sanitation—Targets and global indicators. Retrieved February 5, 2020, from https://www.unwater.org/publications/sdg-6-targets-indicators

GEMI. (2018). The water sustainability tool. Retrieved February 5, 2020, from http://gemi.org/water/module4.htm

Gray, E., Ozment, S., Altamirano, J. C., Feltran-Barbieri, R., & Morales, G. (2019). Green-Gray assessment: How to assess the costs and benefits of green infrastructure for water supply systems. Retrieved February 5, 2020, from https://wriorg.s3.amazonaws.com/s3fs-public/green-gray-assessment.pdf

Jeong, H., Broesicke, O. A., Drew, B., & Crittenden, J. C. (2018). Life cycle assessment of small-scale graywater reclamation systems combined with conventional centralized water systems for the City of Atlanta, Georgia. *Journal of Cleaner Production, 174*, 333–342.

JMP. (2018). JMP 2018 global report on WASH in schools. Retrieved February 5, 2020, from https://www.washdata.org/report/jmp-2018-wash-in-schools-final

JMP. (2019). WASH in health care facilities: Global baseline report 2019. Retrieved February 5, 2020, from https://www.washdata.org/sites/default/files/documents/reports/2019-04/JMP-2019-wash-in-hcf.pdf

Kissel, E., Eyland, D., Lawson, T., Swennen, R., & Carpentier, S. C. (2019). Using growth and transpiration phenotyping under controlled conditions to select water efficient banana genotypes. *Frontiers in Plant Science, 10*, 352.

Lin, Y. P., Petway, J., Lien, W. Y., & Settele, J. (2018). Blockchain with artificial intelligence to efficiently manage water use under climate change. *Environments, 5*, 34. https://doi.org/10.3390/environments5030034

Mahajan, G., Singh, K., Singh, N., Kaur, R., & Chauhan, B. S. (2018). Screening of water-efficient rice genotypes for dry direct seeding in South Asia. *Archives of Agronomy and Soil Science, 64*(1), 103–115.

Malek, K., Adam, J., Stockle, C., Brady, M., & Rajagopalan, K. (2018). When should irrigators invest in more water—Efficient technologies as an adaptation to climate change? *Water Resources Research, 54*(11), 8999–9032.

Mekonnen, M. M., & Hoekstra, A. Y. (2016). Four billion people facing severe water scarcity. *Science Advances, 2*(2), e1500323.

OECD. (2012). OECD environmental outlook to 2050: The consequences of inaction. Paris, OECD publishing. Retrieved February 5, 2020, from http://www.oecd.org/centrodemexico/medios/49912980.pdf https://doi.org/10.1787/9789264122246-en

Perea, R. G., Poyato, E. C., Montesinos, P., & Díaz, J. A. R. (2019). Optimisation of water demand forecasting by artificial intelligence with short data sets. *Biosystems Engineering, 177*, 59–66.

Tavakol-Davani, H., Burian, S. J., Butler, D., Sample, D., Devkota, J., & Apul, D. (2018). Combining hydrologic analysis and life cycle assessment approaches to evaluate sustainability of water infrastructure. *Journal of Irrigation and Drainage Engineering, 144*(11), 05018006.

World Water Council. (2019). Towards a way to improve the situation. Retrieved February 5, 2020, from http://www.worldwatercouncil.org/en/water-crisis

WRI. (2015). *Aqueduct projected water stress country rankings*. World Resources Institute, Washington, DC. Retrieved February 5, 2020, from https://www.wri.org/resources/data-sets/aqueduct-projected-water-stress-country-rankings

WWAP. (2019). *The United Nations world water development report 2019: Leaving no one behind*. Paris: UNESCO.

Zaji, A. H., & Bonakdari, H. (2019). Robustness lake water level prediction using the search heuristic-based artificial intelligence methods. *ISH Journal of Hydraulic Engineering, 25*(3), 316–324.

Chapter 17
Climate Change and Cities

Aldo Alvarez-Risco, Shyla Del-Aguila-Arcentales, and Marc A. Rosen

Abstract Climate change is a significant environmental concern in cities and that needs urgent action through the activities of companies, organizations, and citizens. Intergovernmental policies are increasingly required as are, at the same time, transformations of institutions such as governments and universities.

The generation of sustainable cities that promote the development of citizens who take a sustainable approach is increasingly important. Some suggest that climate change focus should be mandatory to the behavior of the people in the cities, including products and services offered and received. Also, since various global agreements and regulations serve as a guide for planned efforts, cities have an important role as a guide activity for enhancing understanding, mitigation, and adaptation of climate change.

Keywords Climate change · Cities · Environment · Ecology · Environment · Mitigation of climate change

A. Alvarez-Risco (✉)
Facultad de Ciencias Administrativas y Recursos Humanos, Universidad de San Martín de Porres, Lima, Peru
e-mail: aldo.alvarez@redsaf.org

S. Del-Aguila-Arcentales
Escuela Nacional de Marina Mercante "Almirante Miguel Grau", Callao, Peru

Universidad Nacional de la Amazonia Peruana, Iquitos, Peru
e-mail: sdelaguila@enamm.edu.pe

M. A. Rosen
Faculty of Engineering and Applied Science, University of Ontario Institute of Technology, Oshawa, ON, Canada
e-mail: marc.rosen@uoit.ca

A. Alvarez-Risco et al. (eds.), *Building Sustainable Cities*,
https://doi.org/10.1007/978-3-030-45533-0_17

17.1 Introduction

The human activities that occur in cities have changed dramatically since the beginning of the industrial revolution and, over the last 30 years, the manufacture, use, and disposal of an expanding number of objects have increased notably. For example, electronic waste is a growing problem with the increased use of tablets, laptops, smartphones, and other electronic products. Issues like this have even been reported in the mass media (Time, 2019). Also, the increase of cars in cities has generated an increasing amount of environmental and ecological impact. Perhaps such problems will be mitigated somewhat when electric and hydrogen cars are available and widely used, but the time at which a majority will be replaced still seems far off. Many cities have been generating regulations that help reduce polluting activities and increase actions such as recycling and reuse. In addition to these efforts, companies in cities also must play an active role in the mitigation of and adaptation to climate change.

Climate change is expected to have increasing effects on living entities in the future. Among these, various health effects are being examined, such as the effects of global warming and extreme weather events. For example, the pathophysiological effect of heat exposure in persons is well understood, and includes heat stress and heat stroke, acute kidney injury, exacerbation of congestive heart failure (Székely, Carletto, & Garami, 2015), and increased risk of interpersonal violence (Sanz-Barbero et al. (2018) and collective violence (Levy, Sidel, & Patz, 2017). During periods of extreme heat, young children have a greater risk of electrolyte imbalance, fever, respiratory illness, and kidney disease (Xu et al., 2014).

This chapter proposes basic definitions of climate change and greenhouse gases. In addition, global environmental agreements are described that are in force and that form a basis of activities in cities and countries, focused on the care of air, water, soil, and other natural resources. Finally, some general strategies are proposed for combating climate change.

17.2 Climate Change

Climate change is conceptualized as statistically significant variations in weather that are maintained over a long period of time (IPCC, 2007). Climate change includes changes in the magnitude and frequency of rare and extreme weather events. This is accompanied by the gradual increase in the global average temperature of the earth's surface (Houghton et al., 2001). The changes in climate can be observed in various ways, including changes in long-term weather trends, the frequencies and magnitudes of heat or cold waves, occurrences of heavy rains, etc. The greenhouse effect is a process that occurs when gases in Earth's atmosphere trap the Sun's heat, making the Earth much warmer than it would be without an atmosphere. Thus, the greenhouse effect makes Earth livable for life on Earth. But increases in atmospheric greenhouse gas concentrations are raising the amount of solar energy trapped in the Earth's atmosphere, and thereby changing the climate.

A greenhouse gas is a gas that absorbs short wave radiation and emits radiant energy within the thermal infrared range. Various gases contribute the greenhouse effect, and the main one is carbon dioxide (CO_2). Other greenhouse gases include methane (CH_4), nitrous oxide (N_2O), ozone (O_3), and fluorinated gases (National Geographic, 2020). The World Meteorological Organization has measured the emission of greenhouse gases and has been able to register a notable increase from 1750 to 2017, for atmospheric concentrations of the three most important greenhouse gases: methane (257%), carbon dioxide (145%), and nitrous oxide (122%) (World Meteorological Organization, 2017).

17.3 Selected Data on Climate Change

17.3.1 Countries Most Affected by Climate Change

The Global Climate Risk Index (Germanwatch, 2019) is an analysis based on one of the most reliable data sets available on the impacts of extreme weather events and associated socio-economic data. Its aim is to contextualize ongoing climate policy debates—especially the international climate negotiations—looking at real-world impacts. The Global Climate Risk Index analyzes to what extent countries have been affected by the impacts of weather-related loss events (storms, floods, heat waves, etc.). It has listed about the ten countries most affected by climate change in 2017. Five of the countries are from Asia, three from Latin America, and two from Africa. Climate risk indexes for various countries and regions follow (Germanwatch, 2019): Puerto Rico (1.5), Sri Lanka (9), Dominica (9.33), Nepal (10.5), Perú (10.67), Vietnam (13.5), Madagascar (15), Sierra Leone (15.67), Bangladesh (16), and Thailand (16.33). Germanwatch (2019) also indicates the ten countries most affected by climate change over the period 1998–2017. Selected climate risk indexes over this time period follow: Puerto Rico (7.83), Honduras (13.00), Myanmar (13.17), Haiti (15.17), Philippines (19.67), Nicaragua (20.33), Bangladesh (26.67), Pakistan (30.17), Vietnam (31.67), and Dominica (33.00). Also, climate change has effects on health and the lives of people. In the USA, for instance, 650 people die annually from causes related to weather, water, and climate events (NOAA, 2019).

17.3.2 Observations Related to Climate Change

Much data has been reported that provides preliminary evidence of climate change. Some is cited in this section.

17.3.2.1 Increase in Global Temperature of Earth

There has been an increase in the average surface temperature, approximately 0.9 degrees Celsius (1.62 degrees Fahrenheit) since the late nineteenth century (NOAA,

2019). This change has been attributed mainly to the increase in atmospheric concentrations of carbon dioxide (CO_2) and other greenhouse gases resulting from human activity (NOAA, 2019).

17.3.2.2 Warming Oceans

The oceans absorb some of the excess heat related to increasing greenhouse gas emissions, leading to an increase in ocean temperature. Rises in ocean temperatures affect marine species and ecosystems. Increases in ocean temperatures also affect negatively the benefits that humans derive from oceans. For instance, it threatens food security, increases the prevalence of diseases, and causes more extreme weather events and the loss of coastal protection (IUCN, 2019).

17.3.2.3 Shrinking Ice Sheets

The mass of ice in the Greenland Ice Sheet has begun to decline. From 1979 to 2006, the summer melt of the ice sheet has increased by 30 percent, reaching a new record in 2007. At higher elevations, an increase in winter snow accumulation has partially offset the melt. Perhaps, the decline continues to outpace accumulation because warmer temperatures have led to increased melt and faster glacier movement at the island's edges (NSIDC, 2019a).

The effects of climate change on sea ice (i.e., frozen ocean water) are also notable. Sea ice forms, grows, and melts in the ocean. In contrast, icebergs, glaciers, ice sheets, and ice shelves all originate on land and are mainly found in remote polar oceans. On average, sea ice covers approximately 25 million square kilometers of the earth and, although sea ice may not directly affect most people, it is a critical component of the planet because it influences climate, wildlife, and people who live in the Arctic. Also, about 15 percent of the world's oceans are covered by sea ice during part of the year (NSIDC, 2019a, 2019b). Sea ice is important due to its function in regulating global temperature. Since sea ice has a bright surface, causing much of the sunlight incident on it to be reflected back into space, areas covered by sea ice do not absorb a high proportion of the solar energy they receive. However, if gradually warming temperatures melt sea ice over time, there will be fewer bright surfaces available to reflect sunlight into space, allowing more solar energy to be absorbed at the Earth's surface. This will cause temperatures rise further, potentially lead to greater global warming. This effect makes polar regions one of the most sensitive areas to climate change on Earth (NSIDC, 2019a, 2019b).

Sea ice coverage has been declining in recent decades. For example, Fig. 17.1 shows the variation in the extent of sea ice in December in the Northern Hemisphere, from 1980 to 2018.

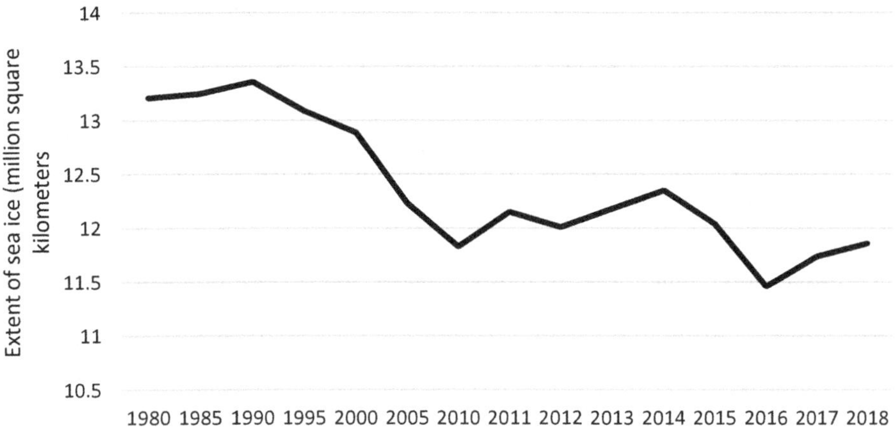

Fig. 17.1 Extent of sea ice in December in the Northern Hemisphere, from 1980 to 2018. (Source: National Snow and Ice Data Center (2019a, 2019b))

17.3.2.4 Glacial Retreat

Glaciers may retreat when their ice melts or ablates more quickly than snowfall can accumulate and form new glacial ice. Higher temperatures and less snowfall have been causing many glaciers around the world to retreat recently (NSIDC, 2019b).

17.3.2.5 Sea Level Rise

Higher sea levels are coinciding with more dangerous hurricanes and typhoons that move more slowly and drop more rain, contributing to more powerful storm surges that can damage objects in their path. One study found that between 1963 and 2012, almost half of all deaths from Atlantic hurricanes were caused by storm surges (National Geographic, 2015).

17.4 Strategies for Addressing Climate Change

17.4.1 Worldwide Efforts for Mitigation of Climate Change

Countries individually as well as various international organizations have been developing initiatives to contribute to climate change mitigation. Some of these are listed and discussed in this section.

Absorb, Anticipate, Reshape (A2R) (UN Climate Resilience Initiative, 2017)
A2R is a global multi-stakeholder initiative that strengthens climate resilience for vulnerable countries and people:

- Anticipate: Capacity to better anticipate and act on climate hazards and stresses through early warning and early action.
- Absorb: Capacity to absorb shocks and stresses by increasing access to climate risk insurance and social protection systems.
- Reshape: Capacity to reshape development pathways by transforming economies to reduce risks and root causes of vulnerabilities and support the sound management of physical infrastructure and ecosystems to foster climate resilience.

Adaptation Fund (Adaptation Fund, 2019)
The Adaptation Fund was established under the Kyoto Protocol of the UN Framework Convention on Climate Change, and since 2010 has committed US$ 720 million to climate adaptation and resilience activities, including supporting 100 concrete adaptation projects. The Adaptation Fund finances projects and programs that help vulnerable communities in developing countries adapt to climate change. Initiatives are based on country needs, views, and priorities.

Africa Renewable Energy Initiative (AREI, 2015)
The Africa Renewable Energy Initiative (AREI) is a transformative, Africa-owned and Africa-led inclusive effort to accelerate and scale up the harnessing of the continent's huge renewable energy potential. Under the mandate of the African Union, and endorsed by African Heads of State and Government on Climate Change (CAHOSCC), the Initiative is set to achieve at least 10 GW of new and additional renewable energy generation capacity by 2020, and mobilize the African potential to generate at least 300 GW by 2030.

Climate and Clean Air Coalition (2019)
In 2012, the governments of Bangladesh, Canada, Ghana, Mexico, Sweden, and the USA, along with the United Nations Environment Programme (UNEP), came together to initiate efforts to treat short-lived climate pollutants as an urgent and collective challenge. Together, they formed the Climate and Clean Air Coalition to support fast action and deliver benefits on several fronts at once: climate, public health, energy efficiency, and food security. Today, the Coalition brings together hundreds of experienced and influential stakeholders from around the world to leverage high-level engagement and catalyze concrete actions in both the public and private sector.

Global Environment Facility (GEF, 2019)
The Global Environment Facility (GEF) was established on the eve of the 1992 Rio Earth Summit to help tackle our planet's most pressing environmental problems. Since then, the GEF has provided close to $20 billion in grants and mobilized an additional $107 billion in co-financing for more than 4700 projects in 170 countries.

GRID-Arendal (GRIDA, 2019)
GRID-Arendal was established in 1989 to support environmentally sustainable development by working with UN Environment and other partners. It communicates environmental knowledge that strengthens management capacity and motivates decision-makers to act. GRID-Arendal also transforms environmental data into credible, science-based information products, delivered through innovative communication tools and capacity building services.

International Carbon Action Partnership (ICAP, 2019)
Founded in 2007 in Lisbon, Portugal, by leaders of more than 15 governments, ICAP provides the opportunity for member jurisdictions to share best practices and discuss emissions trading systems (ETS) design elements with a view to creating a well-functioning global carbon market through linking ETS. ICAP's objectives are:

- Share best practices and learning from each other's experience of ETS.
- Help policymakers recognize ETS design compatibility issues and opportunities for the establishment of an ETS at an early stage.
- Facilitate future linking of trading programs.
- Highlight the key role of emissions trading as an effective climate policy response.
- Build and strengthen partnerships among governments.

International Environmental Technology Centre (IETC, 2019)
According to its vision, the International Environmental Technology Centre (IETC) works with developing countries to implement sustainable solutions to environmental challenges, with focus on holistic waste management and projected towards three directions:

1. Knowledge products on environmentally sound methods and good practice.
2. In-country technical and advisory services for design and implementation of environmentally sound technologies and approaches.
3. Promotion and dissemination of environmentally sound waste management practices and methods.

Kyoto Protocol (UNFCCC, 2019)
The targets for the first commitment period of the Kyoto Protocol cover emissions of the six main greenhouse gases, namely:

- Carbon dioxide (CO_2);
- Methane (CH_4);
- Nitrous oxide (N_2O);
- Hydrofluorocarbons (HFCs);
- Perfluorocarbons (PFCs); and
- Sulfur hexafluoride (SF_6).

The maximum amount of emissions (measured as the equivalent in carbon dioxide) that a Party may emit over a commitment period in order to comply with its emissions target is known as a Party's assigned amount.

Portfolio Decarbonization Coalition (UNEPFI, 2019)
The Portfolio Decarbonization Coalition (PDC) is a multi-stakeholder initiative intended to drive GHG emissions reductions on the ground by mobilizing a critical mass of institutional investors committed to gradually decarbonizing their portfolios.

Partnership for Action on Green Economy (PAGE, 2019)
The Partnership for Action on Green Economy (PAGE) was launched in 2013 as a response to the call at Rio+20 to support those countries wishing to embark on greener and more inclusive growth trajectories. PAGE seeks to put sustainability at

the heart of economic policies and practices to advance the 2030 Agenda for Sustainable Development and supports nations and regions in reframing economic policies and practices around sustainability to foster economic growth, create income and jobs, reduce poverty and inequality, and strengthen the ecological foundations of their economies.

UNESCO Climate Change Initiative (UNESCO, 2009)
This programme uses innovative educational approaches to help a broad audience (with particular focus on youth), understand, address, mitigate, and adapt to the impacts of climate change, encourage the changes in attitudes and behaviors needed to put the world on a more sustainable development path, and build a new generation of climate change-aware citizens. The UNESCO Climate Change Initiative was launched by the Director-General of UNESCO, Ms. Irina Bokova, in 2009. It seeks to reinforce the scientific, mitigation, and adaptation capacities of countries and communities that are most vulnerable to the effects of climate change.

UNEP DTU (2019)
The Danish Ministry of Foreign Affairs, UN Environment, and the Technical University of Denmark (DTU) established UNEP DTU Partnership in 1990. As a UN Environment Collaborating Centre, UNEP DTU Partnership is actively engaged in implementing UN Environment's Climate Change Strategy and Energy Programme. UNEP DTU employs 70 researchers of 26 different nationalities working around the world from offices in UN City, Copenhagen.

UN Climate Change Conference (UNFCCC, 2019)
The UNFCCC secretariat (UN Climate Change) was established in 1992 when countries adopted the United Nations Framework Convention on Climate Change (UNFCCC). With the subsequent adoption of the Kyoto Protocol in 1997 and the Paris Agreement in 2015, Parties to these three agreements have progressively reaffirmed the secretariat's role as the United Nations entity tasked with supporting the global response to the threat of climate change. Since 1995, the secretariat is located in Bonn, Germany.

United Nations Environment Programme Finance Initiative (UNEPFI, 2018)
United Nations Environment Programme Finance Initiative (UNEP FI) is a partnership between UNEP and the global financial sector to mobilize private sector finance for sustainable development. UNEP FI works with more than 300 members—banks, insurers, and investors—and over 100 supporting institutions—to help create a financial sector that serves people and planet while delivering positive impacts. UNEP FI aims to inspire, inform, and enable financial institutions to improve people's quality of life without compromising that of future generations. By leveraging the UN's role, UNEP FI accelerates sustainable finance.

UN Environment Programme World Conservation Monitoring Centre (WCMC, 2019)
The UN Environment Programme World Conservation Monitoring Centre (UNEP-WCMC) works with scientists and policy makers worldwide to place biodiversity at the heart of environment and development decision-making to enable enlightened choices for people and the planet. UNEP-WCMC's 100-strong international team

are recognized leaders in their field and have unrivalled understanding of the institutional landscape surrounding biodiversity policy and ecosystem management. Based in Cambridge, UK, UNEP-WCMC is a collaboration between UN Environment Programme and the UK charity, WCMC.

World Resources Institute (WRI, 2020)
WRI is a global research organization that spans more than 60 countries, with offices in the USA, China, India, Brazil, Indonesia, and more. WRI's more than 1000 experts and staff work closely with leaders to turn big ideas into action to sustain our natural resources—the foundation of economic opportunity and human well-being. The organization's work focuses on seven critical issues at the intersection of environment and development: climate, energy, food, forests, water, cities, and the ocean.

17.4.2 International Environmental Agreements

Numerous global agreements that support environmental protection efforts exist or have existed in the past. A chronological selection of these follow:

1940
Convention on Nature Protection and Wild Life Preservation in the Western Hemisphere
URL: https://www.fws.gov/migratorybirds/pdf/Treaties-Legislation/Treay-WesternHemisphere.pdf

1954
International Convention for the Prevention of Pollution of the Sea by Oil
URL: https://sedac.ciesin.columbia.edu/entri/texts/pollution.of.sea.by.oil.1954.html

1958
Convention on Fishing and Conservation of Living Resources of the High Seas
URL: https://www.gc.noaa.gov/documents/8_1_1958_fishing.pdf

1966
International Convention for the Conservation of Atlantic Tunas
URL: https://www.iccat.int/en/

1969
Boon Agreement
URL: https://www.bonnagreement.org

1973
International Convention for the Prevention of Pollution from Ships
URL: http://www.imo.org/en/About/Conventions/ListOfConventions/Pages/International-Convention-for-the-Prevention-of-Pollution-from-Ships-(MARPOL).aspx

1973
Convention on the International Trade in Endangered Species of Wild Flora and Fauna
URL: https://www.cites.org/eng/disc/what.php

1975
Biological Weapons Convention
URL: https://www.un.org/disarmament/wmd/bio

1977
Working Environment (Air Pollution, Noise and Vibration) Convention
URL: https://www.ilo.org/dyn/normlex/en/f?p=NORMLEXPUB:12100:0::NO::P12100_ILO_CODE:C148

1979
Convention on Long-Range Transboundary Air Pollution
URL: http://www.unece.org/fileadmin//DAM/env/lrtap/welcome.html

1979
Convention on the Conservation of Migratory Species of Wild Animals (CMS)
URL: https://www.cms.int

1981
Convention for the Protection of the Marine Environment and Coastal Area of the South-east Pacific
URL: https://uia.org/s/or/en/1100038714

1981
Convention for Co-operation in the Protection and Development of the Marine and Coastal Environment of the West and Central African Region
URL: http://abidjanconvention.org

1982
United Nations Convention on the Law of the Sea
URL: https://www.un.org/Depts/los/convention_agreements/texts/unclos/UNCLOS-TOC.htm

1982
Berne Convention on the Conservation of European Wildlife and Natural Habitats
URL: https://www.coe.int/en/web/bern-convention

1982
Convention for the Conservation of Antarctic Marine Living Resources
URL: https://www.ccamlr.org/en/organisation/camlr-convention-text

1983
Convention on Certain Conventional Weapons
URL: https://www.armscontrol.org/factsheets/CCW

1984
EMEP Protocol
URL: https://www.unece.org/env/lrtap/emep_h1.html

1985
FAO International Code of Conduct on the Distribution and Use of Pesticides
URL: http://www.fao.org/3/y4544e/y4544e00.htm

1986
Convention for the Protection and Development of the Marine Environment of the Wider Caribbean Region
URL: http://cep.unep.org/cartagena-convention

1986
Nairobi Convention
URL: https://www.unenvironment.org/nairobiconvention

1986
Noumea Convention
URL: https://www.sprep.org/convention-secretariat/noumea-convention

1986
Convention on Early Notification of a Nuclear Accident
URL: https://www.iaea.org/topics/nuclear-safety-conventions/convention-early-notification-nuclear-accident

1986
Convention on Assistance in the Case of a Nuclear Accident or Radiological Emergency
URL: https://www.iaea.org/sites/default/files/infcirc336.pdf

1987
Montreal Protocol on Substances that Deplete the Ozone Layer
URL: https://www.unenvironment.org/ozonaction/who-we-are/about-montreal-protocol

1988
China Australia Migratory Bird Agreement
URL: http://www.austlii.edu.au/au/other/dfat/treaties/1988/22.html

1989
Convention on Civil Liability for Damage Caused during Carriage of Dangerous Goods by Road, Rail, and Inland Navigation Vessels
URL: https://www.unece.org/trans/danger/publi/crtd/crtd_e.html

1991
Espoo Convention on Environmental Impact Assessment in a Transboundary Context
URL: https://cil.nus.edu.sg/databasecil/1991-convention-on-environmental-impact-assessment-in-a-transboundary-context

1991
Convention on the Ban of the Import into Africa and the Control of Transboundary Movements and Management of Hazardous Wastes within Africa
URL: https://au.int/en/treaties/bamako-convention-ban-import-africa-and-control-transboundary-movement-and-management

1991
Protocol on Environmental Protection to the Antarctic Treaty
URL: https://www.ats.aq/e/protocol.html

1992
Convention on Biological Diversity
URL: https://www.cbd.int

1992
Basel Convention
URL: http://www.basel.int

1992
Convention for the Protection of the Marine Environment of the North-east Atlantic (OSPAR Convention)
URL: https://www.ospar.org/convention

1992
Convention on the Protection and Use of Transboundary Watercourses and International Lakes
URL: https://www.unece.org/env/water.html

1992
Convention on the Protection of the Black Sea against Pollution
URL: http://www.blacksea-commission.org/_convention.asp

1992
Convention on the Protection of the Marine Environment of the Baltic Sea Area
URL: https://helcom.fi/about-us/convention

1993
North American Agreement on Environmental Cooperation
URL: http://www.cec.org/about-us/NAAEC

1994
Convention to Combat Desertification
URL: https://www.unbonn.org/UNCCD

1994
Agreement of Agriculture
URL: https://www.wto.org/english/docs_e/legal_e/14-ag_01_e.htm

1995
Alpine Convention
URL: https://www.alpconv.org/en/home/convention/framework-convention

1995
Barcelona Convention
URL: http://web.unep.org/unepmap

1996
Comprehensive Nuclear-Test-Ban Treaty
URL: https://www.un.org/disarmament/wmd/nuclear/ctbt

1997
Chemical Weapons Convention
URL: https://www.opcw.org/chemical-weapons-convention

1998
Aarhus Convention on Access to Information, Public Participation in Decision-making and Access to Justice in Environmental Matters
URL: http://www.unece.org/env/pp/welcome.html

1999
Agreement on the Conservation of African-Eurasian Migratory Waterbirds
URL: https://www.unep-aewa.org

2000
Cartagena Protocol on Biosafety 2000
URL: https://bch.cbd.int/protocol/background

2002
ASEAN Agreement on Transboundary Haze Pollution
URL: https://asean.org/asean-socio-cultural/cop-to-aathp-conference-of-the-parties-to-the-asean-agreement-on-transboundary-haze-pollution/

2003
Framework Convention on the Protection and Sustainable Development of the Carpathians
URL: http://www.carpathianconvention.org

2003
Putrajaya Declaration of Regional Cooperation for the Sustainable Development of the Seas of East Asia
URL: http://pemsea.org/sites/default/files/putrajaya-declaration.pdf

2004
International Treaty on Plant Genetic Resources for Food and Agriculture (ITPGRFA)
URL: http://www.fao.org/in-action/right-to-food-global/global-level/itpgrfa/en

2004
Stockholm Convention on Persistent Organic Pollutants
URL: http://www.pops.int

2004
Rotterdam convention
URL: http://www.pic.int

2004
International Convention for the Control and Management of Ships' Ballast Water and Sediments (BWM)

URL: http://www.imo.org/en/About/Conventions/ListOfConventions/Pages/International-Convention-for-the-Control-and-Management-of-Ships'-Ballast-Water-and-Sediments-(BWM).aspx

2005
Asia-Pacific Partnership on Clean Development and Climate
URL: https://aric.adb.org/initiative/asia-pacific-partnership-on-clean-development-and-climate

2006
Convention on the Prevention of Marine Pollution by Dumping Wastes and Other Matter
URL: http://www.imo.org/en/OurWork/Environment/LCLP/Pages/default.aspx

2006
International Tropical Timber Agreement
URL: https://www.itto.int/council_committees/itta

2006
Energy community
URL: https://energy-community.org/aboutus/whoweare.html

2009
The Hong Kong International Convention for the Safe and Environmentally Sound Recycling of Ships
URL: http://www.imo.org/en/OurWork/Environment/ShipRecycling/Pages/Default.aspx

2010
The Nagoya Protocol on Access to Genetic Resources and Benefit-Sharing
URL: https://www.cbd.int/undb/media/factsheets/undb-factsheet-nagoya-en.pdf

2015
Paris Agreement
URL: https://unfccc.int/process-and-meetings/the-paris-agreement/what-is-the-paris-agreement

2017
Minamata Convention on Mercury
URL: http://www.mercuryconvention.org

17.4.3 Communication to Mitigate Climate Change

All efforts that can be made to contribute to addressing climate change require a communication plan so that each actor knows in detail all the efforts that are made and can actively participate. This is important since it requires a significant commitment to transform societies. This communication requires that all possible means be used, such as curricular plans, websites, and other written material. Likewise, dissemination of successful activities on social media can help promote efforts made throughout societies to contribute to addressing climate change. Engagement is necessary to ensure that the population actively participates in learning and activities.

17.5 Climate Change and Cities

People need to know more about climate change. This includes the general public, as well as especially students from junior education through to universities. This raises the importance of education and awareness about climate change. In order for actions against climate change education to be sustained over time, the contents of the learning curricula need to be enhanced and teachers need to be empowered; they can achieve inform students based on evidence and knowledge. Once the thematic contents related to climate change have been chosen, it is necessary to include an emphasis on those elements that may be easier for citizens to understand, so that they can transform that knowledge into intentions and subsequently into visible and impactful actions. This need for structured educational content also exists for companies during the execution of corporate social responsibility programs, since they require transmitting certain concepts to their employees. One strategy that sometimes has good results is to use recreational activities so that people can have fun and learn at the same time.

Nakayama et al. (2019) mention that culture and education are factors that influence the effect of programs aimed at mitigating climate change. It is important for cultures to accept the responsibility for managing the planet Earth, so that optimal concrete actions to care for the planet are adopted, including measures for climate change mitigation. *Culture* also influences the interest that a person may have in an environmental issue. For example, populations that develop in fundamentally jungle habitats may be more interested in contributing since they are close to the issue of the environment and the necessary care.

Finally, *environmental literacy* is a barrier to education programs because although the contents used are designed to inform, a prior understanding of environmental issues can facilitate understanding by junior school pupils, university students, and managers of firms.

Cities are a crucial part in the fight against climate change. It is noted that half of the world's population lives in cities and the fraction is expected to become higher in the future, reaching 2/3 by 2030 (Ritchie & Roser, 2018). The daily life of a city implies the use of a large amount of energy; as a result, cities emit approximately 70% of greenhouse gases of a country. Currently, high CO_2 emissions are reported by most countries, mainly due to fossil fuel combustion to obtain energy for the development and operation of cities.

One may ask: Is it normal for cities to generate a large amount of CO_2? The answer depends on the environmental policies of a city aimed at achieving the efficient use of resources and on the level of environmental education that has been devoted to cities to motivate environmentally responsible behavior and good environmental stewardship. These aspects are related to planning by all levels of government as well as directly by cities themselves.

Transportation in cities contributes significantly to CO_2 emissions, especially for large cities, and those in which traditional and old vehicles are mainly used (e.g., old cars that are significant polluters) and long distances are traveled by these vehicles. Transportation-related emissions are clearly different when intermodal transport is utilized by cities, as it can be greatly advantageous environmentally. A similar effect is observed in residences as well as commercial and institutional buildings (e.g., recreation centers). In these, fossil fuels are usually the main fuel still used for the generation of the energy necessary for the operation of such buildings. A paradox may be detected at this point, since cities are often the greatest contributors to climate change in a country and at the same time they are often the parts of a country most affected by climate change. Depending on the location of cities, the risks and dangers they face regarding climate change can differ. For instance, cities near the mountains have a significant risk of landslides that can destroy buildings and injure or kill inhabitants. Cities near large water bodies, such as rivers and oceans, are often subject to significant flood risks.

Cities have tools at their disposal to develop sustainable options for addressing climate change. These can involve some or all of the stakeholders, including authorities, public and private environments, citizens (including children and adults), businesses, and universities. Research planning regarding climate change is a commitment that cities need to undertake to achieve short- and long-term solutions to their environmental problems and to avoid complications or reaching irreversible points of environmental damage. Specifically, cities need to change the way they operate and do things. For example, ecologically and environmentally benign forms of transport need to be provided and utilized. Also, companies need to be better weighted regarding organic certifications. Furthermore, cities can use their purchasing power to require suppliers to meet sustainability requirements in their manufacturing, storage, disposal, and other operations. Cities should encourage the production and consumption of sustainable food, ensuring its authenticity.

In cities, buildings (residential, institutional, commercial, etc.) need to achieve more efficient use of energy, utilizing appropriate technologies for heating, cooling, lighting, refrigeration, and operation of appliances and electronic devices. Intelligent lighting systems are helpful as are appliances that are eco-friendly certified. Green buildings, including net-zero and net-positive energy buildings, need to become the norm and construction guides need to reflect this priority, in coming years. To achieve clean energy systems, multidisciplinary research teams are need. An example of the types of proposals in this area is the call for proposals of the Executive Agency for Small and Medium-sized Enterprises (EASME) of the European Commission, in which innovation proposals are solicited in the following areas (EASME, 2019):

1. Research and demonstration of more energy-efficient technologies and solutions.
2. Market uptake measures to remove market and governance barriers by addressing financing, regulations, and the improvement of skills and knowledge. This focuses on six areas:

    a. Consumers
    b. Buildings

c. Public authorities
d. Industry, products, and services
e. Heating and cooling
f. Innovative financing.

Cities are dynamic agents in a world increasingly in need of actions to address climate change. Through the types of global initiatives and agreements described, they should be able to achieve results that ensure the well-being of their citizens. More broadly, it is noted that all of the points raised above can help cities in following the guidelines implicit in the U.N. Sustainable Development Goals for 2015–2030 and achieving outcomes that support these goals.

Closing Remarks

Climate change needs the commitment of institutions, governments, and people to achieve the results necessary to avoid significantly damaging civilization and life on the planet. Agreements on climate change need to be disseminated so that the teaching and development of environmental projects are appropriately oriented towards these agreements, and contribute in measurable ways. Cities must generate policies, regulations, and efforts in relation to climate change. Regardless of policies that arise at the national level, cities have great opportunities to generate concrete actions and changes in the short term. Education is a crucial tool in relation to the dissemination of information regarding climate change.

References

Adaptation Fund. (2019). About the adaptation fund. Retrieved February 5, 2020, from https://www.adaptation-fund.org/about

Africa Renewable Energy Initiative. (2015). About AREI. Retrieved February 5, 2020, from http://www.arei.org/#about

Climate and Clean Air Coalition. (2019). Who we are. Retrieved February 5, 2020, from https://www.ccacoalition.org/en/content/who-we-are

EASME. (2019). 2020 call for proposals. Retrieved February 5, 2020, from https://ec.europa.eu/easme/en/section/horizon-2020-energy-efficiency/2020-call-proposals

GEF. (2019). Global environment facility: About us. Retrieved February 5, 2020, from https://www.thegef.org/about-us

Germanwatch. (2019). Climate risk index of the most affected countries in 2017. Retrieved February 5, 2020, from https://germanwatch.org/sites/germanwatch.org/files/Global%20Climate%20Risk%20Index%202019_2.pdf

GRID-Arendal. (2019). About us. Retrieved February 5, 2020, from http://www.grida.no

Houghton, J. T., Ding, Y., Griggs, D. J., Noguer, M., Linden, P. J., Dai, X., et al. (Eds.). (2001). *Climate change 2001: The scientific basis. A contribution of working group I to the third assessment report of the intergovernmental panel on climate change (IPCC)* (p. 881). Cambridge, UK: Cambridge University Press.

ICAP. (2019). About ICAP. Retrieved February 5, 2020, from https://icapcarbonaction.com/en/partnership/about

IETC. (2019). Who we are. Retrieved February 5, 2020, from https://www.unenvironment.org/ietc/who-we-are

Internet World Stats. (2019). Internet world stats. Retrieved February 5, 2020, from https://internetworldstats.com

IPCC. (2007). Climate change 2007. Climate change impacts, adaptation and vulnerability. *Working Group II*. IPCC, Geneva.

IUCN. (2019). Ocean warming. Retrieved February 5, 2020, from https://www.iucn.org/resources/issues-briefs/ocean-warming

Levy, B. S., Sidel, V. W., & Patz, J. A. (2017). Climate change and collective violence. *Annual Review of Public Health, 38*, 241–257.

Nakayama, M., Taafaki, I., Uakeia, T., Seru, J., McKay, Y., & Lajar, H. (2019). Influence of religion, culture and education on perception of climate change, and its implications. *Journal of Disaster Research, 14*(9), 1297–1302.

National Geographic. (2015). Sea level rise, explained. Retrieved February 5, 2020, from https://www.nationalgeographic.com/environment/global-warming/sea-level-rise

National Geographic. (2020). Greenhouse effect. Retrieved February 5, 2020, from https://www.nationalgeographic.org/article/greenhouse-effect

National Snow and Ice Data Center. (2019a). All about Sea ice. Retrieved February 5, 2020, from https://nsidc.org/cryosphere/seaice/index.html

National Snow and Ice Data Center. (2019b). Sea ice. Retrieved February 5, 2020, from https://nsidc.org/cryosphere/sotc/sea_ice.html

NOAA. (2019). Global climate change indicators. Retrieved February 5, 2020, from https://www.ncdc.noaa.gov/monitoring-references/faq/indicators.php

NSIDC. (2019a). Quick facts on ice sheets. Retrieved February 5, 2020, from https://nsidc.org/cryosphere/quickfacts/icesheets.html

NSIDC. (2019b). Glacier types: Retreating. Retrieved February 5, 2020, from https://nsidc.org/cryosphere/glaciers/gallery/retreating.html

PAGE. (2019). Who are we. Retrieved February 5, 2020, from https://www.un-page.org/about/who-are-we

Ritchie, H., & Roser, M. (2018). Urbanization. Retrieved February 5, 2020, from https://ourworldindata.org/urbanization

Sanz-Barbero, B., Linares, C., Vives-Cases, C., González, J. L., López-Ossorio, J. J., & Díaz, J. (2018). Heat wave and the risk of intimate partner violence. *Science of the Total Environment, 644*, 413–419.

Székely, M., Carletto, L., & Garami, A. (2015). The pathophysiology of heat exposure. *Temperature (Austin), 2*, 452.

Time. (2019). The world has an E-waste problem. Retrieved February 5, 2020, from https://time.com/5594380/world-electronic-waste-problem

UN Climate Resilience Initiative A2R. (2017). A2R. Retrieved February 5, 2020, from http://www.a2rinitiative.org

UNEP DTU. (2019). Who we are. Retrieved February 5, 2020, from https://unepdtu.org/who-we-are

UNEPFI. (2018). About UNEP FI. Retrieved February 5, 2020, from https://www.unepfi.org/about

UNEPFI. (2019). PDC. Retrieved February 5, 2020, from https://unepfi.org/pdc

UNESCO. (2009). The UNESCO Climate Change Initiative (2009). Retrieved February 5, 2020, from www.unesco.org/en/climatechange

UNFCC. (2019). Kyoto protocol. Retrieved February 5, 2020, from https://unfccc.int/process-and-meetings/the-kyoto-protocol/what-is-the-kyoto-protocol/kyoto-protocol-targets-for-the-first-commitment-period

WCMC. (2019). About us. Retrieved February 5, 2020, from https://www.unep-wcmc.org/about-us

World Meteorological Organization. (2017). The State of Greenhouse Gases in the Atmosphere Based on Global Observations through 2017. Retrieved February 5, 2020, from https://library.wmo.int/doc_num.php?explnum_id=5455

World Resources Institute (2020). What we do. Available in https://www.wri.org/our-work. Accessed 5 May, 2020

Xu, Z., Sheffield, P. E., Su, H., Wang, X., Bi, Y., & Tong, S. (2014). The impact of heat waves on children's health: A systematic review. *International Journal of Biometeorology, 58*, 239–247.

Part IV
Economic Sustainability in Cities

Chapter 18
Consumer Debt and Social Sustainability

Aldo Alvarez-Risco, Shyla Del-Aguila-Arcentales, Santiago Diaz-Risco, and M. Chandra Sekar

Abstract Globally, the financial behavior of consumer is very relevant for companies, governments, and of course the consumers themselves. Major problem currently is the lack of sufficient knowledge among consumers about financial planning, known as financial literacy. Lack of such knowledge leads families and citizens to opt for wrong decisions that are financially expensive, resulting in increased debt load and depletion of retirement funds.

Keywords Consumer · Consumer debt · Sustainability · Buyer · Trade

18.1 Management of the Money

Financial education of citizens will result in smart consumers making better financial decisions (Brown, Grigsby, Van Der Klaauw, Wen, & Zafar, 2016; Xiao & O'Neill, 2016) but the effects depend on the content of these programs (Brown et al., 2016); however, Berry, Karlan, and Pradhan (2018) concluded that the financial education lacks short-term effect and suggest that the programs should be designed

A. Alvarez-Risco (✉)
Facultad de Ciencias Administrativas y Recursos Humanos, Universidad de San Martín de Porres, Lima, Peru
e-mail: aldo.alvarez@redsaf.org

S. Del-Aguila-Arcentales
Escuela Nacional de Marina Mercante "Almirante Miguel Grau", Callao, Peru

Universidad Nacional de la Amazonia Peruana, Iquitos, Peru
e-mail: sdelaguila@enamm.edu.pe

S. Diaz-Risco
Centro de Fertilidad Cajamarca, Cajamarca, Peru

M. C. Sekar
University of Findlay, Findlay, OH, USA
e-mail: sekar@findlay.edu

A. Alvarez-Risco et al. (eds.), *Building Sustainable Cities*,
https://doi.org/10.1007/978-3-030-45533-0_18

based on the target population. The banks and other financial organization need to continuously update their information on consumer behaviors that determines how consumers save, borrow, spend, and invest. The same information is also important for consumer protection organization to protect consumers (and finally their families). This will enable the banks, insurance, and credit-card companies from offering products and services that people need (or that they believe they need) in the short-term but that could be "an inconvenient deal" for consumers in the long term. One needs to understand that the economy of a country may stagnate if large number of consumers has financial problems that result in delaying and curtailing of their spending. The global crisis in 2008 and 2013 are good examples of changes in consumer behavior resulting in decreased consumer spending and increased saving—that affected sales of products and services across a broad category around the world. Money management is an essential activity for the survival of an individual and family; without this—some people "make ends meet" without problematic debt, while others exacerbate their economic problems each month. A good management of the money is an essential element to reach the person reach their personal and familiar goals. For example, one person may have a goal to buy a computer for university classes, while the other person may be interested in playing off his credit card. Without appropriate money management skills one cannot meet these goals. Poor people put a lot of effort in solving urgent financial problems (Banerjee, 2016). Surprisingly, many people neither have appropriate training nor interest in personal money management. When it is measured how important are the financial issues, many consumers state that it is really relevant but in the same breath indicate that they do not spend time to read on this subject or follow the advice made by experts in this area. Expressions as "no worry, the magic card pay all" are sadly very common. There are some countries where governments provide free access to basic services like education and health, but then there are many others where no such services are provided. In later countries, consumers have to make their own investments. Global migrations have further burdened many societies and decreasing resources available for public services. Trust is an important determinant of spending, saving, and credit, especially for durable products and services. Consumers with higher trust in the future of the economy tend to use more credit, spend more, and save less. Negative media reports, falling stock prices, and opaque product information also affect trust in banks (Jansen, Mosch, & van der Cruijsen, 2015).

Inflation obviously is one of the main factors that affect consumer's decision. Money illusion happens when one is focusing on the nominal value (the numbers, example: 500 USD) of the cash at hand rather than the real value. Workers who are happy with a salary increase of 1.5%, while inflation is 2%, can be considered to be suffering from "money illusion" as actually their purchasing power has decreased by 0.5%. Similarly, savers may receive 2% interest while inflation is 3%. Again, gap in financial literacy is the cause that people are unable to follow simple financial transactions that affect their well-being. The banks and other financial actors play an important role in both advising consumers and selling those financial services. Trust in the agent dealing with consumer is needed for consumers to evaluate and accept these pieces of advice in transactions such as open a new bank account for his enterprise, buy life insurance for the family, or renew the insurance of the car. Financial behavior is determined by age, gender, education, family income, current debts, and others. Responsibility of payment differs widely across cultures. In some households, all earners pool their incomes and expenses determined by the single leader of the house, whereas in other households the wage earners keep their "own" money and manage their individual budgets. Situational factors often play a role on the type of expenses incurred by a given consumer. For example, a young student basically needs to be in charge of his/her budgets that deal with education (books, academic material, etc.), but when this student becomes a professional will incur new expenses in clothes, transportation, etc. (to buy a car). Probably his/her marital status changes with progression of time and it means more investment in a house/department, beds, kitchen, refrigerator, etc. According to Koschate-Fischer, Hoyer, Stokburger-Sauer, and Engling (2018), consumers can experience major life events (e.g., the birth of a child or retirement) that generate changes in the strategic plans for enterprises in terms of creating new market segments. Prior research has measured individual differences in self-control to predict consumer behavior (e.g., Tangney, Baumeister, & Boone, 2004). However, other research describes two alternative strategies for improving self-control: pre commitment (Ariely & Wertenbroch, 2002) and outcome elaboration (Nenkov, Inman, & Hulland, 2007), and they demonstrate how these strategies differentially affect three categories of consumers described by O'Donoghue and Rabin's (1999) consistent, naïve, and sophisticated individuals.

Citizens, who take financial risks with personal loans, credit, and mortgages and purchasing products with little information, and even accepting insecure jobs, are the ones who are more likely to be affected by financial crises in their country. Specifically, men are more willing to take risks compared to women (Montford & Goldsmith, 2016). Age is another main factor that influences risky behavior: older individuals are more risk averse than younger people (Brooks, Sangiorgi, Hillenbrand, & Money, 2018). Of all factors, lack of financial literacy may be the most critical factor that enables one to buy financial products that do not match with their real needs. The daily money management consists of such acts as paying the school and university bills, putting money in the savings account, getting cash from ATM, paying water and electricity bills. Paying methods may involve cheque, bitcoins, cash money, credit and debit card, and smartphone/Iphone. Now there are different ways for digital payments. Confidence in payment through the Internet has gradually increased; in some cases, even with monthly automatic debit (in the case of electricity or telephone service) or through Web purchases in low-cost companies. Runnemark, Hedman, and Xiao (2015) found that consumer spends more when paying with a credit card, as compared with paying cash. One reason paying with a credit card is a less visible "loss" because of the payment delay till the end of the month. A credit card seems to remove constraints or inhibitions for the consumers because future income can be used for different current payments. It is difficult for consumers to understand and follow successfully the payments that must be made monthly by the credit card. Even consumer paying with an app they have installed on their smartphone may have difficulty understanding the minimum payment amount due and what is the total final amount paid for the product or services. Bank further enhances their profits by imposing multiple types of fines and charges for their services. Societal changes have now made "living in debt" more acceptable, believing that "in order to have things in life, one must live in debt." Then, consumers justify this behavior by making an inventory of the goods they have obtained incurring large debts and they will tell themselves "if we had not gotten into debt, we would not have these goods today." The truth is that social sustainability is based on something more than just buying goods—it is actually the emotional cost paid by the family for buying that good or services, considering that it is bought through a large debt. So, what is the true value paid for a car if the payment leads to a family crisis due to the crippling monthly payment that had to be made over 3 years and the essentials that need to be sacrificed to make those payments? What is fun worth of a holiday trip that was taken to create nice family memory, but ends up creating multiple years of credit card payments and tension in the family?

Students are identified as vulnerable consumers where financial matters are concerned, as they are mostly young with little financial experience and low wage earners (BIS, 2013; Braunsberger, Lucas, & Roach, 2004; FCA, 2015; Rivard, 2002). However, the vulnerability of students whose education is financed by their families to study and the vulnerability of students who work and pay for their studies are quite different. The latter have more experience in financial management, spending carefully and in a planned manner. There are many financial self-help books and videos that provide guidance on money and debt management, but books and videos are only effective when the consumers are ready to modify their behavior. Knowledge alone is not enough to cause behavioral change, with the result that there are consumers with personal libraries on personal finances but they remain in high debt because of their continued fascination for products and services that are really unnecessary.

Social media has a big influence on expenditure and saving. Consumption by the friends shown in Instagram or Facebook creates a social pressure on consumers who think "they must match their friend's expenditures as they are from the same tribe and must enjoy similar things." As we can see, a consumer is led by their culture, their subculture, their social class, their group membership, their family, their personality, their psychological factors, etc. and is influenced by cultural trends as well as their social and societal environment (Ramya & Mohamed Ali, 2016). People with a relatively low income living in a "wealthy" social environment will spend a higher proportion of their income on consumption than people with a relative high income and living in the same social environment. Bankruptcy and divorce are seen as indicators of too high expenditure and too low saving of these households (Frank, Levine, & Dijk, 2014). One relevant concept is conspicuous consumption. Conspicuous consumption refers to the phenomenon where individuals purchase goods for signaling social status, rather than for its inherent functional value (Wu, Eisenegger, Sivanathan, Crockett, & Clark, 2017).

Correia, Kozak, and Reis (2016) have proposed some items for evaluating the conspicuous consumption

- Visit a place where I could maximize my experience
- The holidays are my gift, my way of pampering me
- I like adventures and risk situations
- I like to travel to places to where very few people travel
- When on vacation I like to have a good time without being censored
- When I choose my holiday destination I look for places where I will not be recognized
- I like to visit places with fame and prestige
- To take pictures and show them to others
- I always choose destinations that I have not visited yet
- To gain others' respect
- I like to visit fashionable destinations
- I always bought a T-shirt with the name of the place where I have been
- The holidays are a way of achieving my dreams
- Visiting a destination my family and friends recommend
- Going to places my friends want to go
- I like to visit places where my friends have never been
- I like to talk about my travel experiences with other people
- The holidays are a way of achieving my dreams
- I like to travel to places to where very few people travel
- To take pictures and show them to others
- Developing close friendships
- I always bought a T-shirt with the name of the place where I have been
- I like adventures and risk situations
- I like to visit familiar destinations where I can feel at home
- I always choose destinations that I have not visited yet
- Going to places my friends want to go
- Being entertained
- Visiting a place where I could spend good time with family
- I like to visit destinations recommended by my family and friend.

According to Gandelman (2016), the difference in national saving rates between Latin America and Caribbean economies can mainly be attributed to differences in saving behavior of the citizens and, to a lesser extent, to differences in the distribution of the population by education levels. Also Rocher and Stierle (2015) investigated what can contribute to explaining why household saving rate levels are persistently so dispersed across EU countries, ranging from 10% of household income in Romania to 16% in Germany in 2013. They found that traditional explanatory variables like income levels, age dependency, and uncertainty can explain more than half of the cross-section variance in saving rates; however, large unobserved country fixed effects (e.g., because of institutional differences) appear to be present and affect the outcomes. Beginning in 1990, the gap between the "standard of living" and real disposable incomes went negative with the resultant "gap" filled through the use of debt. Despite the temporary uptick in the savings rate following

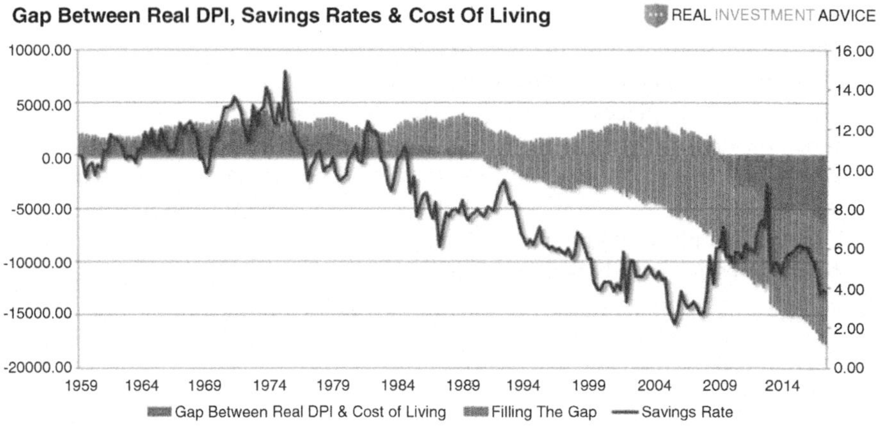

Fig. 18.1 Gap between real disposable personal incomes (DPI), savings rates, and cost of living. (Source: Real Investment Advice)

the 2008 financial crisis, the real cost of living continues to erode the middle class as is described in Fig. 18.1.

As there is great societal pressure to have a "good life," many consumers continue to spend beyond their means on "transient happiness" that in turn increases debts to unsustainable levels. The sustainability of the families, especially those with low-income, is in risk unless savings is promoted, starting from early school years and continuing through college (independently the career students follow).

18.2 Debts and Sustainability

Saving requires the ability to postpone gratification. Saving is not using the present income, wealth, or budget for current expenditures, but refraining from spending this income, wealth, or budget at this moment so it is available at a later date. Uncertainty about the future is the main reason for saving. In this way, saving may be simply not spending a part of the available income during a certain period, because the income is too high for what consumers want to buy or the desired products may not be available yet, but may become available in the future. Not spending all of one's income is called residual saving. It is assumed that most people prefer spending money now (immediate gratification of needs) rather than in the future (gratification of needs later). In this line the marketing companies.

The following types of saving may be recognized:

1. Putting money in a savings box (without interest), as children are taught to do.
2. Residual saving: not spending a residual part of the available budget.
3. Discretionary saving: making a purposeful decision to put money into a savings account.

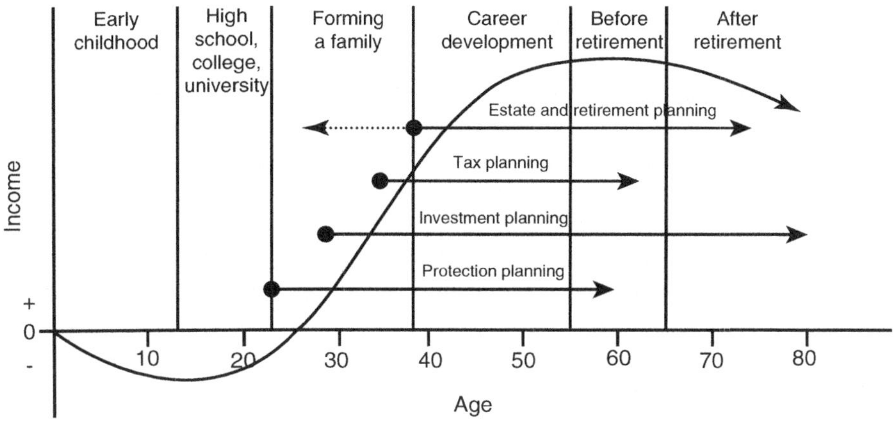

Fig. 18.2 Personal financial life cycle

4. Repaying debt and mortgage: "savings afterwards" in order to decrease debt.
5. Buying products/services on sale, for a temporary lower price, in order to save money (some time prices are not really lower, marketing tricks).
6. Buying more economical services/products that are cheaper in use and maintenance.

It is important to have a good level of financial literacy because all consumers will walk through the personal financial life cycle as it is mentioned in Fig. 18.2.

Globally in the last ten years, consumer credit is accepted as a way to finance consumption. Mau (2015) describes the new culture of indebtedness. This culture is based on the following belief that "if everyone is using credit, it must be okay." Even as possession of certain credit cards gives an illusion of social status to the holder of the card, microcredit refers to small loans extended to impoverished borrowers who usually lack steady employment, lack collateral, and a verifiable credit history. These poor borrowers frequently live in developing countries and have no access to regular credit from big financial organizations. These poor people are usually victims of "loan sharks" (predatory lending). In Latin America, this type of microcredit has been linked to delinquent actors who deliver microcredits and charge weekly. When the person cannot hit they are attacked or even killed. Despite this kind of risk, microcredit is designed to support entrepreneurship, to alleviate poverty, and also to empower people and uplift entire communities.

Recently, effectiveness of microcredit has been evaluated in different scenarios—enterprises in UK (McHugh, Baker, & Donaldson, 2019), farmers in Indonesia (Mariyono, 2019) and China (Qin, Wachenheim, Wang, & Zheng, 2019), non-profit institution in Pakistan (Mahmud & Wahhaj, 2019), and E-business enterprise in China (Zhang, Daim, & Zhang, 2018). Impact of microcredit lending is measured in Bangladesh (Hussain, Nargis, Ashiquzzaman, & Khalil, 2017), self-sustainability of women's group supported by government (Imran, Ślusarczyk, & Haque, 2019), microcredit repayment in Portugal (Mota, Moreira, & Brandão, 2018) and China (Ding & Abdulai, 2018).

18.3 Families and Debts

There is a great risk for families who go into debt to match the standard of living of their peers and neighbors. Is debt bad? Depends on the reason for indebtedness. For example, debt to take a vacation is bad, but debt for an education that will cause a significant increase in lifetime earning is fine. The social sustainability that is expected in a society is, as we have seen, influenced by all the actors (consumers, social environment, banks, and companies) and has its own dynamics. Can consumers do something to keep their spending level under control? Can the government do anything to get its citizens to have a controlled level of debt? Can companies do something that want to ensure that their customers continue to buy more products because they can have a proper management of their debts?

When we talk about sustainability, we often think about environmental issues and also about economic aspects such as the sustainability of companies, but little is said and done to ensure that families can ensure their sustainability, that is, keep healthy emotionally and economically. We have been able to see how much debt can affect the individual and family level. What are the strategies that the countries have proposed to take charge of this social issue?

Different strategies have been reported to increase the level of financial literacy in different populations with different outcomes (Cordero & Pedraja, 2018; Fabris & Lubурić, 2016; Knoote, Partington, & Penner, 2016; Opletalová, 2015; Ribeiro & Soares, 2017; Urban, Schmeiser, Collins, & Brown, 2018). Even there is available historical research of microcredit in Italy (Gatto, 2018) and India (Parekh & Ashta, 2018).

For example, per capita debt in Lima Peru increased five-fold between 2001 and 2017. Second, 80% of Peruvians spend more than their income. Third, 29% of families have debts that consume in monthly installments greater than 30% of their monthly incomes. Fourth, 12% of consumer loans are in default, which means around 590,000 loans. It is unnerving to see a growth rate of 5% annually in personal debt. We need strategies to solve such deep social problems (Gestión, 2018). For that, the SDG 1 called “no poverty” needs to be addressed on many fronts and one main way is to educate the population enabling them to avoid or reduce conspicuous consumption to the maximum. Likewise, the SDG 8 that proposes economic growth can only be had by financially literate citizens, but the higher income generated will lead to greater indebtedness. What else can be done around others SDG linked to social sustainability? There is urgent need to start developing strategic plans to build sustainable cities.

Closing Remarks

Contributing to the fulfillment of Sustainable Development Goals requires planned actions from governments, institutions, and citizens. Ensuring social sustainability is a relevant aspect that should be part of the educational agenda at all levels, starting from primary school to college. Financial education to ensure social sustainability must be a mandatory as without it, we will be training citizens, who in spite of higher incomes spend their lifetime in debt.

References

Ariely, D., & Wertenbroch, K. (2002). Procrastination, deadlines, and performance: Self-control by precommitment. *Psychological Science, 13*(3), 219–224.

Banerjee, M. M. (2016). "We routinely borrow to survive": Exploring the financial capability of income-poor people in India. *Social Work, 61*(4), 349–358.

Berry, J., Karlan, D., & Pradhan, M. (2018). The impact of financial education for youth in Ghana. *World Development, 102*, 71–89.

BIS. (2013). Student income and expenditure survey 2011/12. Business innovation and skills research. Paper Number 115. Retrieved February 5, 2020, from https://www.gov.uk/government/uploads/system/uploads/attachment_data/file/301467/bis-14-723-student-income-expenditure-survey-2011-12.pdf

Braunsberger, K., Lucas, L., & Roach, D. (2004). The effectiveness of credit-card regulation and vulnerable customers. *Journal of Services Marketing, 18*(5), 358–370.

Brooks, C., Sangiorgi, I., Hillenbrand, C., & Money, K. (2018). Why are older investors less willing to take financial risks? *International Review of Financial Analysis, 56*, 52–72.

Brown, M., Grigsby, J., Van Der Klaauw, W., Wen, J., & Zafar, B. (2016). Financial education and the debt behavior of the young. *The Review of Financial Studies, 29*(9), 2490–2522.

Cordero, J. M., & Pedraja, F. (2018). The effect of financial education training on the financial literacy of Spanish students in PISA. *Applied Economics*, 1–15.

Correia, A., Kozak, M., & Reis, H. (2016). Conspicuous consumption of the elite: Social and self-congruity in tourism choices. *Journal of Travel Research, 55*(6), 738–750.

Ding, Z., & Abdulai, A. (2018). Smallholder preferences and willingness-to-pay measures for microcredit: Evidence from Sichuan province in China. *China Agricultural Economic Review, 10*(3), 462–481.

Fabris, N., & Luburić, R. (2016). Financial education of children and youth. *Journal of Central Banking Theory and Practice, 5*(2), 65–79.

FCA. (2015). Consumer vulnerability. Financial conduct authority. Occasional Paper Number 8. Retrieved February 5, 2020, from www.fca.org.uk

Frank, R. H., Levine, A. S., & Dijk, O. (2014). Expenditure cascades. *Review of Behavioral Economics, 1*(1–2), 55–73.

Gandelman, N. (2016). A comparison of saving rates: Microdata evidence from seventeen Latin American and Caribbean countries. *Economia, 16*(2), 201–258.

Gatto, A. (2018). Historical roots of microcredit and usury: The role of Monti di Pietà in Italy and in the kingdom of Naples in XV–XX centuries. *Journal of International Development, 30*(5), 911–914.

Gestión. (2018). Endeudamiento de los hogares y morosidad. Retrieved February 5, 2020, from https://gestion.pe/blog/economiaparatodos/2018/09/endeudamiento-de-los-hogares-y-morosidad.html

Hussain, A. K. M., Nargis, N., Ashiquzzaman, S. M., & Khalil, F. (2017). *The employment impact of microcredit program participation in Bangladesh: Evidence from a longitudinal household survey (no. 59)*. GLO Discussion Paper.

Imran, M., Ślusarczyk, B., & Haque, A. U. (2019). The moderating role of Malaysian government in microcredit organization and quality of women self-sustainability. *Calitatea, 20*(S1), 481.

Jansen, D. J., Mosch, R. H., & van der Cruijsen, C. A. (2015). When does the general public lose trust in banks? *Journal of Financial Services Research, 48*(2), 127–141.

Knoote, F. E., Partington, G., & Penner, J. (2016). Children and youth as economic citizens: Working towards an integrated financial education approach. In *International handbook of financial literacy* (pp. 193–211). Singapore: Springer.

Koschate-Fischer, N., Hoyer, W. D., Stokburger-Sauer, N. E., & Engling, J. (2018). Do life events always lead to change in purchase? The mediating role of change in consumer innovativeness, the variety seeking tendency, and price consciousness. *Journal of the Academy of Marketing Science, 46*(3), 516–536.

Mahmud, M., & Wahhaj, Z. (2019). Charitable giving or signalling? Voluntary contributions by microcredit borrowers in Pakistan. *Journal of Economic Behavior & Organization, 158*, 394–415.

Mariyono, J. (2019). Microcredit and technology adoption: Sustained pathways to improve farmers' prosperity in Indonesia. *Agricultural Finance Review, 79*(1), 85–106.

Mau, S. (2015). The new culture of indebtedness. In *Inequality, marketization and the majority class: Why did the European middle classes accept neo-liberalism?* (pp. 65–72). London: Palgrave Pivot.

McHugh, N., Baker, R., & Donaldson, C. (2019). Microcredit for enterprise in the UK as an 'alternative' economic space. *Geoforum, 100*, 80–88.

Montford, W., & Goldsmith, R. E. (2016). How gender and financial self-efficacy influence investment risk taking. *International Journal of Consumer Studies, 40*(1), 101–106.

Mota, J., Moreira, A. C., & Brandão, C. (2018). Determinants of microcredit repayment in Portugal: Analysis of borrowers, loans and business projects. *Portuguese Economic Journal, 17*(3), 141–171.

Nenkov, G. Y., Inman, J. J., & Hulland, J. (2007). Considering the future: The conceptualization and measurement of elaboration on potential outcomes. *Journal of Consumer Research, 35*(1), 126–141.

O'Donoghue, T., & Rabin, M. (1999). Doing it now or later. *American Economic Review, 89*(1), 103–124.

Opletalová, A. (2015). Financial education and financial literacy in the Czech education system. *Procedia-Social and Behavioral Sciences, 171*, 1176–1184.

Parekh, N., & Ashta, A. (2018). An institutional logics perspective to evolution of Indian microcredit business models. *Strategic Change, 27*(4), 313–327.

Qin, M., Wachenheim, C. J., Wang, Z., & Zheng, S. (2019). Factors affecting Chinese farmers' microcredit participation. *Agricultural Finance Review, 79*(1), 48–59.

Ramya, N., & Mohamed Ali, S. A. (2016). Factors affecting consumer buying behavior. *International Journal of Applied Research, 2*(10), 76–80.

Ribeiro, R. B., & Soares, I. (2017). Insights and directions for sociological approaches to saving: The case of a Financial Education Programme for children in Portugal. *Journal of Consumer Culture, 17*(3), 845–863.

Rivard, N. (2002). Money management 101 needed: Credit card use is up among students; financial literacy down. *University Business, 5*(9), 17.

Rocher, S., & Stierle, M. (2015). *Household saving rates in the EU: Why do they differ so much?* (no. 2015.01).

Runnemark, E., Hedman, J., & Xiao, X. (2015). Do consumers pay more using debit cards than cash? *Electronic Commerce Research and Applications, 14*(5), 285–291.

Tangney, J. P., Baumeister, R. F., & Boone, A. L. (2004). High self-control predicts good adjustment, less pathology, better grades, and interpersonal success. *Journal of Personality, 72*(2), 271–324.

Urban, C., Schmeiser, M., Collins, J. M., & Brown, A. (2018). The effects of high school personal financial education policies on financial behavior. *Economics of Education Review*. https://doi.org/10.1016/j.econedurev.2018.03.006

Wu, Y., Eisenegger, C., Sivanathan, N., Crockett, M. J., & Clark, L. (2017). The role of social status and testosterone in human conspicuous consumption. *Scientific Reports, 7*(1), 11803.

Xiao, J. J., & O'Neill, B. (2016). Consumer financial education and financial capability. *International Journal of Consumer Studies, 40*(6), 712–721.

Zhang, W., Daim, T., & Zhang, Q. (2018). Understanding the disruptive business model innovation of E-business microcredit: A comparative case study in China. *Technology Analysis & Strategic Management, 30*(7), 765–777.

Chapter 19
Entrepreneurship for Sustainable Cities

Aldo Alvarez-Risco, Dennis Lopez-Odar, Raquel Chafloque-Cespedes, and M. Chandra Sekar

Abstract This study analyzes the effect of the university ecosystem, self-efficacy of the students, social entrepreneurial perception of the entrepreneur on the social entrepreneurial intention. Second-generation structural equation method (PLS-SEM) was used to analyze the results, specifically, the SmartPLS 3.2.7 software applied to data on 6425 Peruvian university students. The validity of an explanatory model of entrepreneurial intention was evaluated based on personal and situational determinants. The main contribution of this study is to show that 63% of students had an intention of social entrepreneurship; no significant differences were found in the sex of the students with respect to the intention of entrepreneurship. Only 18% were participating in a social enterprise. It was evidenced that the entrepreneurial support provided by the university ecosystem does not directly influence the social entrepreneurial intention ($\beta = 0.038$, $p = 0.001$), nor in the self-efficacy developed by the students ($\beta = -0.045$, $p = 0.551$). However, it directly influences the entrepreneurial orientation ($\beta = -0.185$, $p = 0.001$) and the positive perception of the social entrepreneur ($\beta = -0.241$, $p = 0.001$). Through these variables (which assume a mediating role), the university ecosystem exerts an indirect influence on self-efficacy and later on entrepreneurial intention. The results are useful to design models of development of social enterprises in the universities taking into account all the factors reported as important by the students.

Keywords Social entrepreneurship · Self-efficacy · Entrepreneurial orientation · PLS · University students

A. Alvarez-Risco (✉) · D. Lopez-Odar · R. Chafloque-Cespedes
Facultad de Ciencias Administrativas y Recursos Humanos, Universidad de San Martín de Porres, Lima, Peru
e-mail: aldo.alvarez@redsaf.org; dlopezo@usmp.pe; mchafloquec@usmp.pe

M. C. Sekar
University of Findlay, Findlay, OH, USA
e-mail: sekar@findlay.edu

A. Alvarez-Risco et al. (eds.), *Building Sustainable Cities*,
https://doi.org/10.1007/978-3-030-45533-0_19

19.1 Entrepreneur

The term entrepreneur was initially popularized by the founder of Ashoka, which is a non-governmental organization that globally supports social entrepreneurship (Drayton, 2002). Current increased interest in this area is reflected by the greater number of conferences held, as well as increased number undergraduate and graduate courses offered at the universities, together with current peer-reviewed publications and academic journals specialized in social entrepreneurship. There are several causes for the rise of social entrepreneurship. According to Doherty, Haugh, and Lyon (2014)—the main cause has been the failure of the role played by philanthropic organizations that make donations. It is also necessary to remember the role of failure of governments (as American scholars point out) and markets (as noted by European academics) in meeting the basic needs of individual groups in areas such employability, health, clean air and drinking water for consumption.

The recent global financial crisis and increased environmental damage caused by the current models of economic development in our societies are a clear case to reevaluate our current system. In the last 10 years, there has been a significant growth in business activities focusing increasingly on achieving social and environmental welfare. On the academic level, social entrepreneurship is not new, since its concepts and importance have been described for over two decades—Shapero and Sokol (1982), Brinckerhoff (2000) and Thompson, Alvy, and Lees (2000).

Forbes (2015) points out five reasons why social entrepreneurship is the new business model to follow:

1. It connects you with your life's purpose
2. It keeps you motivated
3. It brings you lasting happiness
4. It helps others discover their life's purpose
5. It is what today consumers want

It is relevant that 94% of young people today wish to produce a positive impact in the world while generating monetary gains (Achieve, 2014). Due to this, social entrepreneurship is presented as a new formula to achieve success by combining economic gains that can also produce collective benefit to the society. We are talking about for-profit and self-financing companies that also have the mission to address global problems in a local context such as hunger, improve health outcomes, educating their citizens and generate strategies to address climate change. To achieve these goals—these companies can finance specific programs, partner with governmental or private organizations, even combining several of these forms and generating a series of activities at local, national and even international level.

Peru's economy has been growing rapidly achieving annual growth of 4.8% in the period 2001–2006, 6.9% in the period 2006–2011 and 4.2% in the period 2011–2016 (Gestion, 2017). However, this growth has resulted in the depletion of resources and has increased risk of natural disasters from the consequences of climate change, such as the damage caused by the El Niño Phenomenon. Peru is recognized as an entrepreneurial country by the GEM report 2017–2018, which indicates that the entrepreneurship opportunity in Peru in the next 6 months is 56.6% compared to 46.2% of Latin America and the Caribbean.

In order for this business dynamic to produce sustainable changes in relation to social entrepreneurship, it is necessary to have professionals whose university education imbibes them with such orientation and therefore it is crucial to know at present the current opinion of the students regarding social entrepreneurship.

For this reason, in this study, we seek to know the opinion of students of administrative sciences and human resources regarding their orientation and entrepreneurial intention. In addition, we seek to know what is the influence of the university environment on the development of social entrepreneurship by students. Likewise, the aim is to test a conceptual model that has been developed to explain which variables explain the orientation and entrepreneurial intention.

19.2 Entrepreneurship in Universities

19.2.1 University Ecosystem

The education system contribute to the development of entrepreneurs capable of contributing to the economic development of countries through self-employment (Rueda, Fernández, & Herreo, 2012) and by adding social capital the impact can be multiplied. The university environment has been analyzed by the bottom-up approach of the Institutional Theory, which considers that bottom-up evolution can lead to institutional perfection (Easterly, 2008). That is how Universities can promote intentions and behaviors of social entrepreneurship in university students. Institutions by educating on forms of human interaction provide "rules" for operation for that community. Both formal (regulations, rules, contracts, etc.) and informal institutions (attitudes, values, culture, etc.) play a role in this process (North, 1990, page 3). Manolova, Eunni, and Gyoshev (2008) investigated the influence of the institutional environment in the development of enterprises from three dimensions—regulatory, cognitive and normative in emerging countries of Eastern Europe. In a sample size of 254 business students from Hungary, Bulgaria and Latvia, it was evident that in these three countries all of the above three components were unfavorable, as they did not encourage entrepreneurship. In Hungary and Latvia, lack of knowledge and skills necessary to participate in ventures contributed to low entrepreneurship; in Hungary, social attitudes that contributed to preventing supporting policies that promoted entrepreneurial skills, while in Bulgaria, regulations or lack of social approval was the major impediment.

Entrepreneurial education can improve increased awareness of business opportunities and to understand how such ideas that initiate an enterprise are developed. In the study of Hussain, Mohammad, and Ahmed (2016) carried out with 499 senior students from nine universities in Sindh and Pakistan, a positive effect was noted between the business education and the practical exposure of the entrepreneurial initiative from the development of entrepreneurial alert with the absence of mediators ($\beta = 0.48$, $p < 0.01$), as well as the effect of entrepreneurial education in the detection and search of opportunities ($\beta = 0.46$, $p < 0.01$), in the capacity of association and connection ($\beta = 0.54$, $p < 0.01$), and with the skills of evaluation and critical judgment ($\beta = 0.31$, $p < 0.01$). The authors recommend working on business programs that consider the development of skills that allow recognition of business opportunities, their development and evaluation of progress, in order to efficiently utilize the specific human capital of students using their knowledge, skills and experiences in search for opportunities and their ability to gather and process information. Therefore, training courses must develop mental schemata with theoretical and practical insights, in order to maintain ample information capital, to identify, associate and evaluate their importance in the face of the perception of possible opportunities. Likewise, in the study of Yeng, Selvarajah, and Meyer (2011), it was found that the entrepreneurial intention of Malaysian students has a significant relationship with educational factors such as study programs (0.118, $p < 0.01$), perception of entrepreneurship (0.239, $p < 0.01$), the entrepreneurship promoter role of the universities (0.302, $p < 0.01$) and the content of the entrepreneurship curriculum (0.319, $p < 0.01$). Indeed, the authors conclude that university entrepreneurship ecosystems play an important role in promoting the entrepreneurial spirit among students.

19.2.2 Entrepreneurial Orientation

The factors of entrepreneurial universities are incipient in Latin America, while in Europe and the United States the major entrepreneurial universities are already in existence; the importance of their analysis based on the institutional economic theory lies in their favorable or unfavorable effect for growth and development of an entrepreneurial university (Madrigal & Santamaria, 2015). The literature has shown a positive influence of business education on the skills and knowledge of students (Guerrero, Urbano, Cunningham, & Gajón, 2017). A study in the United Kingdom with 2000 students demonstrated that higher quality organizational growth happens, if the graduating students received educational resources, developed skills and experiences to maintain a sustainable growth of companies they are leading (Galloway & Brown, 2002). For the study, the main formal and informal factors of university entrepreneurial environments have been proposed, which according to North (1990) are the rules that can favor or limit the development of a business.

Entrepreneurial orientation corresponds to entrepreneurial culture as a category of entrepreneurship (Ireland& Webb, 2009). Within the university academic context, the role of universities is of great importance and demonstrates universities intention by its offering of specialized programs and courses (Comisión de las Naciones Europeas, 2006). Courses offered provide the foundation for change in attitudes of students compared to traditional educational system, with emphasis on the training of employees and employers (Moriano, Palací, & Morales, 2006a). Entrepreneurial education in universities has a positive influence on the intentions and behaviors of business school students, through the creation of new business, consulting, teaching and research in its thematic area, since the entrepreneurial spirit can be motivated by certain factors through education (Kolvereid & Moen, 1997). Likewise, studies have shown that entrepreneurship education programs improve student learning about entrepreneurial alertness, which can be of great importance in terms of the development of human capital. So, entrepreneurship programs must include in their course design—reflections, readings and discussion on capacity that allow the student to detect and identify information about need in the market to start an entrepreneurship (Chang, Liu, & Chiang, 2014).

Factor influencing "intention to undertake" is dependent on gender. For women—attitude, self-efficacy, subjective norms and having a family business are major influencing factor, while in men only self-efficacy is a significant component (Da Fonseca, 2015). The training for the development of entrepreneurial attitudes is critical, since entrepreneurial spirit is possible and desired in economies with greater economy, since they not only provide a good structure, education and efficiency factors but also by the conception of a positive perception of undertaking by the motivation to know well-known entrepreneurs that come from these economies (Kelley, Bosma, & Amorós, 2011).

19.2.3 Self-Efficacy

Self-efficacy is important in various disciplines that analyze the cognitive behavioral relationship, such as—psychology, education, sociology, communication, political science, business and law (Bandura, 1986) to be considered as a resource that not only favors the initiative to perform a task but also to encourage identification of new opportunities (Navarro, 2014). Albert Bandura developed the term self-efficacy, based on his Theory of social learning—as one of the three constituent functions which guides and leads to the action of a task, where self-efficacy can be a seed of motivation. Therefore, it is necessary to remember that perceived self-efficacy must be evaluated considering a specific performance in a specific activity such as entrepreneurship (Bandura, 2012). Indeed, perceived self-efficacy is not a global personality trait, but a differentiating element from self-beliefs to different areas of functioning (Bandura, 2012).

Consequently, Bandura (2012) developed a guide for the solid conceptual analysis of the types of self-efficacy applied to different spheres of functioning and stated that it is likely that high school students, unlike lower grades, can develop high perceived self-efficacy by the diversity of subjects and curricular areas. Likewise, a difference in levels of self-efficacy of people is observed when developing a direct experience, it is related to a solid development of self-efficacy, assuming a high cost of laboriousness and an expensive trial error (Bandura, 2012). Studying perceived effectiveness is of great importance, due to the possible debilitating or null effects, during the process of testing beliefs in their own capacities, due to the multidetermination and the contingent nature of daily life, which conditions the behavior of people, in fact, these endogenous factors distort the self-confidence of the capacity and the action. Self-efficacy is considered as a positive element for entrepreneurship (Navarro, 2014) since it results in the perception of having the ability to undertake activities that require knowledge, skills and experience. At the global level, according to the GEM Report (2018), Latin America and the Caribbean is the continent with the highest level of self-efficacy with 58.8% and Europe with 43.4%.

19.2.4 Social Entrepreneurial Perception

The perception of social entrepreneurs refers to the favorable or unfavorable attitude assumed by students in relation to the characteristic features or behavior of social entrepreneurs. When considering the expectation-value model of Fishbein and Ajzen (1973)—attitude is the result of the belief regarding the behavior and its consequences, which when evaluated can have a positive or negative value. In the study by Liñán and Chen (2009), the model of entrepreneurial behavior follows a sequence of beliefs, attitudes and intentions, where the intention to undertake is influenced by the attitudes that one has about entrepreneurship and beliefs or perceptions about entrepreneurial behavior (Ajzen, 1985).

Under this perspective, a positive belief in the behavior of social entrepreneurs will have a positive effect on the intention to develop social enterprises, motivation and the strength of the intention of behavior (Da Fonseca, 2015), since a favorable attitude is associated with a greater predisposition to carry out a behavior (Moriano, Palací, & Morales, 2006b). Cultural and social differences can influence perception of people; however, in a study of 519 Taiwanese and Spanish students (Liñán & Chen, 2009) on the perception of entrepreneurial intention, no significant differences were found. Considering that the first step in the mental process starts with the subjective norms as a first filter for the formation of perception on the perceived behavioral Control and personal attitudes on entrepreneurship. For a change of perception of the people, a change in their beliefs is required and at the micro institutional level, the universities must design and carry out entrepreneurship courses to train the capacities and entrepreneurial skills of the students, and a strengthening of the business self-efficacy (Chen, Gene, & Crick, 1998). In effect, people's beliefs about its effectiveness affect the type and level of objectives that were established for themselves and the strength of their commitment to them (Bandura, 2012).

19.2.5 Social Entrepreneurial Intention

Entrepreneurship is developed in productive units of goods and services, while social enterprises are more related to projects on social issues that involve activities in communication, training, medicine, etc. (Duarte & Ruiz, 2009). The intention is influenced by personal and social factors. At a personal level, it is determined by attitudes toward entrepreneurship, influenced by beliefs about entrepreneurship and an evaluation of them. The studies on the entrepreneurial intention within the Theoretical Framework of Planned Action of Ajzen (1991) in different countries have shown how the influence of social and personal factors can explain the intention to undertake. According to the GEM model, external factors such as the opportunities offered by the market or the capacities that potential entrepreneurs believe they possess are enough criteria to realize and ensure the success of a business. These entrepreneurs put a greater emphasis on entrepreneurial spirit through the formation of favorable attitudes in relation to the positive image of entrepreneurs, which can influence the decision to start a new business. Indeed, in the last Report of the GEM, about the intention to undertake in the next 3 years, Africa is the continent with the greatest intention to undertake (41.6%), followed by Latin America and the Caribbean (31.9%), Asia (24.3%). %), North America (12.9%) and Europe (11.9%) (Serida, Guerrero, Alzamora, Borda, & Morales, 2017).

The intention is a predictor of behavior (Ajzen, 1991), but you cannot establish a direct relationship between the two (Da Fonseca, 2015) since there are other factors such as time, by means of which despite having a good attention, the behavior is not specified. Intentions are like orders or instructions that people give themselves, through decisions to perform an action in particular and are identified in statements such as "I try to do …", "I plan to do …" or "I will…" (Triandis, 1979, cited in, Da Fonseca, 2015).

19.3 Measurement of Entrepreneurship: Case Peru

According to the above, the present research proposes a multivariate explanatory model according to Fig.19.1.

19.3.1 Methodology

19.3.1.1 Sample

A sample of 3407 students (55.5%) is analyzed in universities of the coast, 1890 (30%) students in universities of the mountain range and finally 930 (14.5%) students of universities of the forest, with average ages of 20.82 (DE = 2.1), 20.66

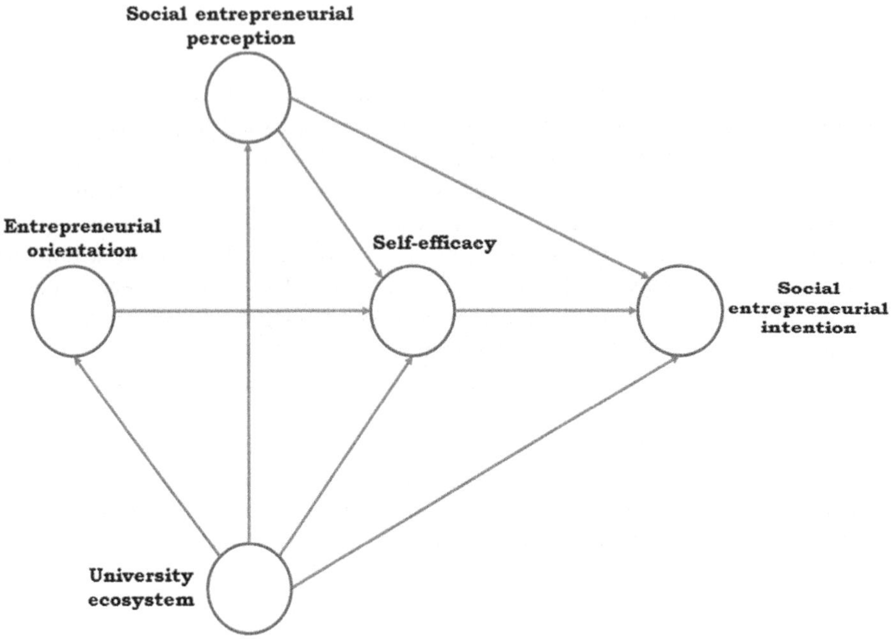

Fig. 19.1 Proposed research model

(SD = 2.2) and 20.83 (SD = 2.0), respectively. The sample was represented by 49.1% of male students and 50.9% of female students.

The sample of the coastal region is represented by 50.9% and 49.1% of male and female students, respectively, where the majority of students entered in 2014 (SD = 1.73), with a weighted average of 13.31 (SD = 1.64), in this region the majority of the sample studied business sciences (73.5%) and in less engineering (26.5%). For the Sierra region, the picture is different from that of the coastal region, the largest number of students is represented by female students (women 53.9% vs. men 46%), the majority of students entered in 2014 and the weighted average of the students is 13.63 (SD = 1.77), the area of knowledge with the highest number of students is business sciences (business sciences 65.8% vs engineering 34.2%). In the jungle region, the sample of students is represented by 49% of male students and 51% of female students, as in the other two regions most of the students surveyed entered in 2014; however, the average weighted of the students ascended to 14.74 (SD = 1.68), and the majority of the students of the jungle region have chosen to study business sciences (79.5%) (Table19.1).

Table 19.1 Reliability and construct validity of the scales using SEM-PLS

Scale and indicators		Weighs	Cronbach's alpha	FCI	AVE
University ecosystem	University social entrepreneurs are provided with financial, business and/or technical advice.	0.761	0.896	0.916	0.554
	It promotes networks of contact between university social entrepreneurs and investors.	0.784			
	A favorable environment for the development of social enterprises is fostered among students and teachers.	0.811			
	Social entrepreneurial ideas are promoted through contests, fairs or contests.	0.783			
	There are support programs for the creation of social enterprises (raising seed capital, incubators, etc.).	0.772			
	There is the support of senior management and authorities to start social enterprises.	0.788			
	There are subjects or courses related to social entrepreneurship.	0.717			
	There is practical training in social enterprises (realization of projects, business plans, etc.).	0.768			
	They have specialized offices that support and provide advice for the development of social enterprises.	0.759			
Self-efficacy	Creating and maintaining a social enterprise is a task that I can carry out.	0.784	0.822	0.883	0.652
	I have the necessary knowledge to develop a social enterprise.	0.804			
	I have enough skills to develop a social enterprise.	0.846			
	I believe that in the future I will be able to develop a successful social enterprise.	0.797			
Social entrepreneurial intention	I plan to develop a business initiative that addresses the social problems of my region or community.	0.844	0.848	0.898	0.687
	I recommend my colleagues develop business initiatives that seek to solve the social problems of a community.	0.828			
	My future initiatives will prioritize social benefits more than financial ones.	0.824			
	If I had the opportunity and the resources, I would carry out a social enterprise.	0.821			

Source: 3407 questionnaires of university students

19.3.1.2 Measurement of Variables

For the university ecosystem, the scale measures the frequency of access to resources and training programs that drive the development of social enterprises in students. For this purpose, a one-dimensional scale has been designed, consisting of nine items adapted from the instruments used by Franke and Lüthje (2004) and Guerrero et al., 2017. The type of response is presented on a Likert scale of 5 points (1 = Strongly disagree to 5 = Strongly agree).

For social entrepreneurial perception, we used a scale composed of six items grouped in two dimensions (innovation and sustainability), which have been developed from the adaptation of the instrument developed by Nga and Shamuganathan (2010). The items are completed using a Likert scale of 5 points, where 1 means totally disagree and 5 completely agrees.

For entrepreneurial orientation, a one-dimensional scale has been designed, consisting of nine items adapted from the instruments used by Franke and Lüthje (2004) and Guerrero et al., 2017. The type of response is presented on a Likert scale of 5 points (1 = Strongly disagree to 5 = Strongly agree).

For self-efficacy, a one-dimensional scale was designed, consisting of four items, adapted from the instruments proposed by Jung, Ehrlich, De Noble, and Baik (2001) and Soria-Barreto, Zúñiga-Jara, and Ruiz Campo (2016) to evaluate behavioral intention. Like the previous scales, the items are completed by responding to a Likert-type response format of 5 points (1 = Totally disagree to 5 = Completely agree). It should be noted that the evaluation of self-efficacy responds to the adoption of Bandura's cognitive social theory (1986) as the basis of the theoretical model proposed in this study. Self-efficacy is defined as the perceived ability to perform a certain activity (Krueger, Reilly, & Carsrud, 2000) and is an important predictor of entrepreneurial behavior (Brockhaus & Horwitz, 1986). For social entrepreneurial intention, a scale of social entrepreneurship was designed on the basis of the instruments developed by Liñán and Chen (2009) and Moriano et al. (2006b) to assess the entrepreneurial intention of business or commercial. The scale is one-dimensional and consists of four items, which assesses the student's perception of the possibility of developing a social enterprise. The respondents answered on a Likert scale of 5 points, where 1 means totally disagree and 5 completely agree.

19.3.2 Results/Findings

19.3.2.1 Analysis of the Measurement Model

Reliability and Validity

To determine the reliability of the scales, their internal consistency was analyzed through Cronbach's alpha coefficient and composite reliability. In the case of scales of entrepreneurial orientation and perception of the social entrepreneur, a first- and second-order analysis was carried out.

Table 19.2 Discriminant validity of scales: social entrepreneurial intention, self-efficacy and university ecosystem

	Self-efficacy	University ecosystem	Social entrepreneurial intention
Self-efficacy	*(0.772)*		
University ecosystem	0.194	*(0.762)*	
Social entrepreneurial intention	0.693	0.190	*(0.775)*

Source: 3407 questionnaires of university students

Reliability and Validity of Scales with Structural Equations of Variance with Partial Least Squares (SEM-PLS)

The validity of convergent and discriminant was analyzed, using the SmartPLS statistical package (Ringle, Wende, & Becker, 2015) to calculate the factorial structure of the indicators, using partial least squares. SEM-PLS aims to predict the latent variables through the estimation of partial least squares (PLS) and principal components analysis (PCA). With the PLS technique, two procedures can be evaluated at the same time (the measurement model and the structural model). For the case of validity, the measurement model is used which involves the reliability analysis of each indicator, the internal consistency of each dimension, the analysis of the average extracted variance and the discriminant validity. In a PLS model, the charges between each indicator and its dimension are valued, accepting loads greater than 0.707. Another measure used to evaluate the fit of the model is the average extracted variance that provides the amount of variance that a construct (dimension) obtains from its indicators in relation to the variance of the error. A good fit requires values greater than 50%.

As shown in Table19.1, the reliability of the scales was checked, since in all cases the composite reliability coefficient exceeded the expected minimum of 0.707. In relation to the validity, it was determined that the factorial loads of the items of all the scales are higher than the expected minimum, with average variance extracted between 55.4% and 89.6%. These values confirm the construct validity of the instrument used (Tables19.2 and 19.3).

For scales of self-efficacy, university ecosystem and social entrepreneurial intention, discriminant validity was also analyzed using the criteria proposed by Fornell and Larcker (1981). For this reason, the independence of each of the scales was established by corroborating that the square roots of the average variances extracted from the constructs are greater than their correlations with the other scales (see Tables19.2 and 19.3).

The discriminant validity of the scales of entrepreneurial orientation and perception of the social entrepreneur was determined by the criterion of Heterotrait Monotrait Ratio (HTMT). This criterion has been used because a model of hierarchical components was analyzed, in which the construct is measured by levels (first and second order).

Table 19.3 Reliability and validity of perception of social entrepreneurship and first order of constructs

Scale and indicators	Weighs	Cronbach's alpha	FCI	AVE
Social entrepreneurial perception (second order)			*0.896*	*0.812*
Social vision (first order)		0.726	0.846	0.646
They are quite focused on social problems.	0.805			
They have a strong commitment to a social vision.	0.830			
They have the facility to identify social needs.	0.776			
Innovation (first order)		0.660	0.815	0.595
They see risks as opportunities to create social value.	0.716			
They are people with a flexible mind.	0.771			
They deal with social opportunities in a novel way.	0.784			
Sustainability (first order)		0.716	0.825	0.540
They can create greater social value than classic entrepreneurs.	0.684			
They are capable of improving the quality of life in the long term.	0.779			
They act by protecting and preserving the environment.	0.719			
They provide solutions to unmet social needs.	0.721			
Entrepreneurial orientation (second order)			*0.887*	*0.724*
Take risk (first order)		0.705	0.835	0.629
I like to take a bold action venturing into the unknown.	0.798			
I can invest a lot of time and money in something that is very satisfying.	0.821			
I tend to act "with courage" in situations where there is a high risk.	0.829			
Innovation (first order)		0.704	0.835	0.628
I prefer to focus on unique and novel projects, more than what is already known.	0.744			
When I learn something, I prefer to do it "my way", not to imitate everybody.	0.769			
When solving problems, I love experimenting and trying original approaches.	0.828			
Anticipation (first order)		0.693	0.830	0.619
In general, I anticipate problems, changes or needs.	0.831			
I tend to plan projects well in advance.	0.816			
I prefer to get involved in the project and make it work, not sit down and watch.	0.843			

Source: 6425 questionnaires of university students

19.3.2.2 Analysis of the Structural Model

Tables 19.4 and 19.5 show that both scales meet the established criteria and have discriminant validity in the first order because the values of the HTMT are below the conservative threshold value (0.90) and the confidence intervals are not including the unit. Predictably, at the second level, the HTMT values exceed the threshold; they include the indicators of the first order. In this case, the confidence intervals should include the unit criterion that is met in both scales.

The entrepreneurial support provided by the university ecosystem is insufficient; only 37.2% of students reported that many times, almost always and always have received some type of institutional support from their universities. This factor is important to promote university-based social enterprises, and its deficiency considerably limits the implementation of entrepreneurial initiatives by students. Another evidence about the association of the factors studied, the intention and social entrepreneurial behavior, is the difference found when comparing the group of students who participate with those who do not participate in social enterprises. As shown in Fig.19.2, students who participate in social enterprises have received greater support from their university ecosystem, perceive more positively social entrepreneurs, have a greater entrepreneurial orientation and have developed greater self-efficacy and entrepreneurial intention. The value of these differences becomes more important because in all cases the differences found are statistically significant.

The direct and indirect influence of university support for social entrepreneurship was analyzed based on a conceptual model and applied structural equations (see Fig.19.2).

The results showed that the entrepreneurial support provided by the university ecosystem does not directly influence the social entrepreneurial intention ($\beta = 0.038$, $p = 0.001$), nor in the self-efficacy developed by the students ($\beta = -0.045$, $p = 0.551$). However, it directly influences the entrepreneurial orientation ($\beta = -0.185$, $p = 0.001$) and the positive perception of the social entrepreneur ($\beta = -0.241$, $p = 0.001$). Through these variables (which assume a mediating role), the university ecosystem exerts an indirect influence on self-efficacy and later on entrepreneurial intention.

Table 19.4 Discriminant validity of perception of social entrepreneurs using the criterion Heterotrait Monotrait Ratio (HTMT)

	Innovation	Social vision	Sustainability
Social vision	0.857 [0.0832; 0.883]		
Sustainability	0.913 [0.886; 0.937]	0.770 [0.832; 0.883]	
Perception of social entrepreneur	1.141 [1.126; 1.58]	1.053 [1.041; 1.064]	1.121 [1.110; 1.132]

Note: Values in parentheses represent the 95% confidence interval bias corrected; these values were obtained with a bootstrapping of 5000 cases. Source: 3407 questionnaires of university students

Table 19.5 Discriminant validity of perception of environmental behavior using the criterion Heterotrait Monotrait Ratio (HTMT)

	Innovation	Take risk	Anticipation
Take risk	0.610 [0.873; 0.922]		
Anticipation	0.850 [0.823; 0.873]	0.738 [0.706; 0.768]	
Entrepreneurial orientation	1.133 [1.119; 1.147]	1.093 [1.006; 1.095]	1.008 [1.06614; 1.095]

Note: Values in parentheses represent the 95% confidence interval bias-corrected; these values were obtained with a bootstrapping of 5000 cases. Source: 3407 questionnaires of university students

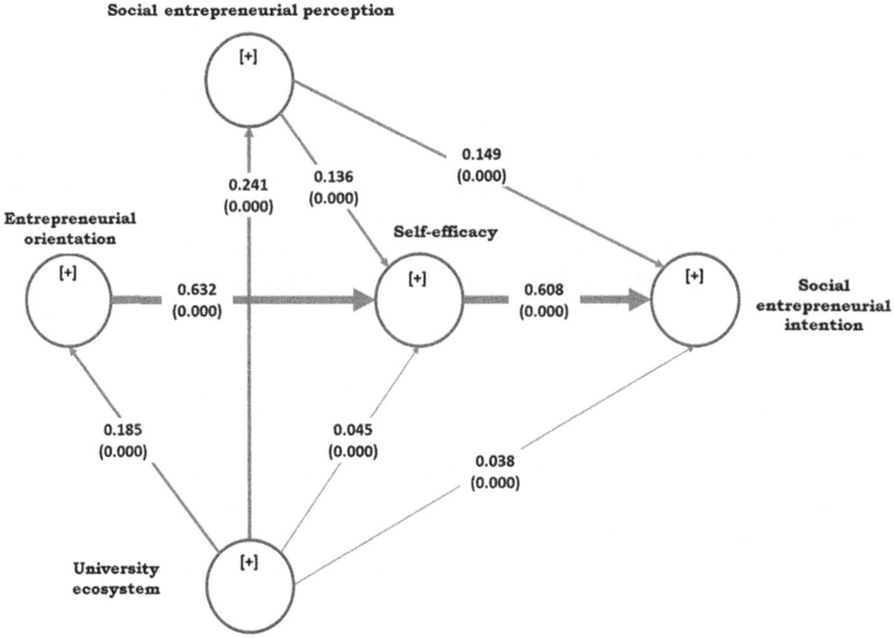

Fig. 19.2 Path analysis of factors that explain the direct and indirect influence of university support for social entrepreneurship. Note: Structural equations of variance using partial least squares. β values and 5% probability (in parentheses). Calculation of "*t*" values using bootstrapping ($n = 5000$). (Source: 6425 questionnaires of university students)

On the other hand, the perception of the social entrepreneur directly and indirectly influences the social entrepreneurial intention ($\beta = 0.149$, $p = 0.001$, indirect effect = 0.083). Likewise, the self-efficacy showed a direct and significant influence on the entrepreneurial intention ($\beta = -608$, $p = 0.001$). The integral evaluation of the model allows observing stable relationships. This aspect is corroborated with the standardized β coefficients, which are significant and similar to the population parameters (Table19.6).

Table 19.6 Significance of the trajectory coefficients (β) between the entrepreneurial intention and the explanatory variables

Relation between dimensions (β values)	Original sample (O)	Mean sample (M)	Standard deviation	*T*-statistic
Self-efficacy → Social entrepreneurial intention	0.608	0.608	0.011	54.611*
University ecosystem → Self-efficacy	0.045	0.045	0.009	4.996*
University ecosystem → Social entrepreneurial intention	0.038	0.038	0.009	4.085*
University ecosystem → Entrepreneurial orientation	0.185	0.186	0.013	14.075*
University ecosystem → Social entrepreneurial perception	0.241	0.242	0.014	17.720*
Entrepreneurial orientation → Self-efficacy	0.632	0.632	0.011	59.133*
Social entrepreneurial perception → Self-efficacy	0.136	0.136	0.012	11.504*
Social entrepreneurial perception → Social entrepreneurial intention	0.149	0.149	0.012	12.230*

*p value = 0.00. Resampling technique bootstrapping (5000 times)

In addition, a path of influence of the institutional support provided by the university ecosystem is identified. The route goes from the university ecosystem to the entrepreneurial orientation (β = 0.185, p = 0.000). These two variables together increase the self-efficacy (β = 0.632, p = 0.000), and finally the three variables increase the entrepreneurial intention (β = 0.608, p = 0.000).

It should be noted that the formulated structural relation model explains the 50% variation of the social entrepreneurial intention of the students. In addition, it is corroborated that personal factors such as entrepreneurial orientation and self-efficacy have a large effect on the social entrepreneurial intention compared to the institutional support received from the university ecosystem, whose impact is indirect (effect size [F2]—Table19.7). On the other hand, the influence of the university ecosystem on the entrepreneurial orientation and the positive perception of the social entrepreneur, as well as the perception of the social entrepreneur on self-efficacy and social entrepreneurial intention, show a medium effect size.

19.3.3 Conclusions of the Study

It is relevant that the present study coincides with the results of previous research as it confirms the growth trend of research in the field of social entrepreneurship. The present investigation has an innovation to have obtained the opinion of the university students of diverse regions of Peru on its intention and practice of social entre-

Table 19.7 Effect size of the β coefficients between the entrepreneurial intention and the explanatory variables

Relation between dimensions	F^2 (O)	F^2 (M)	Standard error	T-statistic	p
Self-efficacy → Social entrepreneurial intentionl	0.545	0.546	0.029	18.973	0.000
University ecosystem → Self-efficacy	0.004	0.004	0.002	2.464	0.014
University ecosystem → Social entrepreneurial intention	0.003	0.003	0.001	2.009	0.045
University ecosystem → Entrepreneurial orientation	0.035	0.036	0.005	6.759	0.000
University ecosystem → Social entrepreneurial perception	0.062	0.062	0.007	8.329	0.000
Entrepreneurial orientation → Self-efficacy	0.579	0.580	0.030	19.562	0.000
Social entrepreneurial perception → Self-efficacy	0.026	0.026	0.005	5.664	0.000
Social entrepreneurial perception → Social entrepreneurial intention	0.032	0.032	0.005	6.107	0.000

Note: Resampling technique bootstrapping (5000 times)

preneurship, which allows to generate diverse strategies to achieve to increase the percentage of students that can be involved and even leading the social entrepreneurship issues that are generated from the universities and then independently. An increase in the percentage of university students involved in social enterprises has been verified. At the same time, it can be verified that there is a growth in the publication of articles that focus on doing meta-analysis in the field of entrepreneurship, such as that published by Schwens et al. (2018) who focused on evaluating the relationship between internationalization and the performance of companies in international entrepreneurship.

On the other hand, Miao, Humphrey, Qian, and Pollack (2018) also conducted a meta-analysis of research that evaluated emotional intelligence and entrepreneurial intention. Another aspect that is also relevant is the evaluation of the characteristics of the entrepreneurial teams and the results of the undertaking undertaken, as analyzed by Jin et al. (2017). As we can see, the progress of research in a more specific way in the field of entrepreneurship continues to grow and diversify, which makes the present research a starting point for the context of Peru and at the same time compels to continue exploring the characteristics of social entrepreneurship at the university level. It is relevant to mention that there is an influence of the family that has previously carried out undertakings on the students. The influence would be given from different aspects: to sell the fear of entrepreneurship (even more so when family businesses have been successful), to have the initial financial support of a successful venture to start the new venture, among others. Having found that work experience is a relevant factor for entrepreneurship, it is useful to think about the incorporation of activities to boost social entrepreneurship in programs aimed at undergraduate training of adults, who due to their greater work experience could have a greater probability of involvement in social entrepreneurship programs. It was reported that 18% of students participated in some social enterprise. While it is true that the percentage may seem low, it is important to consider that students have different stimuli and motivations so you should be optimistic with this finding and evaluate innovative strategies to achieve.

The sex of the students has not been shown to be a differentiating aspect, even though the study by Nicolas Martínez and Rubio Bañón (2012) indicates that social entrepreneurship is mostly carried out by men. According to Nicolas Martínez and Rubio Bañón (2012), the age of starting a social enterprise is variable among countries since, although the average is between 25–34 years, only Spain and Venezuela follow this pattern; thus, in Panama and the Dominican Republic, the average is between 18–24 years, while other countries such as Argentina, Colombia, Chile or Peru average between 35–44 years. Analyzing the limitations that can explain the low percentage of involvement in social ventures, this 63% of student's report that rarely, almost never and never received support from their university. This finding shows two possible realities in universities: the lack of permanent programs to promote student social entrepreneurship and the allocation of few resources to social entrepreneurship programs, which implies that there are activities with students but that their scope is very limited. It has been found that there is a lack of practical training. It is pertinent to mention that practical training requires a standardized procedure, with indicators of learning and results, since it is not only about taking the student to the place where the undertaking should be carried out or only carrying out volunteer activities. Likewise, it was found that there is an absence of specialized advice, which implies that the advice must be constant in the same field of entrepreneurship.

Finally, it was found that networks of contact with investors are lacking and this is clearly a great weakness because when entrepreneurial efforts are not economically supported to scale business, the interest of the student who not only wants to learn but who is also seeking to generate a means of developing own productive activities as an entrepreneur. Social entrepreneurship can be very attractive as an idea and can attract many interested students (63% of interested students have been reported) but it can only be achieved when it has the necessary components described above. The engineering faculties, according to what has been reported, need to increase their attention in the promotion of social entrepreneurship. A concrete strategy may be to take the experience of their business science faculties (if any).

It is necessary to point out that we know the initiatives developed by various universities in relation to social entrepreneurship and even the constant participation in Hult Prize, an initiative supported by Hult International Business School and that seeks to promote ventures that prioritize the resolution of urgent social problems. In relation to professional careers, it is worth mentioning that the vast majority of social enterprises are made up of members with different professional careers, since various competencies are needed, so that students of administrative sciences will have teams of economics students, engineering, health areas, law, agronomy and others, according to the need of social entrepreneurship to be developed. To demonstrate this need for interdisciplinary work, one can review the experience of social entrepreneurship in Ibero-America (De Pablo & Uribe, 2017) in which various initiatives are reported that have teams with different professionals working from their specific competencies.

A limiting aspect so that the initial interest of a student in the realization of a social enterprise becomes a concrete action is the lack of experience and more precisely: the lack of positive experiences of their environment. That is why the present study also showed that a large percentage of students who have the intention or who are already participating in a social enterprise have a history of family members who have developed entrepreneurship. In this way, the students have on the one hand the close motivation of the family member and, also, they could have said family member initially advising the social entrepreneurship. The basis of social entrepreneurship is to achieve social change, which could initially inspire a student but as Haugh and Talwar (2016) point out. Student empowerment is necessary, and this will occur not only due to the information that can be received from the university but also the interrelation of the student with other students who have successfully carried out social ventures and thus empower themselves to achieve the goals and planned results.

Another aspect that must also be taken into account to understand that students may have the intention but then not be materialized is that it is required, as described by Rivera, Santos, Martín-Fernández, Requero, and Cancela (2018) of a service leadership. That is, to train leaders with a service attitude, with characteristics such as listening, caring, becoming aware, conceptualizing, managing and trying to build a sustainable community. The current situation in Peru needs more students focused on developing social entrepreneurship, for which it is required from the universities to incorporate the training and vision of the development of a social enterprise. Thus, volunteering activities, as described by Gaete Quezada (2015), should be the most relevant activity to achieve the development of leadership skills as well as raising awareness among students, so that they can see in social entrepreneurship in its correct magnitude, as mentioned by Ahmad Ashkar, CEO of Hult Prize: Solving the most pressing challenges in the world is not only the right thing to do, it is also a good business.

Is it possible that universities are not working as a programmatic axis generating social entrepreneurship among their students? This could be the explanation why only a third of students have reported that they have support from their universities. The planning from the universities for the development of social entrepreneurship by their students' needs theoretical but mainly practical training, that is, that the student has real experiences that help to dimension the necessary competences to be developed; In addition, teaching should be guided by professionals with current experience in social entrepreneurship and, likewise, universities should be concerned with building collaborative networks, both academic and business.

Although half of the student's report that there are related subjects and that entrepreneurial ideas are even promoted, only a third report that contact networks are promoted and only a fifth report that they received practical training. These results show that all the aforementioned elements must be planned to achieve social undertakings by the students. The intention of entrepreneurial behavior, although present, is a high percentage of students does not seem to have a formative base in universities, which makes it difficult to carry out social enterprises; for this reason, there is a significant challenge from the authorities of the universities. Due to the fact that half of the students reported that their future initiatives would prioritize social benefits over financial ones, the need to generate service leadership among students, who have competences to solve society's problems and that is sustainable, becomes clear, economically in time. Another paradoxical aspect is that there is a high percentage of students reporting that they would develop a successful social enterprise in the future. Will they really do it or is it just an answer to get out of step? The security that students have with the training from the university that they currently receive can be auspicious, pending that the training is increased and improved so that this self-efficacy is really beneficial.

Closing Remarks

Being a social entrepreneur is well seen in society, which has an important influence on students, with half of them reporting a positive opinion. For this reason, as part of the strategies from the universities, it should influence the good reputation that is generated in the participation in social enterprises. In this sense, universities must work on the vision of the Sustainable Development Goals (SDG) so that sustainable development can be specifically promoted based on student innovation proposals. The entrepreneurial orientation has allowed to know how the student looks in relation to the possibility of carrying out a social enterprise. High percentages of entrepreneurial orientation were reported showing that students feel that they have the appropriate profile for entrepreneurship, which is crucial for students to be able to materialize this perception.

Improving entrepreneurial orientation requires the commitment of the team responsible for the planning of social enterprises (van Doorn, Heyden, & Volberda, 2017) through dynamic and interdisciplinary activities. The influence of the university support to social entrepreneurship was analyzed, and it could be shown that it does not directly influence the social entrepreneurial intention, nor in the self-efficacy developed by the students, but does it on the entrepreneurial orientation and the positive perception of the social entrepreneur.

It has been possible to corroborate that personal factors such as entrepreneurial orientation and self-efficacy influence the social entrepreneurial intention in a more significant way compared to the institutional support received from the university ecosystem, whose impact is indirect.

References

Achieve. (2014). The 2014 Millenial impact report. Retrieved February 5, 2020, from http://casefoundation.org/wp-content/uploads/2014/11/MillennialImpactReport-2014.pdf

Ajzen, I. (1985). *From intentions to actions: A theory of planned behavior*. Berlin, Heidelberg: Springer. https://doi.org/10.1007/978-3-642-69746-3_2

Ajzen, I. (1991). The theory of planned behavior. *Organizational Behavior and Human Decision Processes, 50*, 179–211.

Bandura, A. (1986). Social foundations of thought and action: A social cognitive theory. Retrieved February 5, 2020, fromhttp://psycnet.apa.org/record/1985-98423-000

Bandura, A. (2012). On the functional properties of perceived self-efficacy revisited. *Journal of Management, 38*(1), 9–44. https://doi.org/10.1177/0149206311410606

Brinckerhoff, P. C. (2000). Social entrepreneurship: The art of mission-based venture development (Wiley Nonprofit Law, Finance and Management Series).

Brockhaus Sr., R. H., & Horwitz, P. S. (1986). The psychology of the entrepreneur. In D. L. Sexton & R. W. Smilor (Eds.), *The art and science of entrepreneurship* (pp. 25–48). Cambridge: Ballinger.

Chang, W., Liu, W., & Chiang, S. (2014). A study of the relationship between entrepreneurship courses and opportunity identification: An empirical survey. *Asia Pacific Management Review, 19*(1), 1–24. https://doi.org/10.6126/APMR.2014.19.1.01

Chen, C., Gene, P., & Crick, A. (1998). Does entrepreneurial self-efficacy distinguish entrepreneurs form managers? *Journal of Business Venturing, 13*, 295–316. https://doi.org/10.1016/S0883-9026(97)00029-3

Comisión de las Naciones Europeas. (2006). Communication from the Commission to the Council, the European Parliament, the European Economic and Social Committee and the Committee of the Regions. Apply the Lisbon community program: Promote business mentality through education and training [Comunicación de la Comisión al Consejo, al Parlamento Europeo, al Comité Económico y Social Europeo y al Comité de la Regiones. Aplicar el programa comunitario de Lisboa: Fomentar la mentalidad empresarial mediante la educación y la formación]. (pp. 1–12). Bruselas.

DaFonseca, B. (2015). *Study of entrepreneurial intention in Portugal from the perspective of values and gender* [*Estudio de la intención emprendedora en Portugal desde la perspectiva de los valores y de género*]. Doctoral Thesis, Universidad Nacional de Educación a Distancia. 250 pp. Madrid.

DePablo, J., &Uribe, J. (2017). *Experiences of social entrepreneurship in Latin America* [*Experiencias de emprendimiento social en Iberoamérica*] (Vol. 67). Universidad Almería.

Doherty, B., Haugh, H., & Lyon, F. (2014). Social enterprises as hybrid organizations: A review and research agenda. *International Journal of Management Reviews, 16*(4), 417–436.

van Doorn, S., Heyden, M. L., & Volberda, H. W. (2017). Enhancing entrepreneurial orientation in dynamic environments: The interplay between top management team advice-seeking and absorptive capacity. *Long Range Planning, 50*(2), 134–144.

Drayton, W. (2002). The citizen sector: Becoming as entrepreneurial and competitive as business. *California Management Review, 44*(3), 120–132.

Duarte, T., & Tibana, M. R. (2009). Emprendimiento, una opción para el desarrollo. *Scientia et technica, 15*(43), 326–331.

Easterly, W. (2008). Design and reform of institutions in LDCS and transition economies. Institutions: Top down or bottom up? *American Economic Review: Papers & Proceedings, 98*(2), 95–99.

Fishbein, M., & Ajzen, I. (1973). Attribution of responsibility: A theoretical note. *Journal of Experimental Social Psychology, 9*(2), 148–153. https://doi.org/10.1016/0022-1031(73)90006-1

Forbes. (2015). 5 Reasons why social entrepreneurship is the new business model. Retrieved February 5, 2020, fromhttps://www.forbes.com/sites/meimeifox/2016/08/08/5-reasons-why-social-entrepreneurship-is-the-new-business-model/#54af2fd144ca

Fornell, C., & Larcker, D. F. (1981). Structural equation models with unobservable variables and measurement error: Algebra and statistics. *Journal of Marketing Research*, 382–388.

Franke, N., & Lüthje, C. (2004). Entrepreneurial intentions of business students—A benchmarking study. *International Journal of Innovation and Technology Management, 1*(03), 269–288.

Gaete Quezada, R. (2015). University volunteering as a field of service learning and social entrepreneurship: A case study [El voluntariado Universitario como ámbito de aprendizaje servicio y emprendimiento social: Un estudio de caso]. *Ultima década, 23*(43), 235–260.

Galloway, L., & Brown, W. (2002). Entrepreneurship education at university: A driver in the creation of high growth firms? *Education + Training, 44*(8/9), 398–405. https://doi.org/10.1108/00400910210449231

Gestion. (2017). Economic growth in Peru and its presidents [Crecimiento económico en el Perú y sus presidentes]. Retrieved February 5, 2020, fromhttps://gestion.pe/blog/economiaparatodos/2017/09/crecimiento-economico-en-el-peru-y-sus-presidentes.html

GEM (2018). GEM 2018 / 2019 Global Report. Available in https://www.gemconsortium.org/report/gem-2018-2019-global-report. Accessed 10 May, 2020.

Guerrero, M., Urbano, D., Cunningham, J., & Gajón, E. (2017). Determinants of Graduates' start-ups creation across a Multi-Campus Entrepreneurial University: The case of Monterrey Institute of Technology and Higher Education. *Journal of Small Business Management, 56*(1). https://doi.org/10.1111/jsbm.12366

Haugh, H. M., & Talwar, A. (2016). Linking social entrepreneurship and social change: The mediating role of empowerment. *Journal of Business Ethics, 133*(4), 643–658.

Hussain, A., Mohammad, S., & Ahmed, P. (2016). Impact of entrepreneurial education on entrepreneurial alertness. *Journal of Social and Organizational Analysis*, 1–9.

Ireland, R. D., & Webb, J. W. (2009). Crossing the great divide of strategic entrepreneurship: Transitioning between exploration and exploitation. *Business Horizons, 52*(5), 469–479.

Jin, L., Madison, K., Kraiczy, N. D., Kellermanns, F. W., Crook, T. R., & Xi, J. (2017). Entrepreneurial team composition characteristics and new venture performance: A meta-analysis. *Entrepreneurship Theory and Practice, 41*(5), 743–771.

Jung, D. I., Ehrlich, S. B., De Noble, A. F., & Baik, K. B. (2001). Entrepreneurial self-efficacy and its relationship to entrepreneurial action: A comparative study between the US and Korea. *Management International, 6*(1), 41.

Kelley, D., Bosma, N., &Amorós, J. (2011). *Global entrepreneurship monitor 2010 Global report*. Global Entrepreneurship Research Association (GERA). Retrieved February 5, 2020, fromhttp://www.av-asesores.com/upload/479.PDF

Kolvereid, L., & Moen, O. (1997). Entrepreneurship among business graduates: Does a major in entrepreneurship make a difference? *Journal of European Industrial Training, 21*(4), 154–116.

Krueger Jr., N. F., Reilly, M. D., & Carsrud, A. L. (2000). Competing models of entrepreneurial intentions. *Journal of Business Venturing, 15*(5–6), 411–432.

Liñán, F., &Chen, Y. (2009). Development and cross-cultural application of a specific instrument to measure entrepreneurial intentions. *ET&P*, 593–617. Retrieved February 5, 2020, fromhttp://institucional.us.es/vie/documentos/resultados/LinanChen2009.pdf

Madrigal, B.,&Santamaria, C. (2015). Entrepreneurial university, formal and informal factors: Theoretical discussion [Universidad emprendedora, factores formales e informales: discusión teórica]. En Madrigal, B. (Coord.), *Cuatro ejes para conformar una cultura emprendedora*. Prometeo editors.

Manolova, T., Eunni, R., & Gyoshev, B. (2008). Institutional environments for entrepreneurship: Evidence from emerging economies in Eastern Europe. *ET&P*, 203–218. https://doi.org/10.1111/j.1540-652.2007.00222.x

Miao, C., Humphrey, R. H., Qian, S., & Pollack, J. M. (2018). Emotional intelligence and entrepreneurial intentions: An exploratory meta-analysis. *Career Development International*. https://doi.org/10.1108/CDI-01-2018-0019

Moriano, J., Palací, F., & Morales, J. (2006b). The psychosocial profile of the university entrepreneur [El perfil psicosocial del emprendedor universitario]. *Revista de Psicología del Trabajo y de las Organizaciones, 22*(1), 75–99.

Moriano, J., Palací, J., & Morales, J. (2006a). Adaptation and validation in Spain of the entrepreneurship self-efficacy scale [Adaptación y validación en España de la escala de Autoeficacia Emprendedora]. *Revista de Psicología Social, 21*(1), 51–64.

Navarro, M. (2014). *The psychosocial profile of women entrepreneurs* [*El perfil psicosocial de la mujer emprendedora*]. University of Valladolid. Retrieved February 5, 2020, fromhttp://uvadoc.uva.es/bitstream/10324/7059/1/TFG-G%20646.pdf

Nga, J. K. W., & Shamuganathan, G. (2010). The influence of personality traits and demographic factors on social entrepreneurship start up intentions. *Journal of Business Ethics, 95*(2), 259–282. https://doi.org/10.1007/s10551-009-0358-8

Nicolas Martínez, C., & Rubio Bañón, A. M. (2012). Social entrepreneurship: A comparison between Spain and South American countries [El emprendimiento social: una comparativa entre España y países sudamericanos]. *Faedpyme International Review, 1*(1), 38–49.

North, D. (1990). *Institutions, institutional change and economic performance*. Cambridge, MA: Cambridge University Press.

Ringle, C. M., Wende, S., &Becker, J. M. (2015). *SmartPLS 3. Boenningstedt: SmartPLS GmbH*. Retrieved February 5, 2020, fromhttp://www.smartpls.com

Rivera, R. G., Santos, D., Martín-Fernández, M., Requero, B., & Cancela, A. (2018). Predicting attitudes and behavioural intentions towards social entrepreneurship: The role of servant leadership in young people [Predicción de las actitudes y las intenciones conductuales hacia el emprendimiento social: el papel del liderazgo de servicio en los jóvenes]. *Revista de Psicología Social, 33*(3), 650–681.

Rueda, I., Fernández, A., & Herreo, A. (2012). University students and entrepreneurship: Psychological determinants of the intention of creating your own business [Estudiantes universitarios y emprendimiento: determinantes psicológicos de la intención de creación de un negocio propio]. *FAEDPYME International Review, 1*(2), 9–15.

Schwens, C., Zapkau, F. B., Bierwerth, M., Isidor, R., Knight, G., & Kabst, R. (2018). International entrepreneurship: A meta–analysis on the internationalization and performance relationship. *Entrepreneurship Theory and Practice, 42*(5), 734–768.

Serida, J., Guerrero, C., Alzamora, J., Borda, A., & Morales, O. (2017). *Global entrepreneurship monitor: Perú 2016–2017*. Lima: Biblioteca Nacional del Perú.

Shapero, A., &Sokol, L. (1982). The social dimensions of entrepreneurship.

Soria-Barreto, K., Zúñiga-Jara, S., & Ruiz Campo, S. (2016). Determinants of entrepreneurial intention: New evidence [Determinantes de la intención emprendedora: Nueva evidencia]. *Interciencia, 41*(5), 325–329.

Thompson, J., Alvy, G., & Lees, A. (2000). Social entrepreneurship–a new look at the people and the potential. *Management Decision, 38*(5), 328–338.

Yeng, O., Selvarajah, C., & Meyer, D. (2011). Inclination towards entrepreneurship among university students: An empirical study of Malaysian university students. *International Journal of Business and Social Science, 2*(4), 206–222. Retrieved February 5, 2020, fromhttp://www.ijbssnet.com/journals/Vol._2_No._4%3B_March_2011/24.pdf

Chapter 20
Green Consumerism

Sandra Huamán-Pastorelli, Aldo Alvarez-Risco, and Alex Harras

Abstract Green consumerism has expanded quickly throughout the developed world, and it has started to emerge in developing nations as well. Many theories try to explain consumer habits, and in recent years, the importance of consumption has become highlighted primarily because of concerns about the quality of the environment and interest in sustainable development processes. In short, environmentally responsible purchasing is vital as an antidote to the unnecessary purchasing of goods which can severely harm the environment. Over the last decade, many authors have studied different green consumer purchasing behaviors. They have identified several influential factors that help to shape shopping decisions. Such studies explain inconsistencies between green purchasing attitude and actual behavior.

Keywords consumer · consumerism · sustainability · green consumer · ecological buy

20.1 Introduction

According to Joshi and Rahman (2015), today more companies are producing environmentally friendly products, and consumer awareness of these products is increasing. The manufacturers that offer such green products should have a "green thinking" as a part of their culture and mission. Models grounded in social psychology have been used to understand green consumer purchasing behavior. These include the theory of planned behavior (TPB) by Ajzen's (1991) and the theory of reasoned action (TRA) by Fishbein and Ajzen's (1975); however, considering the already

S. Huamán-Pastorelli · A. Alvarez-Risco (✉)
Facultad de Ciencias Administrativas y Recursos Humanos, Universidad de San Martín de Porres, Lima, Peru
e-mail: shuamanp@usmp.pe; aldo.alvarez@redsaf.org

A. Harras
Center for MSP Research, Lima, Peru

A. Alvarez-Risco et al. (eds.), *Building Sustainable Cities*,
https://doi.org/10.1007/978-3-030-45533-0_20

established role of country-context in green consumption, consumers do not likely full or even partial control in green purchases. Thus, the applicability of these models remains to be validated.

20.2 Green Purchasing intention theoretical framework and extended models

20.2.1 Theory of Planned Behavior (TPB)

In 1985, Icek Ajzen developed the theory of planned behavior, which examined attitudes, subjective norms, and perceived behavioral controls. He identified three main components in human behavior: behavioral beliefs (BB), normative beliefs (NB), and control beliefs (CB). Additionally, among the control beliefs were outcomes evaluations (OE) such as attitudes toward the behavior, subjective norms, and perceived behavioral controls. Behavior intention referred to combinations of these elements in concert with one another. For example, behavioral intention is composed of ΣBBiOEi (attitude), ΣNBjMCj (subjective norm), and ΣCBkPPk (perceived behavioral control) (Ajzen, 1991) (Fig. 20.1).

In conclusion, behavioral intention refers that a person's status to perform given behavior. In short, the individual's purpose strengthens when the attitude aligns with it, the subjective norm favors it, and there is a higher perceived control.

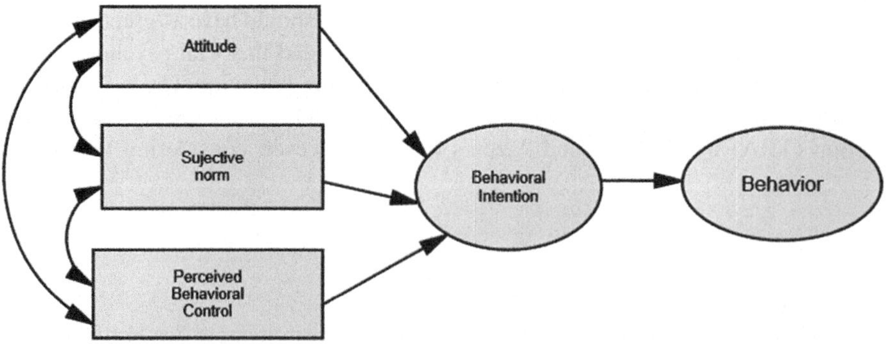

Fig. 20.1 *This represents the following combination: Behavioral Intention = ΣBBiOEi + ΣNBjMCj + ΣCBkPPk. Adapted from TPB complete model (Ajzen, 1985)*

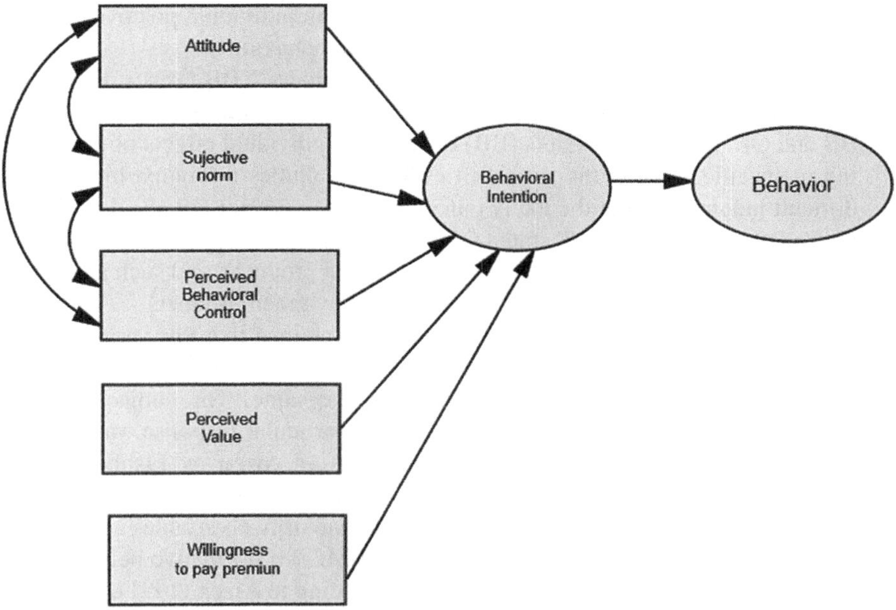

Fig. 20.2 *In the extended theoretical framework, behavioral intention becomes the sum of ΣBBiOEi + ΣNBjMCj + ΣCBkPPk + PV + WPP (*Yadav & Pathak, 2017*).*

20.2.2 Extended Models

Yadav and Pathak (2017) created an extended model of TPB incorporating two more constructs: perceived value (PV) and the willingness to pay a premium (WPP). The authors considered perceived value because it plays a crucial role in green purchasing decisions made by consumers who will not consider only the practical benefits of the product just for the care of the environment. Furthermore, they found that the WPP to be the high price of the eco-friendly product, and they found it to be still an issue for price-sensitive consumers (Fig. 20.2).

20.2.3 The Factors of TPB

a) **Attitude**: Accordingly, Gordon Allport (1935) described attitude as "a mental and neural state of readiness, organized through experience, exerting a directive or dynamic influence upon an individual's response to all objects and situations with which it is related" (p. 810). In addition, Kumar, Manrai, and Manrai (2017) described how attitudes toward environmentally sustainable products arbitrate the direct and positive relationship between environmental knowledge and purchasing intention. For Ajzen and Fishbein (1980), cited in Yadav and Pathak

(2017), this TPB model component is defined as an individual positive or negative evaluation of the performance of a particular behavior.

The corresponding formula is as follows: Attitude (ΣBBiOEi) = behavioral belief (BB)∗outcome evaluation (OE). This means that attitude is the result of BB and OE. Behavioral beliefs (BB) refer to the individual effects of participating in specific conduct though result evaluation alludes to comparing good or difficult judgment about the likely outcomes of the conduct or behavior. To sum up, Yadav and Pathak (2017) stated that behavioral beliefs (∑BBiOEi) positively influence the consumer's attitude toward the green products, and such an attitude significantly alters the consumer's intention to buy green products.

b) **Subjective norm**: Fishbein and Ajzen (1975) explained that this variable could be stated as a form of belief that individuals approve or disapprove specific behavior when undertaking and performing the same. The subjective norm referred to perceive social demand to execute a particular response was found to be very low regarding its effect on purchase intention. Ajzen and Fishbein (1980) stated that this element is characterized as social pressure applied to a person to participate in specific conduct. This is an outcome of two variables as shown in the following formula: Subjective norm (ΣNBjMCj) = normative belief (NB) ∗ motivation to comply (MC). Furthermore, according to Ajzen (1991), normative belief refers to one's personal perception about how other people would like one to behave at a specific time. Motivation to comply refers as the desire of each person to match the others' opinion. In conclusion, according to Kumar et al. (2017), subjective norm has a direct and positive relationship with the purchase intention for environmentally sustainable products. It is likely to positively moderate the relationship between environmental knowledge and attitudes toward ecologically sustainable products.

c) **Perceived behavioral control**: Ajzen and Fishbein (1980) stated that an individual's observed ease or difficulty in performing a particular behavior reflects past experiences and anticipated obstacles. Perceived behavioral control is an outcome of CB and PP as follows: Perceived behavioral control (ΣCBkPPk) = control belief (CB)∗ perceived power (PP).

   Control belief (CB) can be defined as the belief of the individual toward the presence of certain factors that may facilitate or impede the performance of a particular behavior (e.g., time, money, and opportunity). On the other hand, *perceived power* (PP) refers to a personal evaluation of the impact of these factors in facilitating or impeding the particular behavior (Ajzen, 1991). Finally, Yadav and Pathak (2017) concluded that control beliefs (∑CBkPPk) positively influence the perceived behavioral control, and this control significantly influences the consumer's intention to buy green products.

d) **Perceived value**: Zeithmal (1988) defined this factor as "an overall assessment of the utility of the product based on the perception of what is received and what is given." For Chen and Chang (2012), green products are overpriced compared with available alternatives. Consumers are not going to compromise on excellent functionality available with the standard product. When consumers have an alternative to choose between product features and greenness, most probably will select the product attributes over its greenness (Ginsberg & Bloom, 2004). Furthermore, firms should utilize green marketing strategies to enhance the perceived value of their products with respect to consideration for the environment (Chen & Chang, 2012). Finally, Yadav and Pathak (2017) described how perceived value positively influences the intention to buy green products.

e) **Consumer willingness to pay premium (WPP):** Most of the time price is considered as one of the most critical factors that determines the consumer purchasing decision. Understanding the consumers' willingness to pay a premium for socially responsible products is important for organizations as the price is the most important barriers to green consumption. The willingness to pay premium prices for green products may also be considered pro-environmental behavior (Ajzen, 1991). Finally, Yadav and Pathak (2017) stated that consumer willingness to pay the premium (WPP) positively influences their intention to buy green products, and this intention significantly influences actual buying behavior.

20.2.4 Factors from Extended Models

Various extended models contribute additional factors relevant to green purchasing. These are variables specifically related to the individual decision maker. They may be grouped in two categories including those elements related to individual and internal factors and those elements which are situational and external (Fig. 20.3).

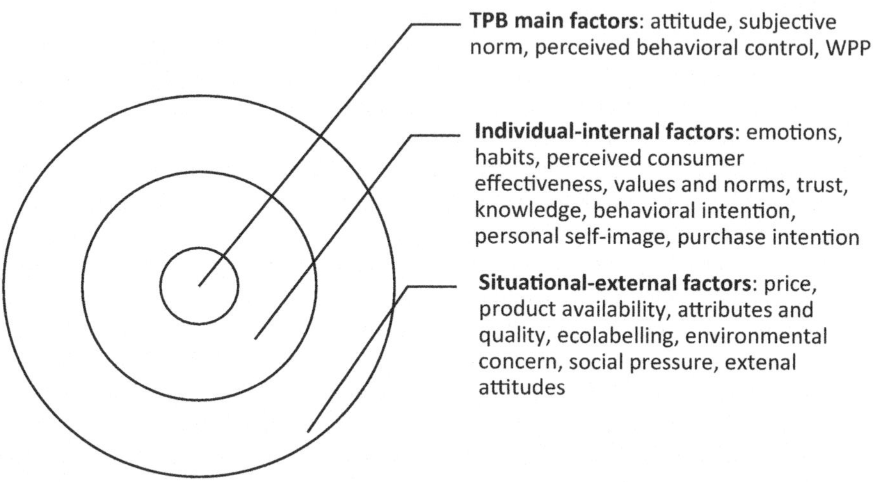

Fig. 20.3 *The constructs defined by several authors. These variables have been divided into TPB main factors, individual-internal factors, and situational-external factors (edited by the authors)*

20.2.4.1 Individual and internal elements

These variables are generally a result of personal life experiences (attitudes, values, personality, etc.) and commonly affect an individual's decision-making process.

- Emotions: According to Joshi and Rahman (2015), consumer emotions, specifically environmental concerns, have a positive and direct impact on green consumer purchasing intention and behavior.
- Habits: Consumer past behavior and habit guide consumer preference and influence present purchasing behavior, often making it difficult to change. Padel and Foster (2005) and Vermeir and Verbeke (2006) reported how habit can have a negative influence on consumer green purchasing behavior. Furthermore, this has been reported as a significant obstacle to green purchasing (Tsakiridou, Boutsouki, Zotos, & Mattas, 2008).
- Perceived consumer effectiveness: This refers to a "consumers' evaluation of the extent to which their consumption can make a difference in the overall problem" (Webster Jr, 1975), and has been extensively studied. According to Joshi and Rahman (2015), seven studies found a positive correlation between perceived consumer effectiveness and purchase intention/adoption of green products. Kang, Liu, and Kim (2013) stated that the perceived consumer effectiveness was also found to indirectly influence consumer purchase intention since it significantly affected consumer attitudes, subjective norms, and perceived behavioral control, which when taken together further determined consumer purchase intention. The research strongly points to a positive correlation between perceived consumer effectiveness and green purchasing intention and behavior.
- Values and personal norms: Joshi and Rahman (2015) examined six studies reporting a positive correlation between the environmental, social, and ethical values of consumers and their subsequent green product purchasing behavior. Values such as benevolence, universalism, and altruism were found to affect purchase intention positively. Such, moral and personal norms were also observed to have a significant influence on purchasing intentions (Arvola et al., 2008; Gleim, Smith, Andrews, & Cronin Jr, 2013).
- Trust: Chen (2010) and Chen and Chang (2013) concerned in the context of green products, trust is defined as a belief or expectation about the environmental performance of specific products and also as consumers' willingness to depend on an object based on the belief or expectation resulting from its credibility, benevolence, and environmental performance. Where the influence of trust on consumer green purchase intention and behavior. These studies found that a lack of consumer trust and confidence in green privileges and characteristics of green products was a significant barrier toward the purchase of green products. In addition, trust is a fundamental factor of long-term consumer attitudes, especially the concept of relationship quality that creates consumer fidelity (Kang & Hur, 2012). In other words, consumers will have a negative attitude toward green products if they do not trust the company or its green products.

- Knowledge: Consumer knowledge has been the most studied variable. Wang, Liu, and Qi (2014) defined this knowledge as the understanding of relevant environmental concepts, environmental problems, and the possible action strategies that may be used for solving such problems. Fryxell and Lo (2003) defined environmental knowledge as a generalized knowledge base which is based on the facts, concepts, and relationships between the natural environment and its significant ecosystems. Many researchers agree that only a small fraction of pro-environmental behavior can be directly linked to environmental knowledge. Joshi and Rahman (2015) examined eighteen papers studying consumers' environmental knowledge. Fifteen papers found that knowledge of environmental issues positively influenced consumer intention and the actual purchase of green products. On the other hand, three studies did not find any relation between environmental knowledge and green purchasing intention. Other studies have explored the role of environmental knowledge as a predictor of environmental awareness between people having a concern about the degradation of the natural environment. MacInnis, Moorman, and Jaworski (1991) mentioned that knowledge has an important role as an antecedent to the receiver's capacity to process the information.
- Behavioral intention: This refers to all indications of an individual's willingness to perform a given behavior. It is assumed to be an immediate antecedent to the actual behavior (Ajzen, 2002). Favorable subjective norms and higher perceived behavioral control were both relevant and influential factors. When individuals have more attitudes toward green behavior then the individual will be more likely to engage in such behavior.
- Purchase intention: According to Kumar et al. (2017), all these variables for influencing environmentally sustainable behavior have direct and positive relationships on the purchasing behavior for specific products. This factor may also diverge depending upon the degree of pro-environmental attitude formed due to the varying levels of consumer environmental knowledge.
- Personal self-image or identity: Environmental (green) self-identity denotes "the extent to which you see yourself as a type of person who acts environmentally-friendly" (Van der Werff, Steg, & Keizer, 2013). Chan (2001) stated that consumer intention to buy environmentally friendly products is affected more by their personal preferences than by social pressure. Thus, self-identity plays an important role in predicting consumer green purchasing intention. This was found to be positively correlated with personal intentions independent of attitude and perceived social norms. This finding indicates that consumption is a form of self-expression. Consumers buy things that are congruent with their values and lifestyles, and therefore people who perceive themselves as "green consumers" or conservationists will be more likely to express those values through their consumption of environmentally sound products (Gatersleben, Murtagh, & Abrahamse, 2014).

20.2.4.2 Situational and External Factors

Situational factors also affect green purchasing decisions. Such external forces either encourage or discourage consumers to adopt green products. The variables included in this category are explained and given below.

- **Price**: Joshi and Rahman (2015) reported ten studies showing higher costs outweigh ethical considerations and expanded the attitude-behavior gap in the purchase of green products.

  In fact, sensitivity to high prices negatively affected green purchasing behavior significantly and consistently. As a consequence, it is clear that high price negatively influences green purchasing intentions and behavior. In addition, lack of economic resources among consumers was found to magnify the effect of price and act as a further barrier to purchase green products (Connell, Fien, Lee, Sykes, & Yencken, 1999); however, the low consumer price sensitivity was found to affect green purchase behavior positively.
- **Product availability**: Studies reported that limited availability of a product had a negative influence on consumer green purchasing intention and behavior (Young, Hwang, McDonald, & Oates, 2010). Inversely, other studies reported that availability of a product had a positive relationship with green purchasing intention and behavior (Tarkiainen & Sundqvist, 2005). Furthermore, the easy availability of the green product positively affected green purchase behavior (Tarkiainen & Sundqvist, 2005; Vermeir & Verbeke, 2006). Consumers generally do not like to spend a lot of time searching for green products. Instead, they prefer products that are easily accessible (Tanner & Wölfing Kast, 2003; Young et al., 2010). In conclusion, the limited availability and inconvenience in procuring products act as barriers and increase the breach between positive attitudes among consumers and the behavior they exhibit toward purchasing green products.
- Product attributes and quality: Some investigations established that product attributes positively influenced the purchase of green products (Chen & Chang, 2012; Young et al., 2010). Likewise, consumers preferred functional attributes of the product fulfilling personal needs and desires over possible ethical characteristics (Chen & Lobo, 2012; Tsakiridou et al., 2008). However, the perceived low quality of green products has a negative influence on consumer green purchase intention and behavior. Therefore, it can be said that functional and sustainable characteristics of products combined with high product quality positively influence consumers' green purchasing behavior.
- Eco-labeling or certification: Young et al. (2010) defined this factor as a variable which informs consumers about the green characteristics of the product and motivates them to purchase green products. Nevertheless, Nittala (2014) stated that consumers do not trust the information provided, and they remain doubtful toward the manufacturing, labeling, and certification procedures of various products. At the same time, eco-labeling can help marketers differentiate their offerings in the minds of consumers. This has a positive effect on consumers' decision

making concerning the purchase of the product, thus increasing the sales (Bougherara & Piguet, 2009).

- **Environmental concern**: Hu, Weng, and Wang (2010) referred to *environmental concerns* (EC) as "the degree to which people are aware of the problems regarding the environment and support efforts to solve them and or indicate the willingness to contribute personally to their solution." Studies explored the consumers' growing attention toward environmental concerns and willingness to pay for sustainable products (Van Doorn & Verhoef, 2011). According to Paul, Modi, and Patel (2016), EC is positively related to attitudes toward the purchase of green products, subjective norms, perceived behavioral control, and green products purchasing intentions.
- **Social pressure and impact**: Joshi and Rahman (2015) described social influence occurring when someone observes the expectations of another or considers the information acquired from another as a symbol of reality. Thus, consumers often purchase green products to show their ecological concern to society. In many studies, social groups as well as subjective or social norms were found to have a positive correlation with purchase intention and the actual purchase of green or other ethically produced products (Eze & Ndubisi, 2013; Vermeir & Verbeke, 2006). In conclusion, people usually adhere to the norms of their society or the groups they belong to because of pressure from influential external sources (Ajzen, 1991).

20.3 Green Consumerism—Tendencies Among Consumers

Several authors have made studies testing various conceptual frameworks related to different green purchasing factors.

20.3.1 Female Consumers

Mobrezi and Khoshtinat (2016) investigated the factors affecting women's willingness to make green purchases based on a model of planned behavior. The study involved 279 women living in the west of Tehran, Iran. The research method was applied, fundamental and developmental in terms of its main objective (see Fig. 20.4) and it is a descriptive-survey in terms of its plan. From this research, came several findings. Firstly, the authors studied that environmental concerns did not affect female consumers' positive attitude. The reasons behind this warrant further investigation. Secondly, the authors analyzed the effect of women's environmental concerns on their willingness to buy green products. The authors' confirmation of this hypothesis suggests that by increasing environmental concerns, female consumers gain increased willingness to buy green products. Thirdly, the subjective

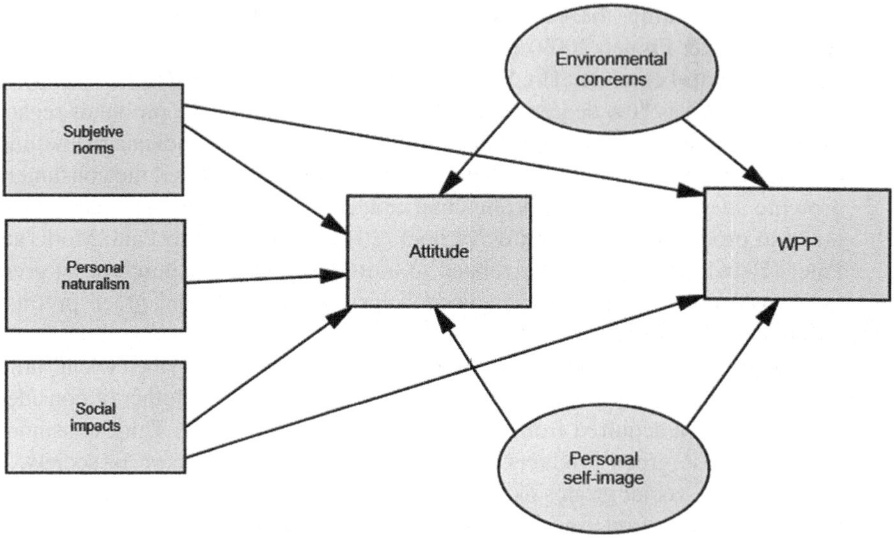

Fig. 20.4 Planned behavior (Mobrezi & Khoshtinat, 2016)

norms of Iranian women do not affect the development of positive attitudes toward green products. An examination of this factor may reveal the root cause: green purchasing may be a part of green consumers' subjective norm. Fourthly, strengthening factors affecting consumers' subjective norm can increase the willingness of Iranian woman to buy green products, Fifthly, the role of social impacts in creating positive attitudes toward green products is highlighted in Iranian women. Therefore, identifying and strengthening social impacts can increase the positive attitude toward green products, but do not cause greater willingness to buy green products among Iranian women. Finally, women's naturalism does not affect their positive attitudes toward green products. Personal self-image does not directly affect women's positive attitudes toward green products.

20.3.2 Young Consumers

According to Paul et al. (2016), an extended *theory of planned behavior* (TPB) includes environmental concern, which is a critical variable in green marketing literature. This study collected data from 521 Indian consumers over 18 years old from India. Using a questionnaire, the authors validated TPB and its extended form called *triple bottom line* (TBL). The findings confirmed the efficacy of an extended TPB as a research model useful for explaining consumers' green product purchase intentions. It specifically validated the claim that when attitude and perceived behavioral control are positive, consumers will be more likely to have purchase

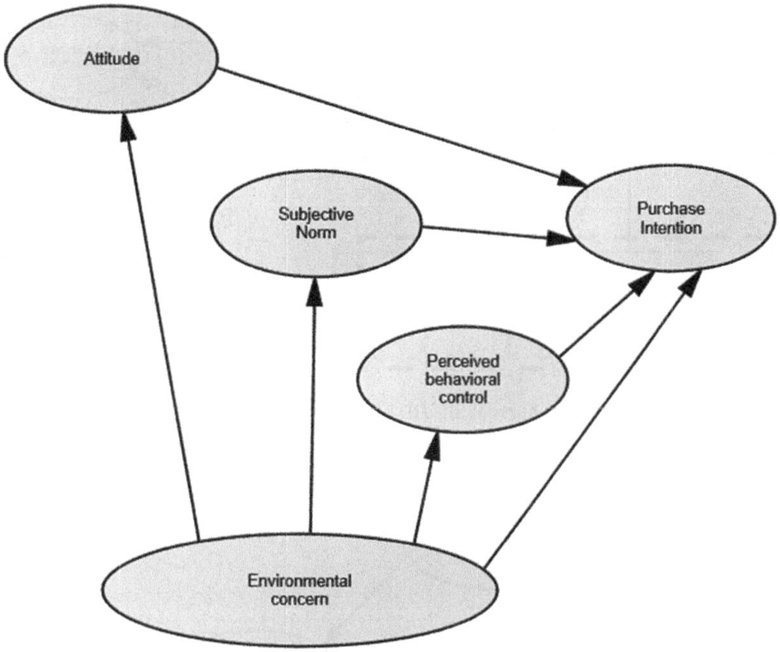

Fig. 20.5 *Proposed research framework (*Paul et al., 2016*)*

intentions for green products. They found that EC was found to be significant and positive for attitude, subjective norm, PBC, and purchase intention for green products. EC was influential indirectly through TPB variables (see Fig. 20.5). EC also influences Indian consumers' perceived behavior control. A plausible reason may be that raised EC motivates consumers to search for sustainable alternatives, thereby yielding greater knowledge about the availability of options. Thus, their searching behavior makes them aware of many green choices which are compatible with their existing brand preferences, thereby reducing the perceived non-availability of green products to some extent.

Another study developed in Portugal by Akehurst, Afonso, and Martins Gonçalves (2012) re-examines the determinants of *ecologically conscious consumer behavior* (ECCB) by analyzing the green consumer profile. Moreover, the study explores the determinants of useful *green purchase behavior* (GPB) considering ECCB and previously evaluated *green purchase intention* (GPI). The target population of this research was composed of individuals living in Portugal, of both sexes, aged over 18 years (186 respondents over 18 years old). The results allow us to conclude that the demographic variables (sex, age, education, and income) are not relevant in explaining environmental conscious consumer behavior. The results reinforce the role of PCE and altruism on ECCB and give support to the attitude-behavior link (Fig. 20.6).

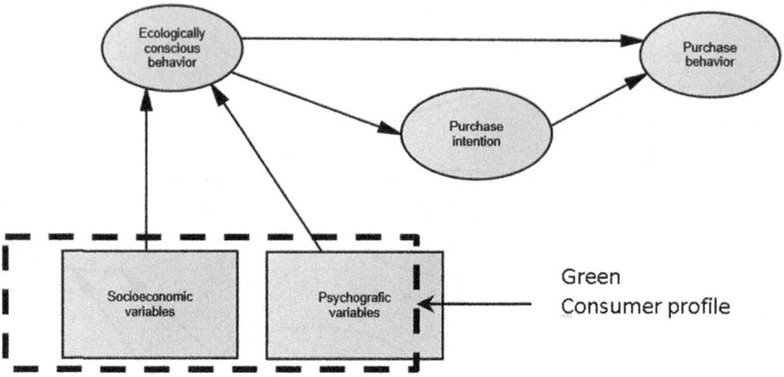

Fig. 20.6 *Research model (*Akehurst et al., 2012*)*

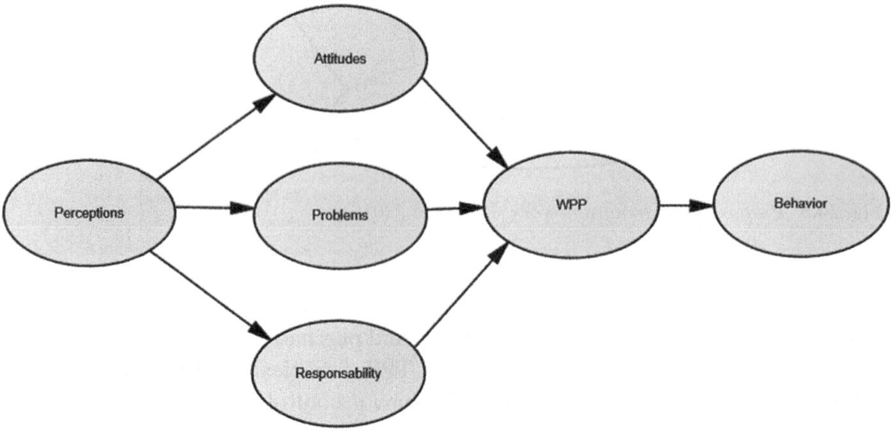

Fig. 20.7 *Structural model (*Lai & Cheng, 2016*)*

20.3.3 Undergraduate Students

Lai and Cheng (2016) examined perceptions of green marketing process among graduate students. It looked at their attitudes toward the environment, including their sense of responsibility and the degree of seriousness represented by such problems. The authors determined whether their behavior and willingness to buy green products will eventually be affected. A sample of 100 undergraduate students in Hong Kong was used. The measurements and structural models (see Fig. 20.7) were examined with the partial least squares approach to structural equation modeling. This research reflects a frugal image about graduate student expectations of green

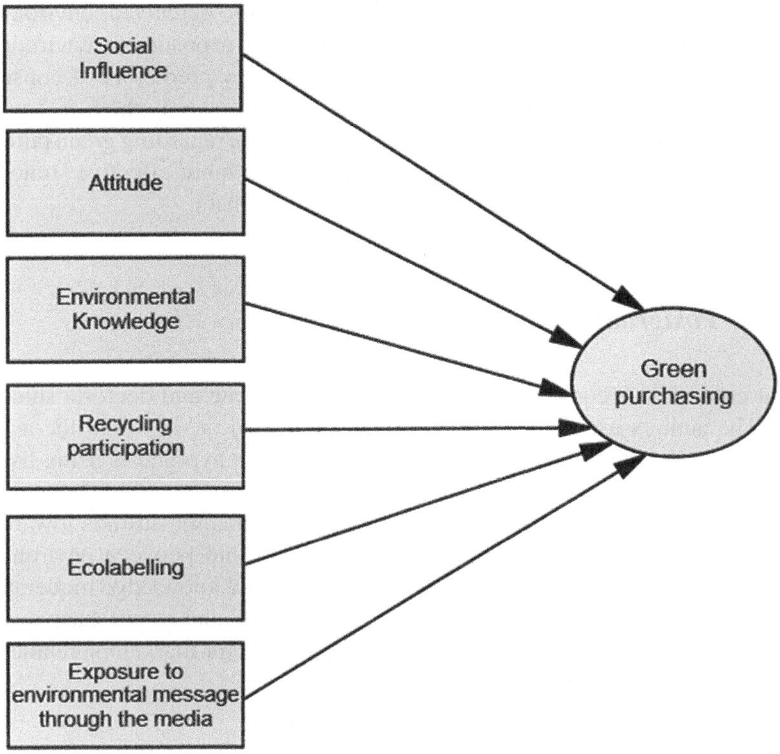

Fig. 20.8 *Conceptual framework (*Joshi & Rahman, 2016*)*

marketing practices, environmental perception, and purchasing behavior. It may provide insights for green marketers formulating strategies to encourage well-educated students to consume additional green products. In conclusion, undergraduate student willingness to purchase green products is more easily activated by rational assessment, can realize their responsibility in environmental protection, and emphasizing the seriousness of the ecological problem may become ineffective in persuading the undergraduate students to have a willingness to purchase green products.

Another study which analyzed undergraduate students was developed by Joshi and Rahman (2015). The purpose was to determine what factors predict green purchase behavior among young, educated consumers in Delhi. A survey was carried out on a sample of 1502 young educated consumers. For each of the variables considered, structural equation modeling was used to assess their predictive power for green purchasing (see Fig. 20.8).

The results indicate that the variables under study predicted green purchase behavior of young, educated consumers of Delhi in the following descending order:

social influence, attitude toward the green purchase, perceived environmental knowledge, recycling participation, eco-labeling, and exposure to environmental messages through the media. The authors identified key predictors of consumers' green purchase behavior, enabling practitioners to understand which factors influence young, educated consumers in their decision making regarding green purchases. This knowledge may help marketing managers design more effective strategies to encourage green purchase behavior among such consumers.

20.3.4 Postgraduate Students

Kumar et al. (2017) considered a sample of 235 graduate and doctoral students in India. The authors used a conceptual framework based on TPB and the available discussion in the literature. From this, they developed a hypothesis using five variables: environmental knowledge, attitude, subjective form, purchase intention, and purchase behavior. The most important findings were that the attitude toward environmentally sustainable products mediates a relationship between environmental knowledge and purchase intention. While environmental knowledge moderates this relationship, the subjective norm is not significantly related to purchase intention. These findings offer valuable information and guidance for marketers, retailers, and green businesses in general (see Fig. 20.9).

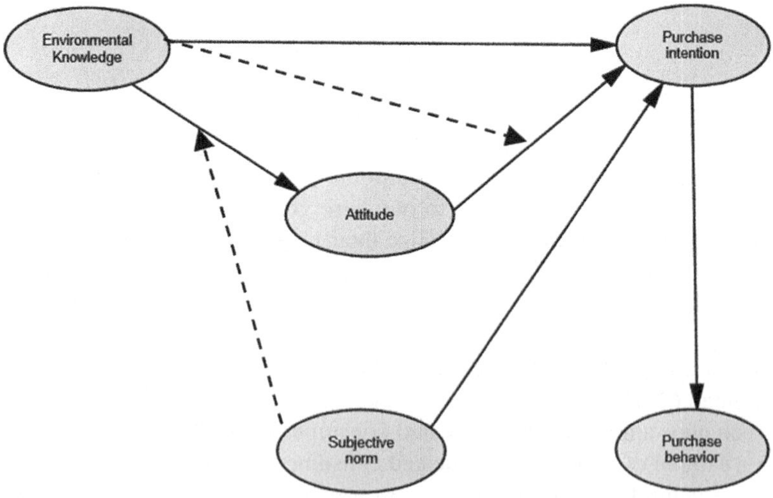

Fig. 20.9 *Conceptual framework (*Kumar et al., 2017*)*

20.4 Green Consumerism—Tendencies by Product Category

20.4.1 Organic Food

Mohd Suki (2016) assessed impacts on green product purchase intention, including the factors of green brand positioning, consumer attitudes toward green brands, and green brand knowledge. The effect of green brand knowledge on consumer attitudes was examined. This specifically included the moderating effect of green brand knowledge on the relationship between green brand positioning and green product purchase intention. Quantitative methods were applied and a structured, self-administered questionnaire was distributed to 350 subjects who practice a green lifestyle. The sample members all had green product purchasing experience, such as purchasing organic vegetables at least once a week at grocery stores and hypermarkets in the Federal Territory of Labuan, Malaysia. The data was analyzed using the *partial least squares* (PLS) method, which is a variance-based technique for the analysis of structural equation modeling (SEM) (see Fig. 20.10).

The authors found several results. First, the green brand positioning has a positive and significant impact on green product purchase intention. Second, the research proves that consumer attitude toward green brands has a significant positive relationship with green product purchase intention. Third, green brand knowledge has a significant effect on green product purchase intention. And finally, green brand knowledge significantly affected consumer attitude toward green brands.

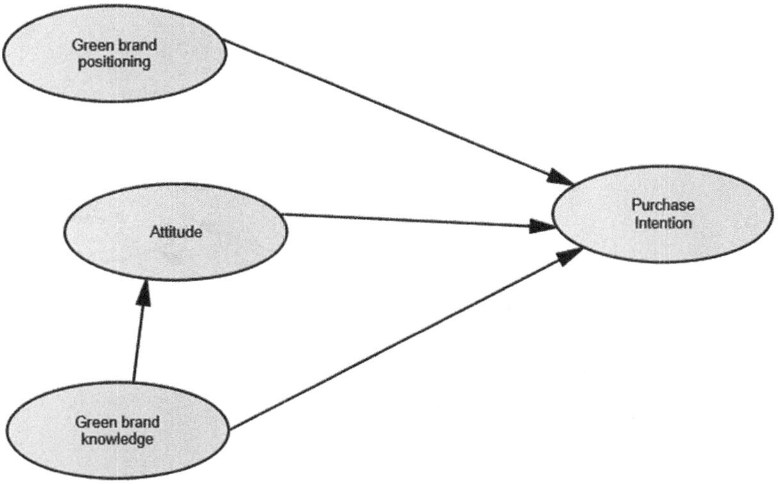

Fig. 20.10 *Proposed theoretical framework (*Mohd Suki, 2016*)*

20.4.2 Household Electrical Appliances

Nguyen, Lobo, and Nguyen (2018) examined some rational, moral, emotional, and self-identity factors. They researched which of these factors may facilitate or impede green purchase behavior, specifically the purchase of energy-efficient appliances in an emerging market, Vietnam. The authors collected data from university students, which yielded a sample of 289 respondents. Multivariate statistics revealed that most factors (i.e., knowledge, attitudes, personal norms, self-identity, and perceived barriers) significantly affected consumer purchase of energy-efficient appliances, except for subjective social norms and warm glow.

From these findings, the authors determined that the environmental self-identity is the most relevant predictor of green purchase behavior among these young consumers. They appeared to associate the purchase of energy-efficient appliances with being environmentally friendly consumers. Another finding showed that knowledge significantly enhances pro-environmental behavior. Additionally, attitude is also a significant predictor of consumer green purchase behavior. Nevertheless, personal norms positively influence green purchase behavior, while the relationship between consumer subjective social norms and their purchase of energy-efficient appliances is not significant. Another study about these products was developed by Chen and Chang (2012). It examined mediation role of green trust. The purpose of this study was to develop an original framework to explore the influences of green perceived value and green perceived risk on green purchase intentions. The research applied four original concepts—green perceived value, green perceived risk, green trust, and green purchase intentions—to develop an integrated model to enhance green purchase intentions. In addition, this research employed an empirical study using the questionnaire survey method to verify the hypotheses and to explore its practical implications. Structural equation modeling (SEM) was also applied to test the research framework (see Fig. 20.11).

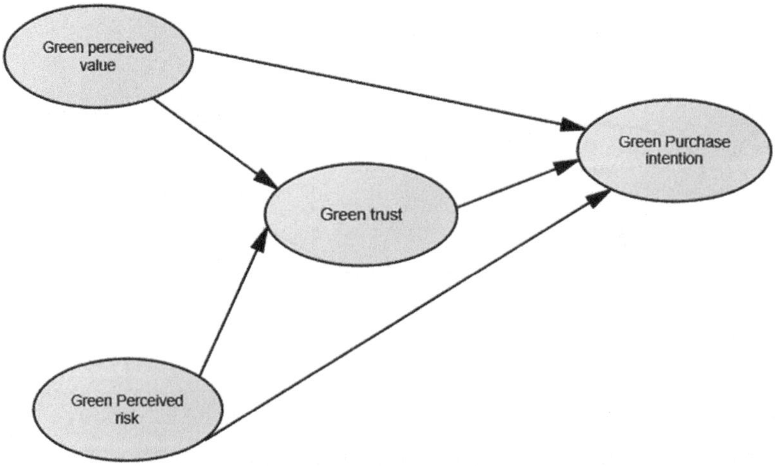

Fig. 20.11 *Research framework* (Chen & Chang, 2012)

The empirical results show that green perceived value would positively affect green trust and green purchase intentions, while green perceived risk would negatively influence both of them. Furthermore, this study demonstrates that the relationships between green purchase intentions and their two antecedents—green perceived value and green perceived risk—are partially mediated by green trust. Hence, investing resources to increase green perceived value and to decrease green perceived risk is helpful to enhance green trust and green purchase intentions. Besides, this study summarizes the literature on green marketing and relationship marketing into a new operational framework of green purchase intentions (Chen & Chang, 2012).

20.4.3 Green Convenience Goods

Chen and Deng (2016) used the theory of planned behavior and examined the moderating effects of product knowledge on the relationships between three independent variables and green purchase intentions. Independent variables included green purchase attitudes, subjective norms, and perceived behavior control. After that, the difference of moderating effects of product knowledge between convenience goods and shopping goods was further analyzed. The data were collected from 306 valid questionnaires. The results of the study showed that product knowledge had a significant moderating effect on the relationship between the three independent variables and green purchase intention, and that the explanatory power of three independent variables would decrease in the context of high product knowledge. Interesting conclusions were reached from the perspective of product classification. This study contributes to the literature in the field of green purchase behavior by treating product knowledge as a moderating variable in the theory of planned behavior and by exploring new research perspective—the perspective of product classification.

20.4.4 Green Personal Care Products and Cosmetics

Wei, Chiang, Kou, and Lee (2017) used cognitive behavior theory to investigate the drivers of green consumer behavior and the missing link in the concern-behavior gap. After collecting 375 valid questionnaires, this study validated the proposed conceptual model using structural equation modeling (see Fig. 20.12). The model indicates that environmental involvement, informational utility, green advertising skepticism, and green trust are antecedent factors influential in consumer attitudes toward green products.

The researchers made several conclusions. First, they confirmed that consumer green purchase behavior is significantly and directly determined by their green purchase intention, which is significantly and directly determined by their attitudes toward green products. Second, environmental involvement significantly and positively affects consumers' attitudes toward green products, confirming a likely

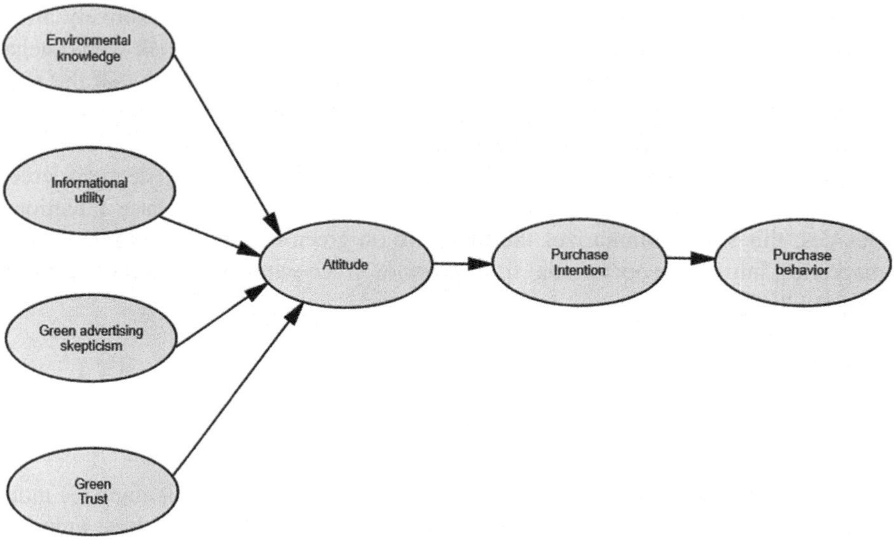

Fig. 20.12 *Proposed model (*Wei et al., 2017*)*

model of persuasion. Third, informational utility and green trust serve as mediating variables in several moderating relationships. Informational service partially mediates the relationships between environmental involvement and green trust, between environmental participation and attitudes toward green products, and between green advertising skepticism and green trust.

Finally, Liobikienė and Bernatonienė (2017) reviewed 80 papers published from 2011 to 2017 on green purchase behavior. This work revealed that most studies on the topic were conducted during the last three years. Moreover, the review showed that authors obtained different results in their analysis of differing categories of green products and the corresponding purchase behavior associated with them. Thus, they suggested that future researchers consider categories of green products since different factors have different influences within these categories. In addition, they proposed a model for the purchasing behavior of green personal care products, including cosmetics (see Fig. 20.13), giving particular attention to health consciousness, as well as with brand and quality as the primary determinants. This study provides insight for other researchers, policy makers, and marketing managers seeking to promote green products purchasing behavior.

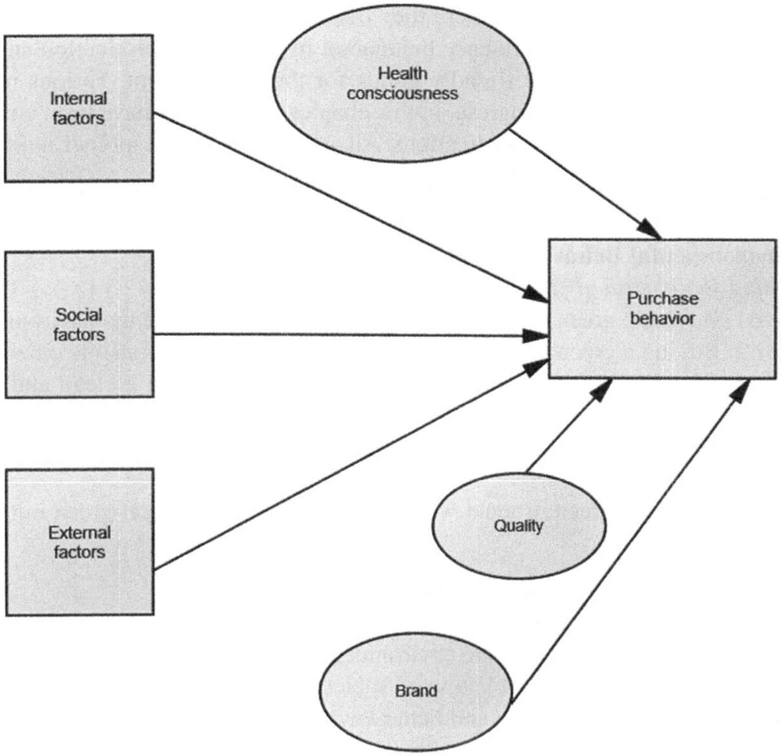

Fig. 20.13 *Model of green personal care and color cosmetics purchase behavior (*Liobikienė & Bernatonienė, 2017*)*

20.5 Evaluation of Determinants of Green Consumption: Case of Peru

Today in Peru, contracting laws are in effect according to legislative decree 1017 and the Regulation of the State Procurement Law issued by supreme decree 184-2008-EF. Both standards limit purchases according to criteria of sustainability. It is important to note that in any contracting, criteria will be applied to guarantee environmental sustainability and avoid adverse impacts on the environment (principle of environmental sustainability). For green purchasing to become a reality in Peru, it is first necessary to gain commitment at the senior-most levels of the Peruvian Government. Thus, the Ministry of the Environment and the Agency Supervisor of State Contracting (OSCE) must lead and commit to this achievement. Only then will practices reflect state policy as outlined in the National Agreement (Palacios, 2006).

According to Foy Valencia (2011), public spending in many states makes a significant contribution to the GDP of those countries and are the primary consumers within the national economies where they belong. For this reason, this purchasing power is an opportunity to influence the market by promoting production and consumption of sustainable and friendly goods for the environment. Various metrics can be used for measuring progress. In this chapter, these items have been collected and can be used for various investigations. All are assessed on a 5-point Likert scale.

I. Attitude = Environmental Behavioral Belief (EBB) + Environmental Outcome Evaluation (OE)

Environmental Behavioral Belief (EBB)

Rating scale: strongly disagree (1) to strongly agree (5)

EBB1: Buying a green product would enable me to help save the environment

EBB2: Buying a green product would enable me to be a responsible citizen

EBB3: Buying a green product would enable me to stay in a clean and better environment

EBB4: Buying a green product would enable me to perform eco-friendly practices

EBB5: Buying a green product would enable me to implement green initiatives in my life.

Environmental Outcome Evaluation (OE)

Rating scale: not at all important (1) to extremely important (5)

OE1: To me helping to save the environment is

OE2: To me being responsible toward society is

OE3: To me staying in clean and better environment is

OE4: To me performing eco-friendly practices is

OE5: To me implementing green initiatives in my life is.

II. Subjective Norm = Normative belief (NB) + Motivation to comply

Normative Belief (NB)

Rating scale: strongly disagree (1) to strongly agree (5)

NB1: My family thinks I should purchase green products in place of conventional non-green products

NB2: My friends think I should purchase green products in place of conventional non-green products

NB3: My colleagues think I should purchase green products in place of conventional non-green products.

Motivation to Comply (MC)

Rating scale: extremely unlikely (1) to extremely likely (5)

MC1: How likely it is for you to do what your family thinks you should do?

MC2: How likely it is for you to do what your friends think you should do?

MC3: How likely it is for you to do what your colleagues think you should do?

III. Perceived Behavioral Control = Control beliefs (CB) + Perceived power (PP).

Control Belief (CB)

Rating scale: since strongly disagree (1) to strongly agree (5)

CB1: While buying the green products, the location needs to be convenient

CB2: Buying green products requires time and effort
CB3: My company/school/others that pay(s) for my expenses encourage(s) me to use green products.

Perceived power (PP)
Rating scale: strongly disagree (1) to strongly agree (5)
PP1: Location is a critical factor while making decision to buy green products
PP2: Time and effort needed to buy is very important while making decision to buy green products
PP3: The expenses available to me is very critical while making decision to buy green products.

IV. Subjective norm (SN)
Rating scale: strongly disagree (1) to strongly agree (5)
SN1: Most people who are important to me would want me to purchase eco-friendly products
SN2: Most people who are important to me would think I should purchase green products
SN3: I think this kind of green products is more in line with my moral values
SN4: I think this type of green products more in line with the wishes of my relatives and friends
SN5: I think this kind of green products is in consistent with the trend of social development
SN6: Most of the people who are important to me think that I should buy energy-efficient appliances
SN7: Most of my acquaintances expect me to buy energy-efficient appliances
SN8: Most of the people who are important to me would support me buying energy-efficient appliances next time
SN9: The people I listen to could influence me buying energy-efficient appliances.

V. Perceived behavioral control (PBC)
Rating scale: strongly disagree (1) to strongly agree (5)
PBC1: Whether or not I buy green product at place of conventional non-green product is completely up to me
PBC2: I have resources, time, and opportunities to buy green product
PBC3: I am confident that if I want to, I can buy green product at place of conventional non-green product
PBC3: It is easy to buy a green product
PBC4: I cannot decide whether to buy this kind of product
PBC5: It is very likely I will choose green products next time
PBC6: Buying habits will have a strong impact on decision making
PBC7: Past purchase experience will have a strong impact on decision-making.

VI. Person's Effects (PE)
Rating scale: strongly disagree (1) to strongly agree (5)
PE1: It frightens me to imagine that many of the products I have are disrupting the environment

PE2: When I think of the way humans are destroying the environment, I get angry and frustrated
PE3: Humans are really abusing the environment
PE4: The balance of nature is easily disrupted, especially by human activity
PE5: We should take responsibility for environmental issues as we are the cause of environmental damage.

VII. Willingness to pay premium (WPP)
Rating scale: strongly disagree (1) to strongly agree (5)
WPP1: I would pay more for a green product that is making efforts to be environmentally sustainable
WPP2: I would be willing to pay this extra percentage on the green products to support the organization's/product efforts to be environmentally sustainable
WPP3: I am willing to buy products that use recycle/recyclable packaging
WPP4: I am willing to buy products that contain no or fewer chemical ingredients
WPP5: I am willing to buy products that support fair community trades
WPP6: I am willing to buy products that are against animal-testing
WPP7: I am willing to buy products that are labeled as environmentally safe
WPP8: I am willing to buy organic products.

VIII. Knowledge (K)
Rating scale: strongly disagree (1) to strongly agree (5)
K1: Practically all of the lead pollution in the atmosphere is caused by humans
K2: Going green products could be a beneficial investment in long-term
K3: Green product's environmental performance meets my expectations
K4: Lack of availability of access is a major reason for low popularity and demand of green products
K5: I purchase green product because it is environmental friendly, has more environmental benefit than other products
K6: I am familiar with this kind of green products
K7: When buying green products, I read the specific information on the label
K8: I believe in testing and identifying green products that are implemented by certification organizations
K9: I am familiar with energy-efficient appliances (EEA)
K10: I am knowledgeable about the environmental impact of EEAs
K11: I am knowledgeable about energy rating labels
K12: I am knowledgeable about EEAs
K13: I know I buy products and packages that are environmentally safe
K14: I know more about recycling than the average person
K15: I know how to select products and packages that reduce the amount of waste ending up in landfills
K16: I understand the environmental phrases and symbols on the product package
K17: I am very knowledgeable about environmental issues.

IX. Intention (I)
Rating scale: strongly disagree (1) to strongly agree (5)

I1: I avoid buying products which are potentially harmful to the environment
I2: I have changed my principal products for ecological reasons
I3: When I have to choose between two similar products, I choose the one that is less harmful to the environment
I4: I make a special effort to buy paper and plastic products that are made from recycled materials
I5: I will not consider environmental issues when making a purchase
I6: I intend to buy green product because of your environmental concern
I7: I expect to purchase green product in the future because of its environmental benefits
I8: Overall, I'm glad to purchase green product because it is environmental friendly
I9: Comparing with ordinary non-green products, I am more willing to buy green products
I10: The next time to buy, the possibility of choosing green products is very high
I11: The next time to buy, the desire to choose green products is not strong.

X. Actual Purchase (AP)

Rating scale: strongly disagree (1) to strongly agree (5)
AP1: I feel that I have played a great part in helping the environment when I use green products
AP2: I feel more comfortable when I use green products rather than normal ones
AP3: There is not much I can do about the environment, and my experience of green products does not change my belief
AP4: I aim to buy green products again after my first purchase
AP5: I would recommend green products to my friends and family.

XI. Attitude (ATT)

Rating scale: strongly disagree (1) to strongly agree (5)
ATT1: I feel that green product's environmental reputation is generally reliable
ATT2: I feel that green product's environmental performance is generally dependable
ATT3: I feel that green product's environmental claims are generally trustworthy
ATT4: Green product's environmental concern meets my expectations
ATT5: Green products keep promises and responsibilities for environmental protection
ATT6: Green purchase brings us more benefits than non-green purchase
ATT7: Buying green energy-saving products will make me happy
ATT8: Buying a product, I will consider how it will affect the environment
ATT9: I am willing to spend a little more money to buy green products
ATT10: Environmental protection is important to me when making purchases of appliances
ATT11: Energy-efficient appliances are important to reduce air pollution
ATT12: Energy-efficient appliances are important to save natural resources that would be used for producing energy, e.g., coal, water

ATT13: If I can choose between energy-efficient and conventional appliances, I prefer energy-efficient
ATT14: Environmental protection is important to me when making purchases of appliances
ATT15: I think environmental protection works are simply a waste of money and resources
ATT16: I think environmental protection is meaningless
ATT17: It is meaningful to invest resources to educate citizens to protect and sustain the environment
ATT18: I think more environmental protection works are needed in your city
ATT19: It is very important to raise environmental awareness among your community
ATT20: Environmental protection should be one of the most important agenda for your government
ATT21: It is important to raise environmental awareness among your community
ATT22: It is meaningful to protect our environment
ATT23: More environmental protection is needed in your country
ATT24: I am concerned about environmental protection.

XII. Perceived value (PV)

Rating scale: strongly disagree (1) to strongly agree (5)
PV1: The green product's environmental functions provide good value to me
PV2: The green product's environmental performance meets my expectations
PV3: I purchase green product because it has more environmental concern than non-green products
PV4: I purchase green product because it is environmental friendly
PV5: I purchase green product because it has more environmental benefit than non-green products
PV6: This product's environmental functions provide very good value for you
PV7: This product's environmental performance meets your expectations
PV8: You purchase this product because it has more environmental concern than other products
PV9: You purchase this product because it is environmental friendly
PV10: You purchase this product because it has more environmental benefit than other products.

XIII. Purchase Intention (PI)

Rating scale: strongly disagree (1) to strongly agree (5)
PI1: I will purchase green products for personal use
PI2: I am willing to purchase green products for personal use
PI3: I will make an effort to purchase green products
PI4: I will consider buying energy-efficient appliances
PI5: I plan to switch to other brands/versions of electrical appliances that are more energy-efficient
PI6: I intend to buy energy-efficient appliances
PI7: I will buy energy-efficient appliances in my next purchase

PI8: You intend to purchase this product because of its environmental concern
PI9: You expect to purchase this product in the future because of its environmental performance
PI10: Overall, you are glad to purchase this product because it is environmental friendly.

XIV. Purchase Behavior (PB)

Rating scale: strongly disagree (1) to strongly agree (5)
PB1: I have been purchasing green products on a regular basis
PB2: I have green purchasing behavior for my daily needs products
PB3: I have green purchasing behavior over the past six months
PB4: When shopping, I deliberately check products for environmentally harmful ingredients
PB5: When shopping, I deliberately choose products with environmentally friendly packaging
PB6: I will choose to buy environment-friendly products, even if they are more expensive than other products
PB7: When shopping, when I consider buying a product, I will look for a certified environmental label
PB8: I often buy products that use recycled/recyclable packaging
PB9: I often buy products that contain no or fewer chemical ingredients
PB10: When I go shopping, I will look for products with certified environmentally safe or organic stamp
PB11: I often buy products that support fair community trades
PB12: I often buy products that are against animal-testing
PB13: I often buy products that are labeled as environmentally safe
PB14: I often buy organic products.

XV. Green Brand Positioning—Eco-labeling (ECO)

Rating scale: strongly disagree (1) to strongly agree (5)
ECO1: Quality and price is important when consumers purchase green products
ECO2: I get to know about green branding through advertisement
ECO3: Green products have matched my personal wants and needs
ECO4: Green product always overpriced
ECO5: I prefer to purchase environmentally green products
ECO6: Whenever possible, I buy products packaged in reusable containers
ECO7: I purchase the green products if they are certified by environmental organization
ECO8: I trust the eco-friendly claims in the advertisements
ECO9: Marketers must advertise the environmental aspects of their products
ECO10: Government must make eco-labeling mandatory
ECO11: Modifying product packaging to minimize harmful to the environment
ECO12: Educating consumers to use products in environmentally friendly manner
ECO13: Manufacturing products through eco-friendly processes

ECO14: Modifying products to make them environmentally friendly
ECO15: Manufacturing eco-friendly products
ECO16: Using green supply chain for procurement and distribution of products
ECO17: Branding products with green labels
ECO18: Promoting products through eco-friendly modes of communication.

XVI. Perceived Risk (PR)

Rating scale: strongly disagree (1) to strongly agree (5)
PR1: There is a chance that there will be something wrong with environmental performance of this product
PR2: There is a chance that this product will not work properly with respect to its environmental design
PR3: There is a chance that you would get environmental penalty or loss if you use this product
PR4: There is a chance that using this product will negatively affect the environment
PR5: Using this product would damage your green reputation or image.

XVII. Trust (T)

Rating scale: strongly disagree (1) to strongly agree (5)
T1: You feel that this product's environmental reputation is generally reliable
T2: You feel that this product's environmental performance is generally dependable
T3: You feel that this product's environmental claims are generally trustworthy
T4: This product's environmental concern meets your expectations
T5: This product keeps promises and commitments for environmental protection.

Validation and use of instruments are needed to measure environmental behavior and purchasing behavior in consumers (Lopez-Odar, Alvarez-Risco, Vara-Horna, Chafloque-Cespedes, & Sekar, 2019).

The characteristics of each market have a direct effect on green consumerism; however, there are very few studies that describe social influence on green behavior. The study by Clark et al (2019) comparing social influence in China versus the USA is one of the few that address this issue.

Closing Remarks

There is still a lot to investigate and more research into the variables impacting green purchasing behavior is required. This should be done in different countries so the socio-economic variables that may be more universal can be seen and be determined and explained. While the price barrier to access different green products is decreasing, it remains only speculation that more consumers prefer green products. Marketing professionals will have to know what attributes current and potential green consumers actually value. They should anticipate and investigate changing preferences and possible increases in green purchasing, as well as learn how to generate targeted advertising strategies for each segment. Several countries already favor ecological consumption. Now it is necessary to evaluate any resulting changes in behavior. This should be accompanied by further efforts to foster greater awareness about green products, provide detailed and correct information, avoid greenwashing, and allow consumers to support or reject such green products based on their personal experience.

References

Ajzen, I. (1991). The theory of planned behavior. *Organizational behavior and human decision Processes, 50*(2), 179–211.

Ajzen, I. (2002). Perceived behavioral control, self-efficacy, locus of control, and the theory of planned behavior. *Journal of Applied Social Psychology, 32*(4), 665–683.

Ajzen, I., & Fishbein, M. (1980). *Understanding attitude and predicting social behavior*. Eaglewood Cliffs: Prentice-Hall.

Akehurst, G., Afonso, C., & Martins Gonçalves, H. (2012). Re-examining green purchase behaviour and the green consumer profile: New evidences. *Management Decision, 50*(5), 972–988.

Allport, F. H. (1935). Attitudes. In C. A. Murchison (Ed.), *Handbook of social psychology* (pp. 1–50). Worcester, MA: Clark University Press.

Arvola, A., Vassallo, M., Dean, M., Lampila, P., Saba, A., Lähteenmäki, L., et al. (2008). Predicting intentions to purchase organic food: The role of affective and moral attitudes in the Theory of Planned Behaviour. *Appetite, 50*(2), 443–454.

Bougherara, D., & Piguet, V. (2009). Market behavior with environmental quality information costs. *Journal of Agricultural & Food Industrial Organization, 7*(2), 22.

Chan, R. Y. (2001). Determinants of Chinese consumers' green purchase behavior. *Psychology & Marketing, 18*(4), 389–413.

Chen, J., & Lobo, A. (2012). Organic food products in China: Determinants of consumers' purchase intentions. *The International Review of Retail, Distribution and Consumer Research, 22*(3), 293–314.

Chen, Y. S., & Chang, C. H. (2013). Greenwash and green trust: The mediation effects of green consumer confusion and green perceived risk. *Journal of Business Ethics, 114*(3), 489–500.

Chen, K., & Deng, T. (2016). Research on the green purchase intentions from the perspective of product knowledge. *Sustainability, 8*(9), 55.

Chen, Y. S. (2010). The drivers of green brand equity: Green brand image, green satisfaction, and green trust. *Journal of Business Ethics, 93*(2), 307–319.

Chen, Y. S., & Chang, C. H. (2012). Enhance green purchase intentions: The roles of green perceived value, green perceived risk, and green trust. *Management Decision, 50*(3), 502–520.

Clark, R. A., Haytko, D. L., Hermans, C. M., & Simmers, C. S. (2019). Social influence on green consumerism: country and gender comparisons between China and the United States. *Journal of International Consumer Marketing, 31*(3), 177–190.

Connell, S., Fien, J., Lee, J., Sykes, H., & Yencken, D. (1999). If it doesn't directly affect you, you don't think about it: A qualitative study of young people's environmental attitudes in two Australian cities. *Environmental Education Research, 5*(1), 95–113.

Eze, U. C., & Ndubisi, N. O. (2013). Green buyer behavior: Evidence from Asia consumers. *Journal of Asian and African Studies, 48*(4), 413–426.

Fishbein, M., & Ajzen, I. (1975). *Beliefs, attitude, intention and behavior: An introduction to theory and research*. Reading, MA.: Addison-Wesley.

Foy Valencia, P. (2011). Consideraciones sobre la contrataci6n pública sostenible ("verde"). Derecho PUCP, (66), 335–350.

Fryxell, G. E., & Lo, C. W. (2003). The influence of environmental knowledge and values on managerial behaviours on behalf of the environment: An empirical examination of managers in China. *Journal of Business Ethics, 46*(1), 45–69.

Gatersleben, B., Murtagh, N., & Abrahamse, W. (2014). Values, identity and pro-environmental behaviour. *Contemporary Social Science, 9*(4), 374–392.

Ginsberg, J. M., & Bloom, P. N. (2004). Choosing the right green marketing strategy: Green marketing has not fulfilled its initial promise, but companies can take a more effective approach if they realize that a one-size-fits-all strategy does not exist. *MIT Sloan Management Review, 46*(1), 79–85.

Gleim, M. R., Smith, J. S., Andrews, D., & Cronin Jr., J. J. (2013). Against the green: A multimethod examination of the barriers to green consumption. *Journal of Retailing, 89*(1), 44–61.

Hu, Q. H., Weng, J. Q., & Wang, J. S. (2010). Sources of anthropogenic radionuclides in the environment: A review. *Journal of Environmental Radioactivity, 101*(6), 426–437.

Joshi, Y., & Rahman, Z. (2015). Factors affecting green purchase behaviour and future research directions. *International Strategic Management Review, 3*(1–2), 128–143.

Joshi, Y., & Rahman, Z. (2016). Predictors of young consumer's green purchase behaviour. *Management of Environmental Quality: An International Journal, 27*(4), 452–472.

Kang, J., Liu, C., & Kim, S. H. (2013). Environmentally sustainable textile and apparel consumption: The role of consumer knowledge, perceived consumer effectiveness and perceived personal relevance. *International Journal of Consumer Studies, 37*(4), 442–452.

Kang, S., & Hur, W. M. (2012). Investigating the antecedents of green brand equity: A sustainable development perspective. *Corporate Social Responsibility and Environmental Management, 19*(5), 306–316.

Kumar, B., Manrai, A., & Manrai, L. (2017). Purchasing behaviour for environmentally sustainable products: A conceptual framework and empirical study. *Journal of Retailing and Consumer Services vol., 34*(C), 1–9.

Lai, C. K. M., & Cheng, E. W. L. (2016). Green purchase behavior of undergraduate students in Hong Kong. *The Social Science Journal, 53*(1), 67–76.

Liobikienė, G., & Bernatonienė, J. (2017). Why determinants of green purchase cannot be treated equally? The case of green cosmetics: Literature review. *Journal of Cleaner Production, 162*, 109–120.

Lopez-Odar, D., Alvarez-Risco, A., Vara-Horna, A., Chafloque-Cespedes, R., & Sekar, M. C. (2019). Validity and reliability of the questionnaire that evaluates factors associated with perceived environmental behavior and perceived ecological purchasing behavior in Peruvian consumers. Social Responsibility Journal. https://doi.org/10.1108/SRJ-08-2018-0201

MacInnis, D. J., Moorman, C., & Jaworski, B. J. (1991). Enhancing and measuring consumers' motivation, opportunity, and ability to process brand information from ads. *Journal of Marketing, 55*, 32–53.

Mobrezi, H., & Khoshtinat, B. (2016). Investigating the Factors Affecting Female Consumers' Willingness toward green purchase based on the model of planned behavior. *Procedia Economics and Finance, 36*, 441–447.

Mohd Suki, N. (2016). Green product purchase intention: Impact of green brands, attitude, and knowledge. *British Food Journal, 118*(12), 2893–2910.

Nguyen, T. N., Lobo, A., & Nguyen, B. K. (2018). Young consumers' green purchase behaviour in an emerging market. *Journal of Strategic Marketing, 26*(7), 583–600.

Nittala, R. (2014). Green consumer behavior of the educated segment in India. *Journal of International Consumer Marketing, 26*(2), 138–152.

Padel, S., & Foster, C. (2005). Exploring the gap between attitudes and behaviour: Understanding why consumers buy or do not buy organic food. *British Food Journal, 107*(8), 606–625.

Palacios, M. (2006). Towards the implementation of environmental sustainability patterns in production and consumption: Sustainable public procurement. [Hacia la Implementación de Patrones de Sostenibilidad Ambiental en la Producción y Consumo: Compras Públicas Sostenibles]. *Derecho & Sociedad, 2*(42), 201–212.

Paul, J., Modi, A., & Patel, J. (2016). Predicting green product consumption using theory of planned behavior and reasoned action. *Journal of Retailing and Consumer Services, 29*, 123–134.

Tanner, C., & Wölfing Kast, S. (2003). Promoting sustainable consumption: Determinants of green purchases by Swiss consumers. *Psychology & Marketing, 20*(10), 883–902.

Tarkiainen, A., & Sundqvist, S. (2005). Subjective norms, attitudes and intentions of Finnish consumers in buying organic food. *British Food Journal, 107*(11), 808–822.

Tsakiridou, E., Boutsouki, C., Zotos, Y., & Mattas, K. (2008). Attitudes and behaviour towards organic products: An exploratory study. *International Journal of Retail & Distribution Management, 36*(2), 158–175.

Van der Werff, E., Steg, L., & Keizer, K. (2013). The value of environmental self-identity: The relationship between biospheric values, environmental self-identity and environmental preferences, intentions and behaviour. *Journal of Environmental Psychology, 34*, 55–63.

Van Doorn, J., & Verhoef, P. C. (2011). Willingness to pay for organic products: Differences between virtue and vice foods. *International Journal of Research in Marketing, 28*(3), 167–180.

Vermeir, I., & Verbeke, W. (2006). Sustainable food consumption: Exploring the consumer "attitude–behavioral intention" gap. *Journal of Agricultural and Environmental Ethics, 19*(2), 169–194.

Wang, P., Liu, Q., & Qi, Y. (2014). Factors influencing sustainable consumption behaviors: A survey of the rural residents in China. *Journal of Cleaner Production, 63*, 152–165.

Webster Jr., F. E. (1975). Determining the characteristics of the socially conscious consumer. *Journal of Consumer Research*, 188–196.

Wei, C., Chiang, C., Kou, T., & Lee, B. (2017). Toward Sustainable Livelihoods: Investigating the Drivers of Purchase Behavior for Green Products. *Business Strategy and the Environment, 26*(5), 626–639.

Yadav, R., & Pathak, G. S. (2017). Determinants of consumers' green purchase behavior in a developing nation: Applying and extending the theory of planned behavior. *Ecological Economics, 134*, 114–122.

Young, W., Hwang, K., McDonald, S., & Oates, C. J. (2010). Sustainable consumption: Green consumer behaviour when purchasing products. *Sustainable Development, 18*(1), 20–31.

Zeithaml, V. (1988). Consumer perceptions of price, quality, and value: A means-end model and synthesis of evidence. *Journal of Marketing, 52*(3), 2–22.

Chapter 21
Sustainable Development Goals and Cities

Aldo Alvarez-Risco, Shyla Del-Aguila-Arcentales, and Marc A. Rosen

Abstract The sustainable development goals (SDGs) proposed by the United Nations seek the transformation of society, in order to achieve sustainable development for all the inhabitants of the Earth. Cities are an important part of this effort, as they are the nucleus for the generation of plans and actions for the fulfillment of many of the SDGs. Including resilience in global plans can help cities succeed in spite of their particular complications. The advantages and disadvantages that artificial intelligence can bring to the SDGs, as well as the various results that have been achieved, are raised. Monitoring of the SDG Index and the Tier Classification for Global SDG Indicators are presented as reference tools for actions oriented towards the SDGs. Finally, progress on SDG compliance is described. The SDGs need coordinated efforts to record and measure the progress made.

Keywords Sustainability · Sustainable development goals · Cities · United Nations

21.1 The Sustainable Development Goals

The sustainable development goals (SDGs), illustrated in Fig. 21.1, are a call for action by all countries—developed and developing—in a global partnership. The SDGs were adopted by all United Nations Member States in 2015, for the 2015–2030

A. Alvarez-Risco (✉)
Facultad de Ciencias Administrativas y Recursos Humanos, Universidad de San Martín de Porres, Lima, Peru
e-mail: aldo.alvarez@redsaf.org

S. Del-Aguila-Arcentales
Escuela Nacional de Marina Mercante "Almirante Miguel Grau", Callao, Peru

Universidad Nacional de la Amazonia Peruana, Iquitos, Peru
e-mail: sdelaguila@enamm.edu.pe

M. A. Rosen
Faculty of Engineering and Applied Science, University of Ontario Institute of Technology, Oshawa, ON, Canada
e-mail: marc.rosen@uoit.ca

A. Alvarez-Risco et al. (eds.), *Building Sustainable Cities*,
https://doi.org/10.1007/978-3-030-45533-0_21

Fig. 21.1 Sustainable development goals adopted at the 70th Session of the UN General Assembly in 2015 (public domain material provided by United Nations at http://www.un.org/sustainabledevelopment/news/communications-material/)

period. 17 goals were established that focused on addressing persistent and profound problems such as poverty, health, and education. The SDGs also encompass actions to reduce inequalities and promote economic growth. At the same time, measures are proposed for tackling climate change and preserving the world's oceans and forests. Finally, the SDGs also encourage states and companies to formulate fair work for the citizens of the world and global collaboration. The SDGs are not only an environmentally focused set of goals, as many think, but rather a multidimensional approach to sustainable development. This approach has been accepted and endorsed at the governmental level in most UN countries, but its practical execution in a given country requires efforts related to the economic-commercial activities of companies and the individual conduct of people.

Cities are where it is expected that the SDGs will be able to achieve the most visible changes. Citizens have an active role in contributing to the SDGs, both in their individual activities and at the group level such as via state institutions, NGOs, companies, and social groups.

21.1.1 How Were the SDGs Developed?

In 2012, the United Nations Rio + 20 summit in Brazil committed governments to create a set of sustainable development goals (SDGs) that would be integrated into the follow-up to the millennium development goals (MDGs) after their 2015 expiry. Taking into account the projected growth of the world population to 9 billion by 2050, it was suggested that the SDGs focus on various aspects of life on earth: addressing poverty, reducing environmental pollution, and enhancing sustainable social/business development.

The resulting SDGs are focused on addressing the many components of sustainability through their 17 objectives, which represent a multidimensional approach to sustainable development. Due to the complex natures of the goals, several of them are complementary to each other. Some may conflict regarding their simultaneous fulfillment. For example, managing hunger through agricultural or marine productivity may lead to problems like deforestation or the disappearance of marine fauna.

21.2 SDGs and Cities

A city is considered to be a complex, multidimensional, and dynamic system in which constant changes occur, physically, socially, economically and, increasingly, environmentally. Given this complexity, cities require planning that supports projects needed by cities to contribute successfully to the sustainable development goals. In line with this need, the participating States in the United Nations (UN) Conference on Housing and Sustainable Urban Development (Habitat III) set a series of objectives that contribute to making cities inclusive, safe and, finally, sustainable.

The concept of urban resilience is increasingly used for the planning and development of public policies as well as the carrying out of activities coordinated by international agencies seeking the sustainable development of cities. The term resilience refers to the ability to cope with borderline situations and, at the same time, to adapt to changing circumstances. From the perspective of urban areas, this concept refers to the ability of cities to resist and overcome the various challenges they face. Some of these are extremely harmful, leading to the destruction of homes, killing people, and causing large financial damages. The poorest people are often the most affected by these problems. The approaches to urban resilience are transversal to what is established by the 17 SDGs, with SDG 1 (poverty), SDG 2 (hunger), SDG 9 (infrastructure), and SDG 11 (sustainable cities) being particularly relevant. In relation to resilience, UN-Habitat has developed various guidelines, indexes, methods, and instruments that allow monitoring of the Resilience Program; likewise, international actors such as the World Bank and the Inter-American Development Bank have developed similar guidelines to achieve urban resilience, especially in cities that have suffered the most damage. Urban growth occurs mostly in emerging economies and developing countries, but a common feature of this growth is that it is unplanned and informally carried out (OECD, 2017). Making the situation more challenging is the report that 1.4 million people move daily to urban areas, and that a growing number of people are subject to wars and climate impact.

In relation to the resilience of cities, it has been recommended that coordinating agencies should (IEG, 2019):

1. Identify and track urban resilience—building efforts
2. Systematically incorporate resilience characteristics in projects
3. Use analytical work to inform support in areas of crime and violence

4. Ensure that urban resilience interventions in a country are complementary and coordinated
5. Articulate long-term, client-oriented financing plans.

One important initiative is the creation of the 100 Resilient Cities (100RC) network. From 2013 to 2019, the network planned and executed technical and financial assistance to cities interested in building resilience and developed the City Resilience Framework (CRF) and the City Resilience Index (CRI). The Cape Town Resilience Strategy was recently launched (Capetown, 2019), composed of five pillars, 20 objectives, and 75 actions. Croese, Green, and Morgan (2020) point out that these are planned to be aligned with the SDGs, as described in Table 21.1.

21.3 SDGs, Cities, and Artificial Intelligence

The amount of data handled in a country, a city, and even a district needs proper management to achieve planning and to execute actions that contribute to the SDGs. Likewise, these data need to be controlled and interconnected, making the use of artificial intelligence (AI) necessary. For example, it has been shown that AI augments the productivity of some workers, and can replace the work done by others. AI will likely transform almost all occupations at least to some degree (Frank et al., 2019), as observed in, for instance, the extensive use of chatbots (Luo, Tong, Fang, & Qu, 2019). AI is expected to affect human resource management (Kreutzer & Sirrenberg, 2020), education (Duong et al., 2019; Mohammed, 2019), entrepreneurship, and other activities. Due to its potential positive impact on human and business activities, AI is expected to play an important role in the development and implementation of the SDGs. This is in part because AI software technology facilitates tasks such as perception (facial recognition), business decision making (purchase of shares of the stock market), prediction (sale of homes in an area of the city), and information evaluation (impact of research published by government research institutes), among others.

When analyzed in greater detail, AI can have a transcendent role in critical aspects of cities such as health. Chen, Joshi, and Ghassemi (2020), for instance, suggest healthcare, as currently practiced, is not equitable, and that disparities that follow the biases of society have been consistently demonstrated. How can AI help correct these disparities? Consciously used, AI can be leveraged as a tool to help highlight and erase the numerous and well-documented inequities in health. However, AI can be advantageous or disadvantageous, since biases existing in the care of minorities can be taken as "normal" and extended in the health system.

But what about the impact of AI on the other SDGs? Vinuesa et al. (2020) describe how AI can act as both an inhibitor and an enabler. Several SDGs have a social impact, which is why we refer to them collectively as social SDGs. Table 21.2 shows the target numbers for each goal and groups them, based on available evidence, into those for which AI is an enabler or an inhibitor.

Several SDGs have an economic impact and are referred to collectively here as economic SDGs. Table 21.3 shows the target numbers for each goal and groups them, based on available evidence, into those for which AI is an enabler or an inhibitor.

Table 21.1 Sustainable development goals and Cape Town Resilience Strategy parallelism

Cape Town's Resilience Strategy	SDGs	
Pillars and goals	*Goals*	*Targets*
Pillar 1: Compassionate, holistically healthy city		
Goal 1.1: Increase awareness of, access to, and uptake of mental health support Goal 1.2: Embrace a more holistic approach to policing and crime prevention to break the cycle of violence and to lower recidivism rates and trauma Goal 1.3: Combat discrimination and build social cohesion Goal 1.4: Promote a culture of health that increases well-being and decreases trauma	SDG 2: Hunger SDG 3: Health SDG 5: Gender SDG 10: Inequality SDG 16: Peace and justice	Target 2.1 Target 3.4–3.5 Target 5.2 Target 10.2–10.3 Target 16.1–16.2
Pillar 2: Connected, climate-adaptive city	*Goals*	*Target*
Goal 2.1: Grow partnerships that strengthen transportation systems and improve mobility. Goal 2.2: Engage communities and the private sector to improve public spaces Goal 2.3: Build climate resilience Goal 2.4: Innovate for improved conditions, service delivery, and well-being in informal settlements	SDG 6: Water SDG 11: Cities SDG 13: Climate action SDG 14: Life below water	Target 6.4–6.5–6.6 Target 11.2–11.3–11.7 Target 13.3 Target 14.2
Pillar 3: Capable, job-creating city	*Goals*	*Target*
Goal 3.1: Foster green economic growth Goal 3.2: Enable enterprise development in the informal economy Goal 3.3: Connect the workforce with a changing economy Goal 3.4: Partner with businesses to achieve a resilient local economy	SDG 4: Education SDG 8: Decent work and economic growth SDG 9: Infrastructure SDG 11: On cities SDG 12: Responsible consumption and production SDG 17: On partnerships	Target 4.4 Target 8.2–8.3–8.5–8.9 Target 9.4 Target 11.6 Target 12.7 Target 17.17
Pillar 4: Collectively, shock-ready city	*Goals*	*Target*
Goal 4.1: Future-proof urban systems Goal 4.2: Strengthen individual, household, and community resilience Goal 4.3: Encourage responsible investment in household and business resilience Goal 4.4: Explore funding mechanisms for shock events	SDG 1: On poverty SDG 6: Water SDG 7: Energy SDG 9: Infrastructure SDG 11: On cities SDG 13: Climate action	Target 1.5 Target 6.b Target 7.2 Target 9.4 Target 11.5–11.b Target 13.3
Pillar 5: Collaborative, forward-looking city	*Goals*	*Target*
Goal 5.1: Develop and approve portfolios of projects that maximize the resilience dividend Goal 5.2: Mainstream resilience in decision-making Goal 5.3: Enhance knowledge management and data-use Goal 5.4: Monitor and evaluate resilience outcomes	SDG 11: On cities	Target 11.b-11.3

Source: Croese et al. (2020)

Table 21.2 Assessment of the impact of AI on social SDGs

Social SDGs	Targets of each goal for which AI acts as an enabler	Targets of each goal for which AI acts as an inhibitor
Goal 1	1.1, 1.2, 1.3, 1.4, 1.5, 1.A, 1.B	1.1, 1.2, 1.3, 1.4, 1.A, 1.B
Goal 2	2.1, 2.2, 2.3, 2.4, 2.A, 2.C	2.3, 2.A
Goal 3	3.1, 3.2, 3.3, 3.4, 3.5, 3.6, 3.9, 3.B, 3.D	3.9
Goal 4	4.1, 4.2, 4.3, 4.4, 4.5, 4.6, 4.7, 4.A, 4.B, 4.C	4.1, 4.2, 4.3, 4.4, 4.5, 4.6, 4C

Source: Vinuesa et al. (2020)

Table 21.3 Assessment of the impact of AI on economic SDGs

Economic SDGs	Targets of each goal for which AI acts as an enabler	Targets of each goal for which AI acts as an inhibitor
Goal 13	13.1, 13.2, 13.3, 13B	13.2
Goal 14	14.1, 14.2, 14.3, 14.4, 14.5, 14.7, 14.A, 14.B, 14.C	14.2, 14.5, 14.7
Goal 15	15.1, 15.2, 15.3, 15.4, 15.5, 15.6, 15.7, 15.8, 15.9, 15.A, 15.B, 15.C	15.2, 15.4, 15.5, 15.7

Source: Vinuesa et al. (2020)

Several SDGs have an environmental impact and are referred to collectively here as environmental SDGs. Table 21.4 shows the target numbers for each goal and groups them, based on available evidence, into those for which AI is an enabler or an inhibitor.

21.4 Monitoring of the SDG Index

The annual SDG Index provides a standardized, quantitative, transparent, and scalable composite measure of SDG baselines for 149 countries, with sufficient data across the goals. According to the Sustainable Development Report 2019 (Sachs, Schmidt-Traub, Kroll, Lafortune, & Fuller, 2019), the SDG Index data leads to seven main findings.

1. *High-level political commitment to the SDGs is falling short of historic promises*

   In September 2019, out of 43 countries surveyed on SDG implementation efforts, 33 have endorsed the SDGs in official statements since January 1, 2018 but only 18 of them claimed that their central budget documents mentioned the SDGs. This gap between rhetoric and action must be closed.

2. *The SDGs can be operationalized through six SDG transformations*

SDG implementation can be organized along the following transformations: a. education, gender, and inequality; b. health, well-being, and demography; c. energy decarbonization and sustainable industry; d. sustainable food, land, water, and oceans; e. sustainable cities and communities; and f. digital revolution for sustainable development.

Table 21.4 Assessment of the impact of AI on environmental SDGs

Environmental SDGs	Targets of each goal for which AI acts as an enabler	Targets of each goal for which AI acts as an inhibitor
Goal 1	1.1, 1.2, 1.3, 1.4, 1.5, 1.A, 1.B	1.1, 1.2, 1.3, 1.4, 1.A, 1.B
Goal 2	2.1, 2.2, 2.3, 2.4, 2.A, 2.C	2.3, 2.A
Goal 3	3.1, 3.2, 3.3, 3.4, 3.5, 3.6, 3.9, 3.B, 3D	3.9
Goal 4	4.1, 4.2, 4.3, 4.4, 4.5, 4.6, 4.7, 4.A, 4.B, 4.C	4.1, 4.2, 4.3, 4.4, 4.5, 4.6, 4C
Goal 5	5.1, 5.2, 5.5, 5.6, 5B	5.1, 5.6, 5B
Goal 6	6.1, 6.2, 6.3, 6.4, 6.5, 6.6, 6.A, 6.B	6.1, 6.4, 6.5, 6A, 6B
Goal 7	7.1, 7.2, 7.3, 7A, 7B	7.1, 7.2
Goal 11	11.1, 11.2, 11.3, 11.4, 11.5, 11.6, 11.7, 11.A, 11.B, 11.C	11.2, 11.3
Goal 16	16.1, 16.2, 16.3, 16.4, 16.5, 16.7, 16.A	16.1, 16.2, 16.3

Source: Vinuesa et al. (2020)

3. *Trends on climate (SDG 13) and biodiversity (SDG 14 and SDG 15) are alarming*
   Countries have not achieved relevant results in SDG 13, even if efforts against climate change appear to be effective. The same applies for SDG 14 and SDG 15, as reported by the Intergovernmental Panel on Climate Change (IPCC, 2019) and the Intergovernmental Science-Policy Platform on Biodiversity and Ecosystem Services (IPBES, 2018).
4. *Sustainable land-use and healthy diets require integrated agriculture, climate, and health policy interventions*
   78% of world nations for which data are available obtain a "red rating" on sustainable nitrogen management. One-third of food is wasted, 800 million people remain undernourished, 2 billion are deficient in micronutrients, and obesity is on the rise.
5. *High-income countries generate high environmental and socioeconomic spill-over effects*
   Domestic implementation of the SDGs should not undermine the ability of other countries to achieve the goals. Tolerance for poor labor standards in international supply chains harms the poor, and particularly women, in many developing countries.
6. *Human rights and freedom of speech are in danger in numerous countries*
   Conflicts in many parts of the world continue to lead to reversals in SDG progress. Modern slavery and the share of unsentenced detainees in prisons remain high, particularly in low-income countries.
7. *Eradicating poverty and strengthening equity remain important policy priorities*
   In middle- and high-income countries, rising income inequalities and persistent gaps in access to services and opportunities by income or territorial areas remain important policy issues. Women in OECD countries continue to spend an average of two hours more than men a day doing unpaid work.

21.5 Tier classification for global SDG indicators (UN, 2019a)

This classification is established to differentiate the indicators according to:

1. If a methodology has been established internationally.
2. If the data is produced continuously by at least 50% of countries and the population.

The tier classification for global SDG indicators can be expressed as follows:

Tier 1: Indicator is conceptually clear, has an internationally established methodology and standards are available, and data are regularly produced by countries for at least 50 percent of countries and of the population in every region where the indicator is relevant.

Tier 2: Indicator is conceptually clear, has an internationally established methodology and standards are available, but data are not regularly produced by countries.

Tier 3: No internationally established methodology or standards are yet available for the indicator, but methodology/standards are being (or will be) developed or tested.

The current classification contains 116 Tier I indicators, 92 Tier II indicators, and 20 Tier III indicators. In addition to these, there are four indicators that have multiple tiers (different components of the indicator are classified into different tiers). Due to the clarity of Tier I indicators, they should be the ones that are mainly addressed and monitored.

The indicators for each of the sustainable development goals classified as Tier I are as follows:

Goal 1. End poverty in all its forms everywhere

1.1.1 Proportion of population below the international poverty line, by sex, age, employment status, and geographical location (urban/rural).
1.2.1 Proportion of population living below the national poverty line, by sex and age.
1.4.1 Proportion of population living in households with access to basic services.

Goal 2. End hunger, achieve food security and improved nutrition, and promote sustainable agriculture

2.1.1 Prevalence of undernourishment.
2.1.2 Prevalence of moderate or severe food insecurity in the population, based on the Food Insecurity Experience Scale (FIES).
2.2.1 Prevalence of stunting (height for age < -2 standard deviation from the median of the World Health Organization (WHO) Child Growth Standards) among children under 5 years of age.

2.2.2 Prevalence of malnutrition (weight for height > +2 or < −2 standard deviation from the median of the WHO Child Growth Standards) among children under 5 years of age, by type (wasting and overweight).
2.5.1 Number of plant and animal genetic resources for food and agriculture secured in either medium- or long-term conservation facilities.
2.a.1 The agriculture orientation index for government expenditures.
2.a.2 Total official flows (official development assistance plus other official flows) to the agriculture sector.
2.b.1 Agricultural export subsidies.

Goal 3. Ensure healthy lives and promote well-being for all at all ages

3.1.1 Maternal mortality ratio.
3.1.2 Proportion of births attended by skilled health personnel.
3.2.1 Under-5 mortality rate.
3.2.2 Neonatal mortality rate.
3.3.1 Number of new HIV infections per 1000 uninfected population, by sex, age and key populations.
3.3.2 Tuberculosis incidence per 100,000 population.
3.3.3 Malaria incidence per 1000 populations.
3.3.4 Hepatitis B incidence per 100,000 populations.
3.3.5 Number of people requiring interventions against neglected tropical diseases.
3.4.1 Mortality rate attributed to cardiovascular disease, cancer, diabetes or chronic respiratory disease.
3.4.2 Suicide mortality rate.
3.5.2 Harmful use of alcohol, defined according to the national context as alcohol per capita consumption (aged 15 years and older) within a calendar year in liters of pure alcohol.
3.6.1 Death rate due to road traffic injuries.
3.7.1 Proportion of women of reproductive age (aged 15–49 years) who have their need for family planning satisfied with modern methods.
3.7.2 Adolescent birth rate (aged 10–14 years; aged 15–19 years) per 1000 women in that age group.
3.8.1 Coverage of essential health services (defined as the average coverage of essential services based on tracer interventions that include reproductive, maternal, newborn and child health, infectious diseases, non-communicable diseases and service capacity and access, among the general and the most disadvantaged population).
3.8.2 Proportion of population with large household expenditures on health as a share of total household expenditure or income.
3.9.1 Mortality rate attributed to household and ambient air pollution.
3.9.2 Mortality rate attributed to unsafe water, unsafe sanitation, and lack of hygiene (exposure to unsafe Water, Sanitation and Hygiene for All (WASH) services).
3.9.3 Mortality rate attributed to unintentional poisoning.

3.a.1 Age-standardized prevalence of current tobacco use among persons aged 15 years and older.
3.b.1 Proportion of the target population covered by all vaccines included in their national programme.
3.b.2 Total net official development assistance to medical research and basic health sectors.
3.c.1 Health worker density and distribution.
3.d.1 International Health Regulations (IHR) capacity and health emergency preparedness.

Goal 4. Ensure inclusive and equitable quality education and promote lifelong learning opportunities for all

4.1.1 Proportion of children and young people (a) in grades 2/3; (b) at the end of primary; and (c) at the end of lower secondary achieving at least a minimum proficiency level in (i) reading and (ii) mathematics, by sex.
4.2.2 Participation rate in organized learning (one year before the official primary entry age), by sex
4.b.1 Volume of official development assistance flows for scholarships by sector and type of study.

Goal 5. Achieve gender equality and empower all women and girls

5.3.1 Proportion of women aged 20–24 years who were married or in a union before age 15 and before age 18.
5.3.2 Proportion of girls and women aged 15–49 years who have undergone female genital mutilation/cutting, by age.
5.5.1 Proportion of seats held by women in (a) national parliaments and (b) local governments.
5.5.2 Proportion of women in managerial positions.

Goal 6. Ensure availability and sustainable management of water and sanitation for all

6.4.1 Change in water-use efficiency over time.
6.4.2 Level of water stress: freshwater withdrawal as a proportion of available freshwater resources.
6.5.1 Degree of integrated water resources management implementation (0–100).
6.5.2 Proportion of transboundary basin area with an operational arrangement for water cooperation.
6.6.1 Change in the extent of water-related ecosystems over time.
6.a.1 Amount of water- and sanitation-related official development assistance that is part of a government-coordinated spending plan.
6.b.1 Proportion of local administrative units with established and operational policies and procedures for participation of local communities in water and sanitation management.

Goal 7. Ensure access to affordable, reliable, sustainable, and modern energy for all

7.1.1 Proportion of population with access to electricity.
7.1.2 Proportion of population with primary reliance on clean fuels and technology.
7.2.1 Renewable energy share in the total final energy consumption.
7.3.1 Energy intensity measured in terms of primary energy and GDP.
7.a.1 International financial flows to developing countries in support of clean energy research and development and renewable energy production, including in hybrid systems.

Goal 8. Promote sustained, inclusive and sustainable economic growth, full and productive employment and decent work for all

8.1.1 Annual growth rate of real GDP per capita.
8.2.1 Annual growth rate of real GDP per employed person.
8.4.2 Domestic material consumption, domestic material consumption per capita, and domestic material consumption per GDP.
8.5.2 Unemployment rate, by sex, age, and persons with disabilities.
8.6.1 Proportion of youth (aged 15–24 years) not in education, employment, or training.
8.10.1 (a) Number of commercial bank branches per 100,000 adults and (b) number of automated teller machines (ATMs) per 100,000 adults.
8.10.2 Proportion of adults (15 years and older) with an account at a bank or other financial institution or with a mobile-money service provider.
8.a.1 Aid for trade commitments and disbursements.

Goal 9. Build resilient infrastructure, promote inclusive and sustainable industrialization, and foster innovation

9.1.2 Passenger and freight volumes, by mode of transport.
9.2.1 Manufacturing value added as a proportion of GDP and per capita.
9.2.2 Manufacturing employment as a proportion of total employment.
9.3.2 Proportion of small-scale industries with a loan or line of credit.
9.4.1 CO_2 emission per unit of value added.
9.5.1 Research and development expenditure as a proportion of GDP.
9.5.2 Researchers (in full-time equivalent) per million inhabitants.
9.a.1 Total official international support (official development assistance plus other official flows) to infrastructure.
9.b.1 Proportion of medium and high-tech industry value added in total value added.
9.c.1 Proportion of population covered by a mobile network, by technology.

Goal 10. Reduce inequality within and among countries

10.5.1 Financial soundness indicators.
10.6.1 Proportion of members and voting rights of developing countries in international organizations.
10.a.1 Proportion of tariff lines applied to imports from least developed countries and developing countries with zero-tariff.
10.c.1 Remittance costs as a proportion of the amount remitted.

Goal 11. Make cities and human settlements inclusive, safe, resilient, and sustainable

11.1.1 Proportion of urban population living in slums, informal settlements, or inadequate housing.
11.6.2 Annual mean levels of fine particulate matter (e.g., PM2.5 and PM10) in cities (population weighted).

Goal 12. Ensure sustainable consumption and production patterns

12.2.2 Domestic material consumption, domestic material consumption per capita, and domestic material consumption per GDP.
12.4.1 Number of parties to international multilateral environmental agreements on hazardous waste, and other chemicals that meet their commitments and obligations in transmitting information as required by each relevant agreement.
12.c.1 Amount of fossil-fuel subsidies per unit of GDP (production and consumption) and as a proportion of total national expenditure on fossil fuels.

Goal 14. Conserve and sustainably use the oceans, seas, and marine resources for sustainable development

14.4.1 Proportion of fish stocks within biologically sustainable levels.
14.5.1 Coverage of protected areas in relation to marine areas.
14.6.1 Degree of implementation of international instruments aiming to combat illegal, unreported, and unregulated fishing.
14.7.1 Sustainable fisheries as a proportion of GDP in small island developing states, least developed countries, and all countries.
14.b.1 Degree of application of a legal/regulatory/policy/institutional framework which recognizes and protects access rights for small-scale fisheries.

Goal 15. Protect, restore, and promote sustainable use of terrestrial ecosystems, sustainably manage forests, combat desertification, and halt and reverse land degradation and halt biodiversity loss

15.1.1 Forest area as a proportion of total land area.
15.1.2 Proportion of important sites for terrestrial and freshwater biodiversity that are covered by protected areas, by ecosystem type.
15.2.1 Progress towards sustainable forest management.
15.3.1 Proportion of land that is degraded over total land area.
15.4.1 Coverage by protected areas of important sites for mountain biodiversity.
15.4.2 Mountain Green Cover Index.
15.5.1 Red List Index.
15.6.1 Number of countries that have adopted legislative, administrative, and policy frameworks to ensure fair and equitable sharing of benefits.

Goal 16. Promote peaceful and inclusive societies for sustainable development, provide access to justice for all, and build effective, accountable and inclusive institutions at all levels

16.1.1 Number of victims of intentional homicide per 100,000 population, by sex and age.

16.3.2 Unsentenced detainees as a proportion of overall prison population.
16.5.2 Proportion of businesses that had at least one contact with a public official and that paid a bribe to a public official, or were asked for a bribe by those public officials during the previous 12 months.
16.8.1 Proportion of members and voting rights of developing countries in international organizations.
16.9.1 Proportion of children under 5 years of age whose births have been registered with a civil authority, by age.
16.10.2 Number of countries that adopt and implement constitutional, statutory, and/or policy guarantees for public access to information.
16.a.1 Existence of independent national human rights institutions in compliance with the Paris Principles.

Goal 17. Strengthen the means of implementation and revitalize the Global Partnership for Sustainable Development

17.1.1 Total government revenue as a proportion of GDP, by source.
17.1.2 Proportion of domestic budget funded by domestic taxes.
17.2.1 Net official development assistance, total and to least developed countries, as a proportion of the Organization for Economic Cooperation and Development (OECD) Development Assistance Committee donors' gross national income (GNI).
17.3.1 Foreign direct investment (FDI), official development assistance, and South-South cooperation as a proportion of total domestic budget.
17.3.2 Volume of remittances (in United States dollars) as a proportion of total GDP.
17.4.1 Debt service as a proportion of exports of goods and services.
17.6.2 Fixed Internet broadband subscriptions per 100 inhabitants, by speed.
17.8.1 Proportion of individuals using the Internet.
17.9.1 Dollar value of financial and technical assistance (including through North-South, South-South and triangular cooperation) committed to developing countries.
17.10.1 Worldwide weighted tariff-average.
17.11.1 Developing countries' and least developed countries' share of global exports.
17.12.1 Average tariffs faced by developing countries, least developed countries, and small island developing states.

21.6 What is the progress of SDG compliance?

According to The Sustainable Development Goals Report 2019 (UN, 2019b), the main progress regarding SDG compliance is as follows:

SDG 1

- 55% of the world's population have no access to social protection.

- 736 million people lived in extreme poverty in 2015, 413 million in sub-Saharan Africa.
- More than 90% of deaths due to disasters occur in low and middle-income countries.

SDG 2

- 821 million were undernourished in 2017, up from 784 million in 2015.
- Two-thirds of undernourished people worldwide live in two regions: sub-Saharan Africa (237 million) and southern Asia (277 million).
- 22% (149 million) of children under 5 are stunted.
- 7.3% (49 million) of children under 5 are affected by wasting.
- 5.9% (40 million) of children under 5 are overweight.

SDG 3

- Under-5 deaths dropped from 9.8 million in 2000 to 5.4 million in 2017.
- The tuberculosis incidence rate declined by 21% between 2000 and 2017; nonetheless 10 million people developed tuberculosis in 2017.
- Vaccinations resulted in an 80% drop in measles deaths between 2000 and 2017.
- The incidence of HIV among adults 15–49 years old in sub-Saharan Africa declined by 37% between 2010 and 2017.
- There were an estimated 3.5 million more malaria cases in the ten highest-burden African countries in 2017 compared to 2016.

SDG4

- 617 million children and adolescents lack minimum proficiency in reading and mathematics.
- 750 million adults still remain illiterate. Two-thirds of them are women.
- More than half of the schools in sub-Saharan Africa do not have access to:
- Basic drinking water
- Handwashing facilities
- The Internet
- Computers.
- 1 out of 5 children between 6 and 17 years are not attending school.
- In Central Asia, 27% more girls than boys of primary school age are not attending school.

SDG 5

- 18% of ever-partnered women and girls aged 15–49 years have experienced physical and/or sexual partner violence in the previous 12 months.
- 24% of national parliamentarians are women 24%, an increase from 19% (2010).
- In Southern Asia, a girl's risk of marrying in childhood has decreased by 40% since 2000.
- 30% of women aged 20–24 years were married before age 18 (2018).
- At least 200 million girls and women have been subjected to female genital mutilation, half of them in west Africa.

SDG6

- 785 million people remain without even basic drinking water services (2017).
- 2 out of 5 people worldwide do not have basic handwashing facility with soap and water at home (2017).
- 1 out of 4 health-care facilities worldwide lack basic drinking water services (2016).
- By 2030, 700 million people could be displaced by intense water scarcity.
- 673 million people (9% of the global population) still practice open defecation (2017). The majority of them are in southern Asia.

SDG 7

- 9 out of 10 people worldwide have access to electricity.
- On average, 2.3% less energy was needed to create a unit of economic output each year (2010–2016).
- 87% of the 840 million people without electricity live in rural areas.
- 17.5% of total final energy consumption comes from renewable energy.
- 3 billion people lack clean cooking fuels and technology.

SDG 8

- Real GDP grew by 4.8% annually in the least developed countries (LDC) (2010–2017), well less than the 7% SDG target.
- In 2018, labor productivity increased by 2.1% from 2017, the highest annual growth since 2010.
- The median hourly pay of men is 12% higher than that of women.
- The global unemployment rate is 5% (2018).
- One fifth of young people have education, employment, or training

SDG 9

- Medium-high and high-tech sectors account for 45% of the global manufacturing value added (2016), but the share is only 15% in sub-Saharan Africa.
- Global investment in research and development is $2 trillion (2016), up from $739 billion (2000).
- 90% of people live within range of a 3G or higher quality mobile network (2018), but not all can afford to use it.

SDG 10

- In more than half of the 92 countries having data, income of the bottom 40% of the population grew faster than the national average (2011–2016).
- Most countries have policies to facilitate safe and orderly migration, but more work remains to be done to protect migrants' rights and socioeconomic well-being.
- 66% of products exported from the LDCs receive duty-free treatment (2017), compared to 51% for developing regions.

SDG 11

- 2 billion people do not have access to waste collection services.

- 1 out of 4 urban residents live in slum-like conditions (2018).
- Only half (53%) of urban residents have convenient access to public transport (2018).
- 9 out of 10 urban residents breathe polluted air.
- 150 countries have developed national urban plans, with almost half of them in the implementation phase.

SDG 12

- The material footprint per capita in high-income countries is 60% higher than in upper-middle-income countries and more than 13 times the level of low-income countries.
- Developed countries use one-fifth of natural resources to produce the same amount of economic output as developing countries.
- Nearly 100 countries are actively adopting policies and measures to promote sustainable consumption and production.

SDG 13

- The global mean temperature in 2018 is approximately 1 °C above the pre-industrial baseline.
- Climate-related and geophysical disasters claimed an estimated 1.3 million lives between 1998 and 2017.
- 186 Parties have ratified the Paris Agreement.
- To limit global warming to 1.5 °C, global carbon emissions need to fall to 55% of 2010 levels by 2030 and continue a steep decline to zero net emissions by 2050.
- The previous point occurred despite an increase in global climate finance flows of 17% (2015–2016), compared with 2013–2014.
- Investment in fossil fuels continues to be higher than investment in climate activities.

SDG 14

- Ocean acidity has increased by 26% since pre-industrial times.
- Ocean acidity is expected to increase rapidly, by 100–150%, by 2100.
- 104,220 coastal regions improved their coastal water quality (2012–2018).
- The proportion of fish stocks within biologically sustainable levels declined from 90% (1974) to 67% (2015).
- 87 countries signed the Agreement on Port State Measures, the first binding international agreement on illegal, unreported, and unregulated fishing.
- 17% of waters under national jurisdiction are covered by protected areas, more than double the 2010 coverage level.

SDG 15

- Red List Index: Biodiversity loss is happening at an accelerated rate. The risk of species extinction has worsened by almost 10% over the last 25 years.

- 116 Parties have ratified the Nagoya Protocol, which addresses access to genetic resources and their fair and equitable use.
- Land degradation is affecting one-fifth of the Earth's land area and the lives of 1 billion people.

SDG 16

- Men make up around 80% of homicide victims overall. But women constitute 64% of homicide victims of intimate partner/family-related homicide.
- 70% of detected victims of human trafficking are women and girls, most of whom are trafficked for sexual exploitation.
- The UN recorded and verified 397 additional killings of human rights defenders, journalists and trade unionists across 41 countries (Jan–Oct, 2018). 91 journalists and bloggers were among the victims.

SDG 17

- Net official development assistance (ODA) totalled $149 billion in 2018, down by 2.7% from 2017.
- In 2018, bilateral ODA to the LDCs fell by 3% in real terms from 2017.
- Aid to Africa fell by 4%.

21.7 Steps for SDG implementation

The steps for SDG implementation are (UNGC, 2019):

1. Understand the SDGs and their targets.
2. Conduct principled prioritization of SDG targets.
3. Define your SDG-related report content.
4. Set business objectives.
5. Select appropriate disclosures.
6. Collect and analyze data.
7. Consider general features of good practice when reporting on the SDGs.
8. Consider data users' information needs.
9. Report and implement change.

Closing Remarks
The United Nations Sustainable Development Goals have been developed to motivate and help countries to make sincere efforts to contribute to the sustainable development of the world, with consequent benefits to their citizens, economically, socially, and environmentally. The indicators are increasingly leading to actions that contribute to the sustainable development. It is important to be able to monitor and assess at the city level the level of SDG compliance, and to identify and demonstrate the corresponding effects of actions. Cities need to able to become more sustainable, and this can be achieved by the relevant authorities in a manner framed by the SDGs.

Various tools are available to assist SDG monitoring. However, knowing in detail the damage that is generated by inaction helps raise awareness of the pending planning and execution. The activities that people, companies, and governments carry out to address the SDGs are important and require indexes or scales through which the effort of all can be reflected. A balanced score card is needed for micro and macro sustainability management at global, regional, local, and individual levels.

References

Capetown. (2019). Cape Town resilience strategy. Retrieved February 5, 2020, from http://resource.capetown.gov.za/documentcentre/Documents/City%20strategies%2C%20plans%20and%20frameworks/Resilience_Strategy.pdf

Chen, I. Y., Joshi, S., & Ghassemi, M. (2020). Treating health disparities with artificial intelligence. *Nature Medicine, 26*(1), 16–17.

Croese, S., Green, C., & Morgan, G. (2020). Localizing the sustainable development goals through the lens of urban resilience: Lessons and Learnings from 100 resilient cities and Cape Town. *Sustainability, 12*(2), 550.

Duong, M. T., Rauschecker, A. M., Rudie, J. D., Chen, P. H., Cook, T. S., Bryan, R. N., et al. (2019). Artificial intelligence for precision education in radiology. *The British Journal of Radiology, 92*(1103). https://doi.org/10.1259/bjr.20190389

Frank, M. R., Author, D., Bessen, J. E., Brynjolfsson, E., Cebrian, M., Deming, D. J., et al. (2019). Toward understanding the impact of artificial intelligence on labor. *Proceedings of the National Academy of Sciences, 116*(14), 6531–6539.

IEG. (2019). Building urban resilience. Retrieved February 5, 2020, from https://ieg.worldbankgroup.org/sites/default/files/Data/Evaluation/files/UrbanResilience.pdf

IPBES. (2018). The assessment report on land degradation and restoration. Retrieved February 5, 2020, from https://ipbes.net/sites/default/files/2018_ldr_full_report_book_v4_pages.pdf

IPCC. (2019). AR6 climate change 2021: Mitigation of climate change. Retrieved February 5, 2020, from https://www.ipcc.ch/report/sixth-assessment-report-working-group-3

Kreutzer, R. T., & Sirrenberg, M. (2020). Fields of application of artificial intelligence – Health care, education and human resource management. In *Understanding artificial intelligence* (pp. 167–193). Cham: Springer.

Luo, X., Tong, S., Fang, Z., & Qu, Z. (2019). Frontiers: Machines vs. humans: The impact of artificial intelligence chatbot disclosure on customer purchases. *Marketing Science, 38*(6), 937–947.

Mohammed, P. S. (2019). Towards inclusive education in the age of artificial intelligence: Perspectives, challenges, and opportunities. In *Artificial intelligence and inclusive education* (pp. 17–37). Singapore: Springer.

OECD. (2017). Resilient cities. Retrieved February 5, 2020, from http://www.oecd.org/cfe/regional-policy/resilient-cities.htm

Sachs, J., Schmidt-Traub, G., Kroll, C., Lafortune, G., & Fuller, G. (2019). *Sustainable development report 2019.* New York: Bertelsmann Stiftung and Sustainable Development Solutions Network (SDSN).

UN. (2019a). Tier classification for global SDG indicators. Retrieved February 5, 2020, from https://unstats.un.org/sdgs/iaeg-sdgs/tier-classification

UN. (2019b). The sustainable development goals report 2019. Retrieved February 5, 2020, from https://unstats.un.org/sdgs/report/2019/The-Sustainable-Development-Goals-Report-2019.pdf

UNGC. (2019). Integrating the SDG into corporate reporting: A practical guide. Retrieved February 5, 2020, from https://www.globalreporting.org/resourcelibrary/GRI_UNGC_Reporting-on-SDGs_Practical_Guide.pdf

Vinuesa, R., Azizpour, H., Leite, I., Balaam, M., Dignum, V., Domisch, S., et al. (2020). The role of artificial intelligence in achieving the sustainable development goals. *Nature Communications, 11*(1), 1–10.

Chapter 22
Closing Remarks

Aldo Alvarez-Risco, Marc A. Rosen, Shyla Del-Aguila-Arcentales, and Dora Marinova

Throughout the book, sustainability is approached from economic, environmental, and social perspectives, in largely because this multidimensional topic requires a comprehensive approach, at governmental, corporate, and individual levels. Various tools available and used to monitor activities that seek to impact sustainability are presented. Energy sustainability is a fundamental pillar of sustainable development that helps ensure that the activities of society are carried out. This is important as in many cities, energy resources are being used without having a defined long-term plan. Hence, it is necessary to select properly the energy resources and their usage for the various human activities in a city. In this book, an emphasis is placed on efficient energy management.

It is possible to analyze the history of urbanization, describing changes in the formation of cities. Such an understanding can help understand current cities, so that sustainability measures can be generated for adapting to economic, social, and environmental needs.

A. Alvarez-Risco (✉)
Facultad de Ciencias Administrativas y Recursos Humanos, Universidad de San Martín de Porres, Lima, Peru
e-mail: aldo.alvarez@redsaf.org

M. A. Rosen
Faculty of Engineering and Applied Science, University of Ontario Institute of Technology, Oshawa, ON, Canada
e-mail: marc.rosen@uoit.ca

S. Del-Aguila-Arcentales
Escuela Nacional de Marina Mercante "Almirante Miguel Grau", Callao, Peru

Universidad Nacional de la Amazonia Peruana, Iquitos, Peru
e-mail: sdelaguila@enamm.edu.pe

D. Marinova
Curtin University Sustainability Policy (CUSP) Institute,
Curtin University, Perth, WA, Australia
e-mail: d.marinova@curtin.edu.au

A. Alvarez-Risco et al. (eds.), *Building Sustainable Cities*,
https://doi.org/10.1007/978-3-030-45533-0_22

Sustainability is evident in cities. Various terminologies and global initiatives to achieve sustainable cities are described in the book. Note that all initiatives that seek to improve cities are relevant and require support, but it is also necessary that they be articulated clearly to ensure that cities can become sustainable, smart, resilient, and green. In general, the United Nations Sustainable Development Goals (SDGs) can be monitored with precision at the city level, which can help changes that are to be made on a strategic scale.

We also discuss the characteristics of informality and describe both positive and negative aspects of it. It is important to evaluate completely informal jobs for understanding and to help create successful strategies to promote the formalization of work in cities. Another factor that is sometimes omitted in sustainability analyses is health. This includes the effective and safe use of medicines, which if not implemented properly can leave populations vulnerable. Hence, results are presented in this book that can be used for the generation of public policies and specific programs to address chronic diseases and their high economic and social costs.

Social equity is important and addressed here. Social equity can be improved through the use of social equity factors, which can help in the formulation of policies and programs that encourage its implementation. Equity is not only about generating welfare but also about empowering populations so that they can create areas of social/business development and have autonomy for their healthy growth.

Food insecurity is a problem that has been addressed increasingly since 2000 and the Millennium Development Goals and is now more strongly factored into the Sustainable Development Goals. This has led to increasing number of studies and efforts and has led to the Food Insecurity Experience Scale being used as a tool for measuring food and nutritional insecurity, in general and in a number of current government interventions. More specific tools also are needed to better shape the local understanding of food insecurity.

Special attention is given in the book to violence against women as the cause of a series of negative social and economic impacts. Such violence is not only a problem on a personal or family level, but also is a more complex problem since it impacts the physical and mental health of women and men as well as the development of productive companies and countries. For this reason, the need to generate prevention programs with the associated benefits for the various actors in society is proposed.

Citizens have the right to expect to live in a city planned in terms of physical construction following the SDGs. Aesthetics are also an issue that can complement a structured approach. Sustainable construction is an important option for cities. As a consequence, innovative proposals are required for urban designs that are efficient, eco-friendly, and sustainable. Similarly, transportation in a city requires planning designed for both the present and the future, focusing on intermodal transportation connections, which are absent in many cities of the world. Transportation also needs to account for and monitor the environmental impact of transport alternatives, taking into account factors ranging from product design to waste disposal and recycling. Municipal planning should involve the generation of ecological transport alternatives, which are accessible in terms of location and cost for a large part of the population. Greater use of bicycles and cars that operate on clean energy can also assist in moving towards sustainability. Transportation efforts can also exploit big data tools to help manage the large amounts of transportation data and to develop solution alternatives in the short term.

The construction that is carried out in the cities is increasingly evaluated from a sustainable viewpoint. Thinking globally, it is important that construction in places in the world having the greatest poverty make sustainability as a priority. The development of sustainable energy in cities is increasingly required by governments and companies, although little is done in many locations to address it. As a result, we discuss the generation and management of energy, especially the sustainable management of energy usage in homes and firms. There is a broad path for transformation. Business management for a sustainable use of energy is important, especially in light of agreements that have been signed recently (e.g., the Paris Agreement), to understand the commitment needed by the world to develop policies aimed at ecological energy management. Additionally, waste management is an important focus of global initiatives. It engages public and private organizations as well as citizens. The available indicators help in understanding the level of progress of the cities and in planning programs of greater scope. Prudent management is also required to be environmentally friendly and economically sustainable so that the activities carried out by companies and governments are compatible with the needs of the population and the planet. Social networks can assist in the dissemination and generation of environmental commitment for waste management as well as in education.

Water initiatives are increasing and they need to be coordinated legislatively so that they cease to be only isolated efforts and become changes that are part of the usual activities of organizations and cities. The price of water is a key factor for the greater or lesser impact a population has in relation to use and misuse. Water restrictions, used in various cities, can help generate changes in the consciousness of citizens, although this takes years, so short-term measures are needed to shape the behavior. Artificial intelligence has proven useful in generating forecasts about water scarcity problems up to 20 years in the future; the results indicate that it is worthwhile to initiate changes in water management today.

The book has a strong focus on climate change, including the factors that cause it and its most significant impacts. Actions that are being developed to understand and mitigate climate change are also described. It is shown that short-term measures need commitment from governments and/or companies, and often response times are slow. There is a need to propose in bilateral or multilateral trade agreements between countries that business transactions should take into account climate change. Such efforts can be developed within an environmental sustainability envelope.

Financial literacy remains a gap in education at all levels, leaving many populations vulnerable. To contribute to social sustainability, financial education needs to be enhanced, so that the general population and small entrepreneurs can be empowered in the management of money. This can allow them to make efficient use of funds and to improve their living conditions. Such initiatives can complement learning between parents and children, helping families achieve full social development based on the sustainable management of their financial resources. Entrepreneurship is monitored fairly extensively in many countries, so the level of progress that is being made in the various countries is understood to some extent. However, limited accurate data are available, explicitly for cities of a country. Such data are needed to assist efforts to support entrepreneurs in large cities. In many countries, universities take a leading role in the generation of enterprises, but other institutions of the state and companies also promote entrepreneurship. Corporate ventures in which an entity in a company also acts as an entrepreneur are a new and challenging concept that has significant potential.

Green consumption is a fertile field of research that is becoming increasingly understood. The relationship between green products/services and consumers is affected by numerous variables, such as the cultural aspects of consumers. Another factor requiring improved understanding is whether or not green behavior influences green buying. Companies do not presently know if it is profitable to invest in encouraging potential buyers to develop environmental behaviors so they are willing to pay for green products and services. Companies require this information in a segmented way to know where to invest with respect to sustainable products and services.

The Sustainable Development Goals of the United Nations provide the global reference framework for 2015–2030 for planning and execution of measures to support sustainability. As part of this framework, a range of sustainability indicators are proposed. The collection of data for many indicators is lacking or problematic, so one challenge is to increase the availability of data to measure and monitor well sustainability actions and their impacts. The SDGs are transforming business management in many countries and the generational replacement in management positions is expected to facilitate implementation. It is increasingly understood that SDGs are not only environmental in nature, but form a broad and comprehensive approach to sustainable development that can help improve the lives of all people and improve the health of the natural environment.

There are many challenges that require a change in the current activities of the societies across the globe towards new ways of doing things under the sustainable development concept. The growing interest at global, national, regional, and local levels in sustainable development bodes well for the future of civilization and the cities where people live. It is essential to restore not only the hope in better opportunities for urban dwellers but also the ecological environment that supports the functioning of the cities and provides a meaningful optimism for a sustainable future.

Index

A. Alvarez-Risco et al. (eds.), *Building Sustainable Cities*,
https://doi.org/10.1007/978-3-030-45533-0

C

The manufacturer's authorised representative in the EU is Springer Nature Customer Service Centre GmbH, Europaplatz 3, 69115 Heidelberg, Germany. If you have any concerns regarding our products, please contact ProductSafety@springernature.com

Printed and bound by CPI Group (UK) Ltd, Croydon, CR0 4YY
30/09/2026
02225558-0014